ITER
Physics

ITER Physics

C. Wendell Horton, Jr.
University of Texas at Austin, USA

Sadruddin Benkadda
Aix Marseille University–CNRS, France

World Scientific

NEW JERSEY · LONDON · SINGAPORE · BEIJING · SHANGHAI · HONG KONG · TAIPEI · CHENNAI

Published by

World Scientific Publishing Co. Pte. Ltd.
5 Toh Tuck Link, Singapore 596224
USA office: 27 Warren Street, Suite 401-402, Hackensack, NJ 07601
UK office: 57 Shelton Street, Covent Garden, London WC2H 9HE

Library of Congress Cataloging-in-Publication Data
Horton, C. W. (Claude Wendell), 1942–   author.
    ITER physics / C. Wendell Horton, Jr. (University of Texas at Austin, USA), Sadruddin Benkadda
(Aix Marseille University, France).
      pages cm
    Includes bibliographical references and index.
    ISBN 978-9814678667 (hard cover : alk. paper)
    1. Fusion reactors. 2. Magnetohydrodynamics. 3. Plasma turbulence. I. Benkadda, S., author. II. Title.
III. Title: International Thermonuclear Experimental Reactor physics.
    TK9204.H65 2015
    621.48'40153--dc23

                                                                                    2015005912

British Library Cataloguing-in-Publication Data
A catalogue record for this book is available from the British Library.

Typeset by Stallion Press
Email: enquiries@stallionpress.com

Printed in Singapore

Prologue

ITER is a large international *nuclear fusion* science and engineering project under construction in Europe, as the world's largest experimental *tokamak nuclear fusion reactor* (`https://www.euro-fusion.org/iter-2/`). In Latin language, ITER means "the way," which, as explained in this book, conveys well that the ITER physics and engineering experiments are important to future generations. Originally, the machine name ITER came from the *acronym* for International Thermonuclear Experimental Reactor. The Latin name reminds one that the way the Earth exists is from the power produced in the Sun from a self-sustained nuclear "burning" or nuclear-reacting ball of hydrogen gas compressed by the pull of gravity. While hydrogen is the dominant gas, the solar plasma has a spectrum of minor elements including C, N, O, Ne, Fe and Ni which correspond closely to the impurities studied in tokamaks, as described in Chapter 4.

Helium (He), the major minor constituent of the solar plasma, is the reaction product of the thermonuclear fusion burning of the hydrogen plasma. The spent fuel from fusion is the element Helium discovered in 1868 by French astronomer Pierre Janssen who found a yellow line (487 nm) in the solar spectrum of the chromosphere while observing light during a total solar eclipse. Janssen observed with a spectroscope of his invention for that purpose as recorded in *The Solar Atomic Emissions*. As described here in Chapters 8 and 9, plasma physicists use modern derivatives of these solar techniques to study all types of emissions from laboratory plasmas hotter than the core of the sun. The atomic line emissions are used to measure the transport of the elements, including helium, that enter the plasma from the walls and the thermal insulating coatings applied to the huge confinement walls of the ITER vessel.

Laboratory fusion plasmas also have a dominant minority of the fully stripped ion of helium He atoms called alpha particles born in the core plasma reaction with a high kinetic energy [3.5 MeV] received from the nuclear reaction of deuterium D and tritium nuclei produced by the collision D+T$\rightarrow$ He[3.5 MeV] + n[14.1 MeV]. This 3.5 MeV helium nucleus, or alpha particle, has a large orbit due to the high energy and larger mass. The alpha particle physics of the ITER plasma is a key area of research and described in Chapters 2 and 3. The high energy [14.1 MeV] neutrons penetrate the metal walls and are slowed down and absorbed in a thick blanket of boron and lithium. To create electric power from fusion reactions, the neutron blanket is designed to absorb the kinetic energy of the neutrons, thus creating a huge thermal power generator. The blanket is laced with lithium that undergoes reactions with neutrons that generate the tritium to fuel the fusion reaction. In this way the fusion machine regenerates the fuel needed to maintain the reaction. Deuterium is abundant in water, and the tritium isotopes of hydrogen are produced in the neutron-lithium reaction in the blanket.

The high energy neutron flux from the "burning plasma" is useful for a wide range of applications, including making pharmaceutical radio isotopes for medical diagnostics and for the burning up of waste products from fission reactors, as described in Chapter 10.

The thermal energy will be used to drive a steam turbine generating electricity free of the carbon (oil, gas, and coal) fuels that are currently used in electric power generation stations. The fusion plasma energy source has no carbon in the fuel cycle and thus will be a "GREEN" electric power source. Other "green" power generators are from solar, wind, and tidal power.

The design of the ITER fusion reactor has evolved from more than fifty years of searching for a device to confine the hydrogen plasma at the solar core temperatures required to reproduce the steady nuclear fusion reactions. The principal ideas on how to approach this project evolved from several research teams in the 1950s shortly after the demonstration of explosive nuclear fusion devices. In 1958 an International Conference was organized by the United Nations in Geneva that opened previously classified research projects attempting to create controlled fusion reactions in laboratories in the United States, the Soviet Union and Great Britain. Sharing of the progress and obstacles from these independent attempts lead to rapid future progress and an international involvement for the future experiments in nuclear fusion.

In the 1960s as a result of numerous experiments with various geometries, it was shown that the toroidal magnetic confinement geometry produced the best results. In particular, the experiments in a toroidal chamber with a large plasma current created by using the plasma itself as the secondary winding in an electrical transformer showed the highest temperature and the longest confinement times. The device constructed in the Kurchatov Laboratory in Moscow by a Soviet team showed a clear advantage over the other machines in the 1960s and continues to lead to the best fusion performance today. The toroidal device is known by the Russian acronym for a "toroidal chamber with a current", which is Tokamak. The tokamak became the internationally accepted name for this type of plasma confinement system with many variations explored in the past fifty years. The ITER design includes all the best results from these design variations. An encyclopedic source for physics and engineering in this toroidal system is found in J. A. Wesson, *Tokamaks*, 3rd Ed. (Oxford University Press, 2004) ISBN:0198509227.

France was an early and steadfast builder of tokamaks. First it was at the laboratory at Fontenay-aux-Roses in the suburb of Paris with a tokamak machine called Tokamak Fontenay-aux-Roses (TFR). The TFR machine showed the efficacy of using high energy neutral atomic beam accelerators to heat the plasma ions to temperatures of tens of thousands of degrees centigrade. The TFR experiments were the first to clearly show the anomalous thermal transport of heat (ion temperature 10^4 K equivalent to 1 KeV of kinetic energy) by the turbulent transport of the hydrogen ions, as explained in Chapter 4.

Core plasmas now reach temperatures ten to twenty times higher than those in the first neutral beam heated tokamak plasmas, exceeding the temperature of the core of the Sun. The nuclear fusion of the hydrogen isotopes becomes fast at these high temperatures, yielding the high energy alpha particle and the 14.1 MeV neutron fluxes that could not be explained by the use of the collisional transport of thermal energy that was used in the 1960s and 1970s, as explained in Chapter 4. The Deuterium-Tritium (D-T) (nuclear fusion reaction) experiments on TFTR at the Princeton Plasma Physics Laboratory clearly ruled out the use of collisional transport models. The confinement data is consistent with the plasma temperature gradient-driven turbulent transport.

Increasing the heating power with externally-accelerated beams of neutral deuterium atoms produce the required tens of megawatts of heating power and is the primary tool used to produced the burning plasma in ITER as described in Chapters 5 and 6. With increased heating power the turbulent thermal transport becomes stronger as described in detail in W. Horton, *Turbulent Transport in Magnetized Plasmas* (World Scientific, 2012), ISBN:978-981-4383-53-0.

The level of auxiliary heating power both from neutral beams and from radio frequency (RF) power has increased steadily over the decades since the 1980s. The neutral beam injection (NBI) power P_{nbi} planned for ITER experiment is 33 to 50 MW. Other heating powers from RF antennas on the inside vessel wall are shown in Table 1.1 for the ITER machine. For comparison the values are 20 MW in the largest fusion power tokamak operating today shown for the Joint European Torus called JET and the previous Japanese machine JT-60U described in Chapter 6.

Subsequently, from the mid-1900s onward, many laboratories around the world reproduced outstandingly successful results in the tokamak fusion experiments. Thus, there is an extensive international database for the performance of the tokamak.

Contents

Chapter 1

Machine Architecture and Objectives

1.1 Beginning of the ITER Project

ITER began in 1985 as collaboration between four countries: Russia, the European Union (through the EURATOM organization), the USA, and Japan. Conceptual and engineering design phases were carried out under the auspices of the International Atomic Energy Agency (IAEA). By 2000 a design acceptable to all parties was completed. Subsequently, these four parties were joined by the People's Republic of China and the Republic of South Korea. India became the seventh ITER partner in December 2005. The first Director General was Professor Kaname Ikeda and the second Director General was Professor Osamu Motojima. From 2015, the ITER Director is General Bernard Bigot from Commissariat Energie Atomic in France [http://www.cea.fr].

The first organized design activities (1980-1990) for a tokamak fusion reactor design were known as the INTOR project. Those first design activities also looked toward the second phase of a demonstration machine to follow the ITER proof-of-principle for fusion power. With the ITER project, the design efforts took a new direction. The earlier INTOR design activity is described by W. M. Stacey in *The Quest for a Fusion Energy Reactor*, Oxford, ISBN:978-0-19-973384-2.

In June 2005, the partners officially announced that ITER would be built in the European Union in Southern France. The negotiations that led to this decision ended in a compromise between the EU and Japan, in which Japan was promised directorship of 20% of the research staff at the French location of ITER, as well as having the Director of the administrative body of ITER. In addition, another research facility for the project was built in Japan, and the European Union contributes about 50% of the costs of this new Japanese-based institution located in Amori at the north end of Honshu. In November 2006 an international consortium signed a formal agreement to build the toroidal fusion machine. In September 2007, the People's Republic of China became the seventh country to send their ITER Agreement to the IAEA. By October 2007, the ITER Agreement was established and the ITER Organization legally came into existence as a Treaty between the participating countries.

The initial design in the 1990s from the four partners is shown in Fig. 1.1 as presented by Aymar, *et al.* (1998). This design, called ITER-FDR, was for a major

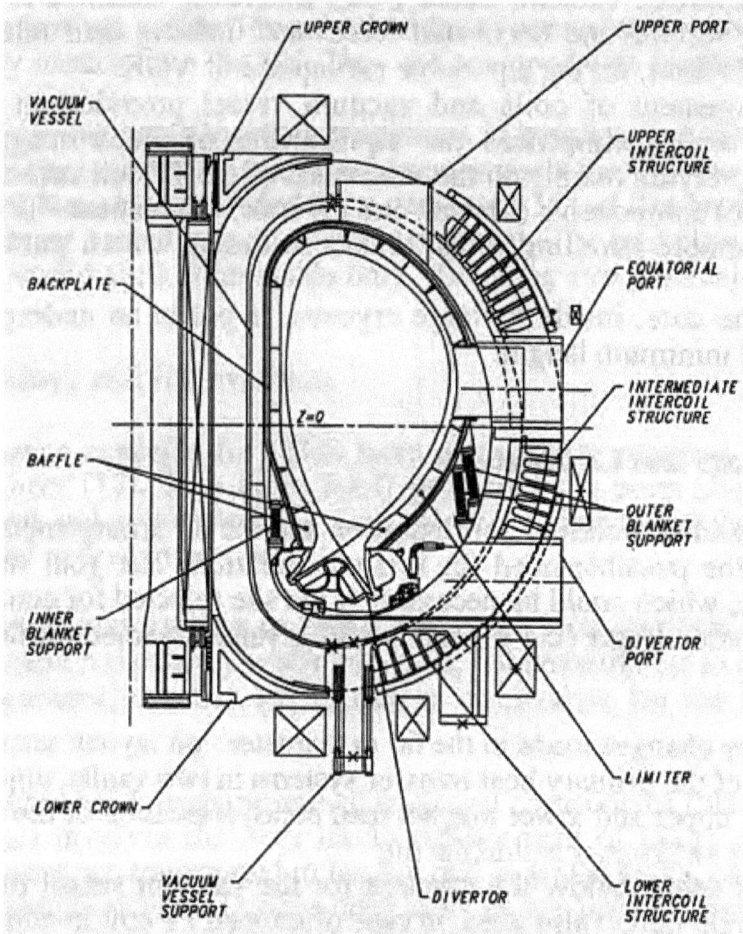

Fig. 1.1   Cross-section of the original 1996 ITER-FDR architecture with $R/a = 8.1\,\text{m}/2.80\,\text{m}$ and $B/I_p = 5.7\,\text{T}/20\,\text{MA}$ designed for fusion power of $1.5\,\text{GW}$. Here FDR refers to the Final Design Report from the Physics Expert Groups published in 1999. The international development path of this design is described in Stacey (2010). Subsequently, the design was charged to that shown in Fig. 1.2 with plasma current reduced to $15\,\text{MA}$ in view of lower-transport regimes discovered in the 2000s. Note the X-point of the separatrix line just above the dome of the divertor in the bottom of the inner chamber [Aymar, *et al.* (1998)]. The subsequent progress in improved plasma confinement described in Chapters 4-7 lead to the ITER-FEAT design [Campbell (2001)] with major radius $R = 6.1\,\text{m}$ as shown in Fig. 1.2.

radius of 8 meters with the aim of achieving ignition and running at $1.5\,\text{GW}$ of fusion power. By 2000 the perspective had changed, owing to new results described in Chapter 6.

Construction of the ITER complex began in 2007 at a new site next to the Cadarache nuclear laboratory supported by the Commissaries Energie Atomique (CEA). The assembly of the tokamak building and machine was started in 2013-2015. The first components finished were the neutral beam injectors built in Japan.

The construction of the ITER device and the supporting facilities is now (2015) well under way with the latest news available at `https://www.iter.org`.

1.2 Architecture of ITER

The architecture of the ITER fusion reactor is shown in Fig. 1.2. The objectives of the engineering architecture are to design and build within a ten-year period, at reasonable cost, a tokamak capable of producing $500\,\mathrm{MW}$ of output fusion power from $50\,\mathrm{MW}$ of input power into a mixture of tritium and deuterium plasma. Achieving this power amplification factor $Q_{\mathrm{DT}} = 10$ is considered more assured than the goal of reaching a self-sustained nuclear fusion from ignition, although the machine allows for the possibility of ignition if the thermal energy confinement proves to be sufficient. For a steady-state power system, the power amplification factor is traditionally defined as the Q of the system. ITER construction and research activities are constantly updated on the comprehensive website `https://www.iter.org`.

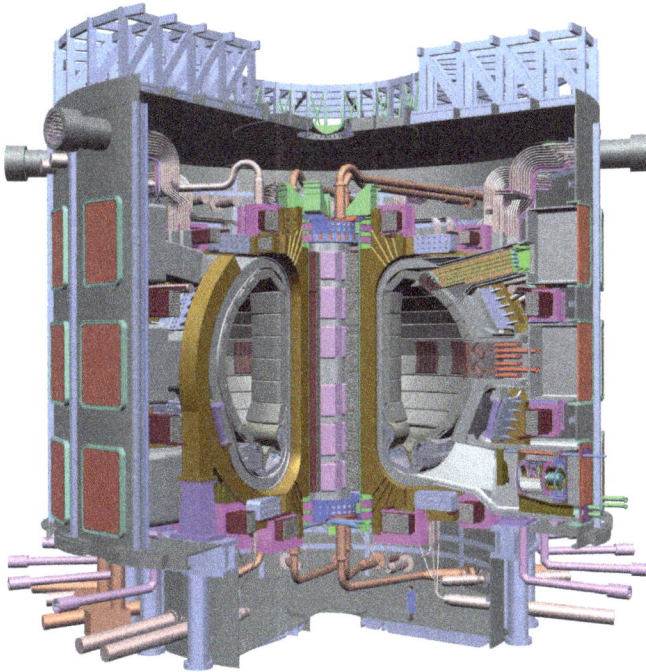

Fig. 1.2   ITER (International Toroidal Experimental Reactor) architecture with $R/a = 6.1\,\mathrm{m}/2\,\mathrm{m}$ with $B_0 = 5.3\,\mathrm{T}$ for confining plasma currents up to $I_p = 15\,\mathrm{MA}$ as given in Table 1.1. The table shows the change from the current European tokamak JET with new ITER fusion experiment.

The reference parameters for ITER, in comparison with the currently-operating Joint European Torus (JET) parameters, are shown in Table 1.1 from Shimada,

Table 1.1   JET and ITER comparison of main parameters from [Shimada, *et al.* (2007)] in *Physics Basis of ITER*. In the left column are the best JET parameters as of 2015 and in the right column are the designed values for ITER.

	Maximum values achieved on JET separately	ITER Design values
R and a	3 m 1.25 m	6.2 m 2 m
Elongation	1.8	1.7
Plasma volume	100 m^3	840 m^3
Magnetic field on axis	4 T	5.3 T
Plasma current in D-shaped plasma	7 MA	16 MA
Plateau current	1 MA for 60 s	15 MA for 500 s
Modes of operation	L, H and ELMy	ELMy
Plasma contact	carbon/beryllium limiters-pumped divertor	beryllium pumped divertor
Neutral injection to the plasma	22 MW	33-50 MW
Coupled ICRH	22 MW	20-40 MW
ECRH	0	20-40 MW
Current drive	3 MA (LH) (5 MA)	15 MA
Solenoid	external	275 V.s
Central density	2×10^{20} m^{-3}	10^{20} m^{-3}
Electron temperature	20 keV	21 keV
Ion temperature	40 keV	18 keV
Q value in DT plasma	0.6 (0.9 net)	10
Fusion power	16 MW	500 MW
Fusion energy	22 MJ in 4 s	120 GJ in 200 s

et al. (2007).   The JET deuterium-tritium fusion experiments in the 1997 time frame produced a power amplification factor of $Q \sim 2/3$ for pulse of about one second for 24 MW injected and a power amplification of $Q \sim 1/3$ from a longer four-second pulse driven by 12 MW of auxiliary heating power. Both the short and long pulses produced record amounts of fusion power, or equivalently a record number of neutrons from the DT nuclear fusion reactions. Thus, ITER may still be the first fusion confinement machine to show fusion power amplification factors greater than unity [$Q > 1$] for significant periods of time, meaning for time intervals of order tens of seconds. Reaching such net power amplification for a period time of order 50 to 100 energy confinement times is the key objective of ITER. Following this success a larger machine designed on what is learned from ITER will be used to develop the electric power producing fusion reactor called DEMO.

The ITER machine is designed to produce a long DT fusion power pulse of up to 300 s with a $Q = 10$. The initial ITER design was given to the international fusion community [Aymar, *et al.* (1998)]. The machine architecture and the tokamak building and pit presented by Aymar and his design team are shown in Figs. 1.1. Note the components labeled divertor, divertor port, limiter, vacuum vessel, and blanket. These components are common to all fusion reactors. Following the improved confinement results obtained with internal transport barriers in the Japanese tokamak JT-60U reported by Fujita, *et al.* (1998, 1999) the ITER parameters were fixed with radii $R/a = 6.2\,\text{m}\ 2.0\,\text{m}$ and $B_T = 5.3\,\text{T}$ as given in the second column of Table 1.1. The second major ITER group publication is Campbell, *et al.* (2001).

Campbell (2001) describes the final design of the ITER machine with major radius $R/a = 6.2\,\text{m}/2.0\,\text{m}$ and minor radius $a = 2\,\text{m}$ and the mean toroidal magnetic field $B = 5.3\,\text{T}$. The expectation is to achieve a plasma current $I_p = 15\,\text{MA}$ and a fusion power amplification of $Q = 10$ from fusion power of $500\,\text{MW}$.

The historical background of the design of ITER begins from the definitive fusion power experiments on the Tokamak Fusion Test Reactor that delivered the first deuterium-tritium experiments in 1993 and ran more than 840 DT discharges. The TFTR experiments were concluded in 1997. The first tritium experimental results in this relatively simple circular cross-section tokamak were reported by Bell, *et al.* (1989) at the International Atomic Energy Agency Meeting in Nice, France. A comprehensive review of the achievements of the TFTR experiments is given by Hawryluk (1998). The DT experiments were successful, producing a fusion power $Q_\text{fus} = 0.5 \pm 0.13$ for pulses lasting a few energy confinement times. The TFTR plasmas reached record ion temperatures of $32\,\text{keV}$ and thus high values of the fusion triple product of $n_e(0)\tau_E T_i(0) = 4 \times 10^{20}\,\text{m}^3\text{s}\,\text{keV}$ by using high-power neutral beam injection [Hawryluk (1998)]. These first DT experiments were performed in a hot ion mode called supershots as described in Chapter 6. The TFTR discharges, or "shots" as referred to in the laboratory, produced 10-12 MW of fusion power from injection of 40 MW of NBI power over period of half a second. The experiments were important for showing that in the fusion plasma there are alpha particle-driven MHD modes, or instabilities, that are excited by the high-energy (3.5 MeV) products of the fusion reactions [Nazikian, *et al.* (1997)]. Plasma instabilities excited by the 3.5 MeV alpha particles released in the fusion reactions place significant constraints on the ITER system as described in Chapter 2. The instabilities in the core plasma show up as structures called "fishbones" and "sawteeth", as explained in Chapter 3. The results of the TFTR experiments were largely incorporated in the design of the Joint European Tokamak or JET. JET began operation in 1991 [Jet Team (1992)] and achieved record fusion power gain Q_DT described in Keilhacker, *et al.* (1999).

A non-technical review of the history of the International Thermonuclear Experimental Reactor (ITER) is given by McCray (2010). McCray emphasizes three aspects of the project's history, focusing largely on the European research community's perspective. First, McCray explores how European scientists and science managers constructed a trans-national research community around fusion energy projects as part of Europe's larger technological integration and development. McCray (2010) expands on Gabrielle Hecht's concept of 'technopolitics' to the larger international dimension and explores how the political environments of the Cold War and the post-9/11 era helped shape ITER's history, sometimes in ways not entirely within researchers' control. The essay considers ITER as a technological project that gradually became globalized. At various stages in the project national borders became less important, while social, economic, legal and technological link-ages created a shared societal space for fusion research on an expanding scale. Presently, ITER is an international treaty with seven countries working together. The project headquarters and building are on a plateau north of Aix en Provence adjacent to a the French nuclear laboratory called Cadarache with complementary nuclear facilities in Japan.

A popular account of scientists' 50-year quest to harness nuclear fusion as an energy source is described by Fowler (1997), who was leader of the fusion program at the Lawrence Livermore National Laboratory (LLNL). From the 1960-1985 the fusion research at LLNL centered on developing an alternative approach to fusion power based on the mirror confinement or mirror machine. This concept is still fol-lowed in two major nuclear laboratories, one in Tsukba, Japan, and the other at the Budker Institute for Nuclear Research in Novosibirsk, Russia. While reaching the fusion reactor power production with plasmas in a mirror machine is now considered as unlikely, the machine can be used as a compact source of fusion neutrons. These neutrons sources can be used for testing the lifetime of materials and for produc-ing radio-pharmaceuticals for the diagnosis and treatment of cancers. Alternative approaches to fusion power are still actively pursued on small-scale machines.

At the 1997 time of Fowler, the four partners collaborating to build ITER were the European Union (EU), the USA, Japan and Russia. Fowler's evocative assess-ment: *'The sun never sets on ITER'*, became all the more salient after China, India, and South Korea joined the effort. By 2005, the mega-project represented over half of the world's population.

There is a qualitatively large extrapolation from JET to ITER. The Japanese Atomic Energy Research Institute (JAERI) designed and showed qualitatively new results in 1994 with knowledge that a design with an elliptical-shaped magnetic chamber, with two separate parts of the magnetic field configuration, would give better fusion engineering design features [Koide and JT-60U Team (1997)]. Koide introduced the term internal transport barriers to describe the new profiles of tem-perature and density with steep radial gradients outside the partial barrier zone discovered in Ishida, *et al.* (1997) and Koide, *et al.* (1998). The physics of these internal transport barriers is explained in Chapters 5 and 6.

Thus, both the JT-60U and JET machines were designed with plasma shaping poloidal field coils as shown in Figs. 1.1 and 1.2 that produce closed magnetic surfaces up to a certain radius of the plasma confinement region followed by an open magnetic field region between the metal chamber walls and the plasma confined inside the closed magnetic surfaces. This change in the magnetic field structure introduces what is called the magnetic separatrix (SX). A surface inside of which the magnetic field lines wind forever, in principle, around the magnetic axis in the core hot plasma region, and outside of which the twisting magnetic field lines terminate on the walls of the chamber, referred to as the divertor chamber. This structure allows the hot escaping plasma to be diverted into a special chamber designed to withstand high temperatures and intense particle bombardment. These surfaces, called the Plasma Facing Components (PFC) are described in Chapters 9 and 10 that deal with several critical components of ITER and any other fusion reactor designs. A key uncertainty in magnetically confined fusion reactor designs is the response and lifetime of the first metallic walls to the bombardment of the energetic alpha particles and the fusion neutrons. The neutrons pass through the metal walls and are then slowed down in a "blanket" designed to convert their kinetic energy into thermal energy to drive the electric power generators as discussed in Chapter 10 on the "Broader Approach" recently launched as a bilateral program between EU and Japan.

Note the region at the lower right of Fig. 1.1 marked as "Divertor Port". This divertor chamber is analogous to "exhaust manifold and exhaust pipe" in gasoline engine designs. The divertor is designed to take the escaping hot plasma away from the first wall of the machine. This technique was developed on the ASDEX machine at the Max Planck Institüt für Plasma Physics outside of Munich by Karl Lackner's team. The new design gave an unexpected improvement in the plasma confinement that is described in Chapter 6. The new higher-confinement regime was named the H-mode and the previous conventional tokamak regime is now called the L-mode. In overly simple terms, H denotes "high confinement" and L denotes "low confinement". To divert the plasma to the "divertor port" the magnetic surface is opened up at what is called an X-point creating surfaces outside the critical magnetic surface defined by the X-point that flow into the divertor chamber. The plasma outside the separatrix on the field lines that flow to the divertor chamber has a considerably lower plasma temperature and pressure than the plasma inside the separatrix.

A large Japanese tokamak was designed to incorporate the magnetic separatrix and the divertor chamber. The first configuration called JT-60 had the divertor on the equatorial plane of the machine, while the second and better design, JT-60U, moved the divertor region to the bottom of the machine, as used in JET and planned for ITER. The JT-60U machine demonstrated confinement at high-plasma pressure and at high ion temperatures. The machine, decommissioned in 2010, never ran with tritium so the comparison with JET and TFTR is difficult. With deuterium injected into a deuterium plasma, one extrapolates from the low-fusion neutron fluxes from DD reactions to estimate what the power would be in an equivalent DT experiment. The extrapolation has numerous transport uncertainties and involves a factor greater than 200. Nevertheless, the JAERI team announced [Koide and JT-60U Team (1997); Kamada (1999)] the extrapolated fusion neutron production level would translate to break even with $Q_{DT} \gtrsim 1$ in a corresponding tritium plasma. The record deuterium plasma was created with reversed magnetic shear at high-plasma current $I_p = 2.6\,\mathrm{MA}$ in Kamada, *et al.* (2001). Further progress in JT-60U is reported in Fujita, *et al.* (2001). The international fusion community, however, generally has reservations about the errors surrounding this difficult extrapolation from the DD plasma to the DT plasma. The clearest result from JT-60U was that the position and shaping of the divertor chamber is of critical importance in achieving the highest tokamak plasma performance. The best one can state is that with the modeling carried out under reasonable assumptions, the DD experiments would extrapolate to a DT experiment that would have produced $Q_{DT} \approx 1$. However, there are questions about the difficult extrapolation to the different fuel with a factor of 200 lower nuclear reactivity and the fact that the turbulence energy transport rates are known to be dependent on the mass of ion components in the plasma as described in Chapter 4.

These three large machines: TFTR, and JET, plus confinement data from approximately a hundred earlier smaller tokamaks, provide a large database used to validate computer modeling and to extrapolate parameters for the expected performance for ITER. Some of the many articles making these extrapolations are Cordey, *et al.* (1999), Hawryluk (1998), Bateman, *et al.* (1998), and a series of articles in a volume with the "Final Design Report" (1998), devoted to the basis for ITER. In setting the parameters, the ITER design teams used the extensive International Tokamak Database in the design of ITER [Kaye, *et al.* (1997)]. Advanced computer simulations make predictions for some aspect of the dynamics, but as in weather systems and magnetospheric simulations, the range of scales and interactions of the complex dynamical sub-systems is too complex to allow for full system predictive simulations. The simulations codes are discussed in Chapter 5.

Thus, the production of a net fusion power output P_{fus} significantly greater than the injected power P_{aux} seems assured for the ITER machine. Amplification of fusion power for extended periods of time by the controlled fusion process in a laboratory is something that has not yet been achieved with previous fusion systems. ITER is designed to achieve this important demonstration in a "Big Science" physics experiment. With successful completion of the ITER fusion power production, the plan is to build a larger tokamak reactor called DEMO, which would output electric power from the nuclear fusion reactions. Plans for the DEMO nuclear fusion power reactor can be completed only after the knowledge gained from the ITER experiments. Successful demonstration of the generation of controlled fusion power in the laboratory will be the climax of a long running dream to create a miniature model of the power of the stars in the laboratory.

At the 1996 Fusion Energy Conference the JT-60U Team (here 60 denotes that the volume of the plasma chamber was approximately 60 cubic meters), presented the upgrade JT-60U in which the divertor was moved from the outboard side to the bottom of the toroidal chamber. The team showed that the neutron emission reached a record for the deuterium beams injected into the deuterium plasma. The discharge plasma achieved a high-beta H-mode plasma in which the core ions reached the temperature $T_i(0) = 45\,\text{keV}$. The researchers interpreted the data to be an equivalent of $Q_{DT} = 1.27$ in a Physical Review Letter [Ishida, *et al.* (1997)] and

[Koide and JT-60U Team (1997)].

Construction of the ITER facility began in 2007, and the first fusion plasma is expected to be produced in 2020. ITER will be the world's largest magnetic confinement plasma physics experiment greatly surpassing the Joint European Torus (`http://en.wikipedia.org/wiki/Magnetic_confinement_fusion`). Plans for the first commercial demonstration fusion power plant, named DEMO, are under development now in Europe, anticipating that the results from ITER will enable engineers to bring fusion energy to the commercial market.

The physics path that led to the design of ITER is a long, complicated one. Few fusion experts even know the full story of physics principles in all the areas from plasma physics, atomic and molecular physics to radio frequency heating and control. There is a complex set of high-technology engineering problems that were overcome in designing the ITER tokamak fusion system. Some countries justify their participation in the fusion program partly on the basis of the development of the high-technology capabilities in areas complimentary to those of space science and defense technologies.

In view of these facts, the authors working together at Aix Marseilles University, decided in 2013 that it was an appropriate time to collect the basic ITER physics material in a single book to archive for a wide range of scientists and engineers the vast information used in the design and construction of one of the largest basic science experiments the world community has undertaken to date. The book is designed to unify the plasma and engineering considerations provided in the vast literature on the physics and engineering work of the past generations, leading to the construction of this large, hugely-important basic science experiment named ITER.

1.2.1 *Operational regimes documented by tokamak experiments*

After several years of operation, it was reported at the 1988 International Atomic Energy Agency (IAEA) conference in Nice, France, that the Tokamak Fusion Test Reactor (TFTR) [Bell, *et al.* (1989); Hawryluk (1998)], with 30 MW of balanced deuterium neutral beam heating, achieved an ion temperature of 32 keV and the triple product of $n_e^{(0)} \tau_E T_i^{(0)} = 4 \times 10^{20} \, \text{m}^3 \, \text{s keV}$. Since the electron temperature remained at about 8-10 keV, this type plasma confinement regime was defined as the "hot ion supershot" regime, which remains as one of the promising approaches to fusion power. The Bell, *et al.* (1989) TFTR article describes how the deuterium-beams of neutral atoms were injected into the chamber to create the hot tritium-deuterium plasma generating the measured neutron flux of $3 \times 10^{16}/\text{s}$ into what a future reactor would use to produce electric power. In 1998 experiments with tritium in TFTR were concluded. The highest fusion power amplification achieved was $Q_{\text{fus}} = 0.5 \pm 0.13$. This world record stood until the JET tritium experiments. The TFTR geometry was simpler with circular cross-section of the plasma without the magnetic separatrix shown in the figures of the ITER machine. Nevertheless, an enhanced confined regime called the hot-ion mode was produced for short periods of time.

Measurements of plasma turbulence and comparisons with theory show that one of the most serious limiting factors for the success of ITER, and similar fusion machines, is the control of drift-wave turbulent transport. The seriousness of the turbulence becomes more evident as machines with increasing injected power P_{aux} reach core power densities of $1\,\text{MW/m}^3$. With increasing heating power, the plasma temperature rises only as a fractional power law, approximately as $P_{\text{aux}}^{1/2}$, owing to the associated, or concomitant, increase in the plasma turbulence levels with the higher core plasma temperatures. The energy confinement time, τ_E, decreases even though the plasma temperature increases. Thus, the triple-fusion-power product may remain essentially constant as more external power is injected into the magnetically confined plasma. The triple-fusion-power product is $n_e^{(0)} \tau_E T_i^{(0)}$ and originates from considering the Lawson product of $n_e^{(0)} \tau_E$ and the rate of fusion power production from the core ion temperature $T_i^{(0)}$ dependence of the nuclear reactivity of the deuterium-tritium plasma. Here the superscript (0) denotes the value of the field at the core of the plasma.

The database from twelve tokamaks showing the dependence of the energy confinement time τ_E on the machine and plasma parameters is shown in Fig. 1.3. This data gives the standard L-mode energy confinement time law called the Troyon formula for the L-mode plasma confinement regime. The horizontal axis gives the complex empirical formula used to organize the data and the extrapolation needed to reach the energy confinement time of two seconds for ITER. Current ITER construction and research activities are found at `https://www.iter.org`.

The data organizing empirical formula in Fig. 1.3 is

$$\tau_{\text{ITERL96}} = 0.023\, R^{1.83}\, I_p^{0.96}\, n^{0.40}\, M^{0.20}\, B^{0.03} (R/a)^{0.06}\, P^{-0.73}. \tag{1.1}$$

giving the energy confinement parameterization derived from twelve tokamaks listed in Fig. 1.3.

From the graph in Fig. 1.3 one sees that the confinement time varies strongly with the plasma current I_p and weakly with the value of the toroidal magnetic field B_T. One also sees that increasing the injected heating power P_{aux} lowers the confinement time τ_E with other parameters fixed. This dependence of $\tau_E \sim 1/P_{\text{aux}}^{0.73}$ is due to the increasing level of plasma turbulence as the heating power P_{aux} increases the temperature gradient driven plasma turbulence. The physics of these temperature gradient driven turbulent transport is described in Chapter 4.

A consequence of this stronger heating power, P_{aux}, is that the turbulence is amplified, which shortens the plasma energy confinement time, $\tau_E = W/(P_{\text{aux}} + P_{\text{ohm}})$, in agreement with the ITER database [Kaye, et al. (1997); L-H Mode Database (1994)]. Here $W(t)$ is the total energy stored in the thermal plasma density and temperature. Here $W(t)$ is the total energy stored in the plasma from the measured

L-Mode scaling

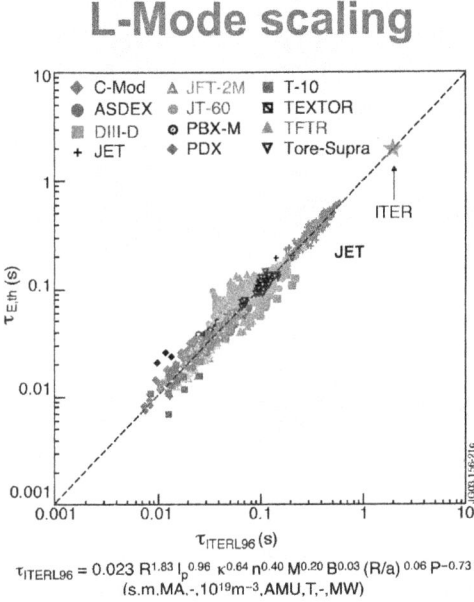

$$\tau_{ITERL96} = 0.023\ R^{1.83}\ I_p^{0.96}\ \kappa^{0.96}\ n^{0.40}\ M^{0.20}\ B^{0.03}\ (R/a)^{0.06}\ P^{-0.73}$$
$$(s.m.MA.\text{-}.,10^{19}m^{-3},AMU,T,\text{-},MW)$$

Fig. 1.3 The standard tokamak energy confinement time τ_E versus the experiment data for twelve machines with widely different configurations and sizes. The scaling law shows the strong dependence of increasing confinement time with increasing toroidal current I_p and decreasing confinement time with injected auxiliary heating power P_{aux}.

density and temperature profiles and the diamagnetic signal diagnostics described in Chapter 7. The drift-wave turbulence has explained the empirical database laws for both helical systems [Yamada, *et al.* (2005)] and Tokamaks [Kaye, *et al.* (1997)], showing that $\tau_E = \tau_E^0 (P_0/P_{aux})^{2/3}$. Thus, doubling the injected power P_{aux} shortens the confinement time, $\tau_E \rightarrow \tau_E^0/1.59 = 0.64\,\tau_E^0$. For example, the reference design for ITER (see Table 1.1) has $P_{ICRF} = P_{ECH} = 20\,\mathrm{MW}$ and $P_{NBI} = 16\,\mathrm{MW}$ injected into the plasma volume of $V = 2\pi^2 R_{ab} \simeq 800\,\mathrm{m}^3$. The average injected power density is then $0.07\,\mathrm{MW/m}^3$. To reach the power density of $1\,\mathrm{MW/m}^3$ at the core, as in current large tokamaks, the ICRH and electron cyclotron emission (ECE) power needs to be focused in the central core volume of $40\,\mathrm{m}^3$, corresponding to the core plasma radius $r_c = 0.6\,\mathrm{m}$.

Following the fusion literature, we introduce the symbols ICRF = ion cyclotron radio frequency heating, ECH = electron cyclotron heating, and NBI = neutral beam injection heating using the acronyms found in the plasma physics literature.

1.2.2 *Why is ITER so important for future generations?*

Carbon-based fuels grow increasingly scarce (http://en.wikipedia.org/wiki/Peak_oil) in the face of the ever-growing demand and therefore we look for new, more sustainable sources of energy. The authors argue that nuclear power provides the clearest alternative to carbon-based fuels for electric power in the coming centuries. Fission nuclear power is currently the only realistic non-carbon-based source of electric power. France, for example, derives about 70% of its electric power from fission nuclear reactors and China (3%), Japan (50%), and the US (30%) have substantial noncarbon electric power from fission reactors. Currently, new methods of extracting butane, propane and other hydrocarbon fuel gases from shale rock in Texas are making significant contributions to the energy supply of the United States [Mealer (2013)]. Whether the sequestration of the carbon-oxygen emitted exhaust compounds from the hydrocarbon fuels can be successfully controlled is a field of current research.

The next natural step, from the science perspective, in search of an abundant non-carbon based electric power is from nuclear fusion. In reality, all the carbon-based fuels owe their existence to the energy from the sun, which is our total inspiration for nuclear fusion research. Can society produce an endless supply of clean energy by using controlled nuclear fusion? This is the science and engineering question being addressed by the international ITER research project. The carbon-based fuels of oil and gas come from plants grown in the Jurassic and Permian periods that were subsequently compressed and heated as the layers of earth were sub-ducted deep below the surface of the earth in the Cretaceous period. Without fusion energy released by the nuclear burning of the sun there would be no life or earth and thus no carbon fuels.

Since the 1970s the research communities, especially those in petroleum engineering, have forecast looming shortages of gas and oil supplies. The consensus is that we must find new ways to meet global energy needs and reduce our dependence on energy sources by adopting a different way of life on Earth. Only fusion power has the potential to provide sufficient energy to satisfy more demand, and to do so sustainably, with a relatively small impact on the environment.

Nuclear fusion power has many potential attractions. First, its hydrogen isotope fuels are relatively abundant, one of the necessary isotopes, deuterium, can be extracted from seawater, while the other fuel, tritium, would be created using neutrons produced in the fusion reaction itself [Bell, *et al.* (1989)]. Furthermore, a fusion reactor would produce virtually no CO_2 or other atmospheric pollutants. The radioactive waste products would be very short-lived compared to those produced by conventional nuclear reactors.

Design Agreement Reached in 2006 by Seven Partners

In 2006-2007, the seven international participants formally agreed to fund the creation of a nuclear fusion reactor. The program is anticipated to last for 30 years: 10 years for construction and 20 years of operation. ITER was originally expected to cost approximately five to ten billion dollars, but the rising price of raw materials and changes in the initial design have seen the estimated cost get pushed up from the initial 16 billion to now (2015) estimated at 20 billion dollars. The reactor is expected to take 10 years to build with completion scheduled for 2020 [Bucalossi (2011)]. Site preparation is now finished. Procurement of the largest machine components has been completed at the time of writing (2015).

ITER is designed to produce approximately 500 MW of fusion power sustained for up to 1,000 seconds (compared to JET's peak of 16 MW for less than a second) by the fusion of about 0.5 g of deuterium/tritium mixture in its approximately 840 m^3 reactor chamber. Although ITER is expected to produce (in the form of heat) 10 times more energy than the amount consumed to heat-up the plasma to the fusion temperatures, the generated heat will not be used to produce electric power [Doyle, *et al.* (2007)].

The origin of ITER concept can be traced to 1985, and perhaps earlier. ITER was originally an acronym for *International Thermonuclear Experimental Reactor*, but that title was eventually replaced, perhaps partly due to the negative connotations of the word "thermonuclear". The word "ITER" in Latin means the "journey", the "direction" or the "way". Thus, the choice was made to declare the old acronym ITER as the new name of the experiment with its Latin origin as meaning of the "way".

Objectives of the ITER Experiments

ITER's mission is to demonstrate the feasibility of fusion power and prove that the tokamak type of magnetic confinement device can work without negative impact [Friedberg (2007)]. Specifically, the project aims are to:

- Produce ten times more thermal energy from fusion power than is supplied by auxiliary heating (a Q value of 10) for a period of a few minutes.
- Produce a steady-state plasma with a Q value greater than 5 in a long pulse for about 480 seconds or 8 minutes.
- Develop the advanced technologies and processes needed for a fusion power plant.
- Verify the rates of tritium breeding from lithium.
- Find the life-time of the vessel's walls from the neutron bombardment and the heating and erosion by the high-temperature plasma.

In view of the complexity of the ITER project, the effort is multi-generational as well as multi-national. The authors recognize the importance of leaving detailed public documents on the physics and engineering of the project for those who will inherit the job of finishing the work started here in Provence in early 2000. This book will provide the record of the early ITER research and design activity. The authors have organized a series of International AIP 2008 ITER Summer Schools in Aix-en Provence and the AIP Conference 2012 ITER Summer School in Austin, Texas. Now the summer schools are held in all the ITER partner countries. Scholarships from the Principality of Monaco for young scientists and engineers have laid

Table 1.2   Proceedings of the International ITER Summer Schools

First ITER International Summer School (2007)	Ed. Sadruddin Benkadda AIP Conference Proceedings, no. 1013, 2007, ISBN:978-0-7354-0534-9
Second ITER International Summer School (2008)	Eds. Sanae Itoh, Shigeru Inagaki, Masatoshi Yagi, Confinement (from device to plasma), AIP Conference Proceedings, no. 1095, 2009, ISBN:978-0-7354-0628-5
Third ITER International Summer School (2009)	Ed. Sadruddin Benkadda, AIP Conference Proceedings, no. 1237, 2009, ISBN:978-0-7354-0781-7
Fourth ITER International Summer School (2010)	Magnetohydrodynamics and Plasma Control in Magnetic Fusion Devices Special Issue, Fusion Science and Technology, vol. 59, no. 3, 2010
Fifth ITER International Summer School (2011)	Eds. Sadruddin Benkadda, Nicolas Dubuit, Zwinglio Guimareaes-Filho, AIP Conference Proceedings, no. 1478, 2012, ISBN:978-0-7354-1087-9
Sixth ITER International Summer School (2012)	RF Heating and Current Drives in Plasmas, Fusion Science and Technology, vol. 65, no. 1, 2014

a foundation for future fusion scientists and engineers. The proceedings of these ITER Summer Schools are listed in a table of the International Summer School Books (ed. S. Benkadda, AIP, vols. 1-10). The Summer School volumes provide more details of the physics and engineering presented in the following chapters.

1.2.3   *Elementary description of the plasma turbulence*

In toroidal magnetic geometry (tokamak or large helical device) there needs to be mechanisms for stabilizing the plasma from interchange motions. These are plasma motions that are analogous to the interchange instability that occurs when a dense liquid or gas is supported above a light, less-dense gas or liquid. This instability is called the Rayleigh-Taylor instability in hydrodynamics. In plasmas the analogous instability is called the magnetic Rayleigh-Taylor instability. The effective plasma acceleration g_{eff} arises from electric forces created by the separation of the ion and electron charges induced by the toroidal magnetic geometry. In the elementary plasma books these are the guiding center drifts produced by the nonuniform and curved magnetic fields from the incomplete closure of the circular cyclotron orbits. Positive and negative charges drift in opposite directions owing to their reversed directions in the cyclotron orbit. The detailed description of the plasma interchange modes is complicated and will be met again Chapters 4 and 5.

The direction of the effective gravity g_{eff} in the plasma-fluid analog is horizontal and directed outward from the axis of symmetry. Thus, for a peaked plasma mass density profile, there is a strong acceleration determined by the sound speed c_s squared divided by the major radius R of the torus as described in Fig. 1.4. This configuration is unstable to the onset of an interchange of plasma from the higher density core to the lower-density outer regions. The methods of maintaining the stability are described in Chapters 5 and 6. There it is shown that the tension in the magnetic field attached to the plasma provides one restoring force. This balance between the magnetic tension force and the acceleration from g_{eff} gives the fundamental limit on the maximum plasma pressure that a torus with a specified magnetic field B_φ and major radius R can contain. This pressure limit is called the beta limit and is given by an expression called the Troyon formula [Troyon, *et al.* (1984)]. The limit is thoroughly validated by the extensive tokamak database accumulated from the many well-documented tokamak experiments performed over the past forty years. The details are found in Chapters 5 and 6. Figure 1.4 gives a few of the key equations that will be derived in Chapter 5.

Conditions for DT Fusion Plasmas Power

Ultimately the condition for the onset of fusion power is set by the values of the cross-sections σ for the atomic scattering and the nuclear fusion reactions. Atomic scattering cross-sections σ are of order 10^{-16}cm^2 and the nuclear cross-sections are of order $10^{-24}\,\text{cm}^2$. The nearly elastic scattering of the hot electrons from the ions results in the intrinsic loss of electron energy by radiation of high-frequency electromagnetic waves in the process called Bremsstrahlung radiation [Jackson (1999)]. Both processes increase with the square of the plasma density so that the ignition temperature is set by a temperature around 77 million degrees centigrade (equivalent to 7000 electron volts) for the fusion power collisions to dominate the atomic-electron collisions producing the radiative power losses.

As with all modern science and engineering fields, fusion plasma physics has many specialized terms and acronyms. A good glossary that defines most of the terms used in this book can be found at `http://www.iter.org/glossary`. A widely-used handbook for the formulas and constants used in plasma physics can be found at `http://www.ppd.nrl.navy/nrlformulary`. The books on fusion power physics [Horton (2012); Stacey (2010)], contain frequently-used physical constants, fusion power formulas and acronyms. The plasma literature has a mixture of cgs and MKS units owing to the mixture of atomic physics and engineering. Reference values for the parameters in the International Thermonuclear Experimental Reactor (ITER) device are given in Table 1.1. There is now a broad level of recognition in the fusion community that an understanding of the chaos and coherent structures created by plasma drift wave instabilities is required

Pressure Gradient Interchange   **Rayleigh-Taylor**

$$p' \equiv -\frac{p}{L_p} \quad \text{and} \quad \boldsymbol{E}_\perp = -\boldsymbol{v} \times \boldsymbol{B}$$

gravity g and
density gradient

$$\gamma_{MHD}^2 = -\frac{p'}{\rho}\frac{2}{R}$$

$$\gamma_{RT}^2 = -\frac{g}{\rho}\frac{dp}{dr}$$

blobs accelerate out

$$= -\frac{\boldsymbol{g} \cdot \nabla \rho}{\rho}$$

$$g_{eff} = 2c_s^2/R$$
$$\sim 2(3 \times 10^5 \text{m/s})^2/(3\text{m})$$
$$\sim 6 \times 10^{10} \text{m/s}^2$$

over outside of torus

g_{eff} sets limits on three scales:
MHD p', ρ_i-ITG and ρ_e-ETG.

Fig. 1.4   Comparison of the instability in a magnetized toroidal plasma on the left with the classical neutral fluid gravitation instability in an inverted mass density profile on the right. Key equations for the magnetized plasma instability are on the left and the neutral fluid analog on the right. The role of gravity in the plasma is played by an electric field arising from the charged particles (negative electrons and positively charged ions) drifting to the top or bottom of the torus owing to the curved magnetic field lines in the torus. The positive and negative charges drift in opposite directions setting up an electric field in plasma. The role of the mass density gradient for the neutral fluid is played by the plasma pressure gradient. The outer half of a torus has an unstable stratification of the pressure gradient, and the instability operates through the drift waves on the micro-scales of the ion and electron gyroradius. As shown on the left, the effective value of the equivalent gravity g_{eff} acting in the plasma is nine to ten orders of magnitude larger than the Earth's gravitational acceleration of a mass [Troyon, *et al.* (1984)].

to reach practical fusion energy producing regimes [Horton and Ichikawa (1996); Horton (2012)].

From the results of a variety of tokamak plasma experiments and particularly from the TFTR, JT-60U, and JET machines, there is a broad expectation that conditions for producing fusion power in the ITER machine will be met. This means that the new machine expects to increase the plasma energy confinement time τ_E to exceed $\tau_E \sim 1\,\text{s}$ and product of the plasma density and the energy confinement time $n_e\tau_E \sim 6 \times 10^{13}$ to $10^{14}\text{cm}^{-3}\text{s}$. The Lawson condition for ignition requires still higher values of $n_e\tau_E \sim 3 \times 10^{14}$ and higher. In retrospect, it took over ten years to complete the experiments in TFTR and the highest Lawson product achieved was $n_e\tau_E \sim 2 \times 10^{13}\text{cm}^{-3}\,\text{s}$. The reason is now clear that the confinement time degrades with increasing plasma temperature and does not continue to increase linearly with plasma density as first found in the earliest tokamak experiments in the 1970s. The degradation with increasing temperature is a consequence of the increasing drift wave turbulence that increases the thermal diffusivity. In other words, the plasma "boiling" increases with the increasing temperature gradient determined by the core plasma temperature, compared to the vessel wall temperature that must be maintained below the melting point of the material. Tungsten has the highest melting temperature of suitable wall materials at about 3700°C. The Tokamak Fusion Test Reactor (TFTR) operated at the Princeton Plasma Physics Laboratory (PPPL) from 1982 to 1997 set records for the core plasma temperature with the core ion temperature $T_i^{(0)} = 45\,\text{keV}$ equivalent to 500 million degrees centigrade [Hawryluk (1998)]. While not achieving breakeven as originally planned, the TFTR tokamak achieved all of its hardware design goals and thus made substantial contributions in many areas of fusion technology development (`http://www.pppl.gov//tokamakfusiontestreactor.cfm`).

The parameters of key tokamaks discussed in some detail in this book are given in Table 1.1. An overview of the history of the progress towards the design of the ITER machine from the forerunner design committee for the INTOR device is found in Stacey (2010). In an appendix Stacey gives a list of the parameters of some hundred tokamaks starting with the T-1 device in the USSR in 1957, and going through to ITER with expected operation in 2020. The list includes EAST and KSTAR, the new superconducting tokamaks in China and South Korea.

JET record fusion power experiments

The Joint European Tokamak called JET was designed in the 1970s and has produced significant fusion power from deuterium-tritium plasmas. The JET experiment, approved in 1975, was a major step up from earlier tokamaks [Jacquinot, *et al.* (1993)]. The large JET chamber allowed plasma currents to increase from $I_p \simeq 300\,\text{kA}$ and volumes of order $10\,\text{m}^3$ to plasma currents $I_p = 4\,\text{MA}$. The plasma chamber was designed to have a small aspect ratio $R/a \le 3$ and height $b/a = 1.7$ allowing a large plasma current I_p for the given magnetic field $B_T \le 4\,\text{T}$ and machine size $R = 3\,\text{m}, a = 1.15\,\text{m}$. The plasma volume is up to $100\,\text{m}^3$.

Eventually, with an upgraded plasma heating system with $22\,\text{MW}$ of neutral beam injection (NBI), $22\,\text{MW}$ of ion cyclotron wave heating (ICRH), and $7.3\,\text{MW}$ of lower-hybrid current device (LHCD) the plasma current reached $7\,\text{MA}$ which is approximately one half the designed current value ($15\,\text{MA}$) of the ITER device currently under construction.

The JET machine was designed with a poloidal divertor and the divertor configuration has gone through numerous changes since 1980. The principle confinement modes of plasma confinement states, the L-mode, the H-mode, the ELMY-H-mode and the reversed magnetic field have been achieved in JET.

The JET tokamak with $R/a = 2.88\,\text{m}/0.9\,\text{m}$ with $B/I_p = 4\,\text{T}/4.5\,\text{MA}$ produced record fusion power plasma performance discharges in 1995-1999 with what has become a standard regime of the ITER design called the H-mode for the high-confinement time τ_E regime. During the 1997 campaign with DT experiments in JET, the current record results were obtained in the H-mode regime with an edge transport barrier or an edge pedestal [Cordey, *et al.* (1999); Gibson (1998)]. A new DT campaign for JET is being planned at the time of writing.

1.3   Plasma Heaters NBI, ICRH, LHCD and ECRH

Chapters 5, 6, and 7 describe the regimes of plasma heating and the types of confinement modes obtained in tokamaks. For the large ITER plasma volume the high-powered radio frequency RF heating will play a dominant role. While TFTR, JT-60U, and JET have RF heating, they relied heavily on the neutral-beam injection (NBI) heating system. This NBI method is less effective in the large volume of the ITER, which would require that the beam injectors have considerably higher energy. Higher energy beams are less efficient and more costly.

Electron cyclotron heating (ECH) is a universally-applicable method of heating the electrons in plasmas and controlling the electron density and electron temperature profiles. The intrinsic tokamak toroidal current, called the bootstrap current, is driven by the electron density and temperature radial gradients when the temperature is sufficiently high, that a significant fraction of the electrons are trapped by the mirror ratio $R_M = B_{\max}/B_{\min}$ from the intrinsic variation of the strength of the magnetic field at the particle orbit as it travels along the magnetic field line. This bootstrap current arises from the variation of the magnetic field experience by these electrons and ions as they travel along the magnetic field lines in the local radial gradient of the electron density. The bootstrap (BS) current density J_{BS} is localized to a radial region $[\rho_1, \rho_2]$ where there is significant trapping of the electrons. At higher collisionality (lower temperatures) there remains an electric current parallel to the magnetic field lines, called the Pfirsch-Schlüter current in tokamak physics and the field-aligned current in space physics. This parallel current is intrinsic to all confined plasma and electrical conductors in a steady state where the electric currents must form closed loops. (The proof is by contradiction. If the currents were not closed, then charges would accumulate, producing sufficiently strong electric fields to force the charges to move to eliminate the unbalanced charge distributions.) Optimizing the self-generated bootstrap current is a key element in the design of the steady state fusion reactor. Along with radio frequency (RF) waves launched so as to "push" the electrons forward, the bootstrap current and RF driven current is the method designed to run a steady-state fusion tokamak reactor.

A recent example of toroidal plasma experiments with high-power electron cyclotron resonance heating (ECH) is shown in Fig. 1.5. Here the parallel bootstrap current has been verified to confirm to theoretical expectations. The bootstrap current theory was developed by Bickerton, *et al.* (1971) and is explained in detail in Friedberg (2012). The TCV tokamak experiment shown in Fig. 1.5 validates the prediction from the computer simulation codes for the current driven by the ECH heating as J_{CD} and the bootstrap current J_{BS} driven by the radial gradients of the electron density and electron temperature profiles.

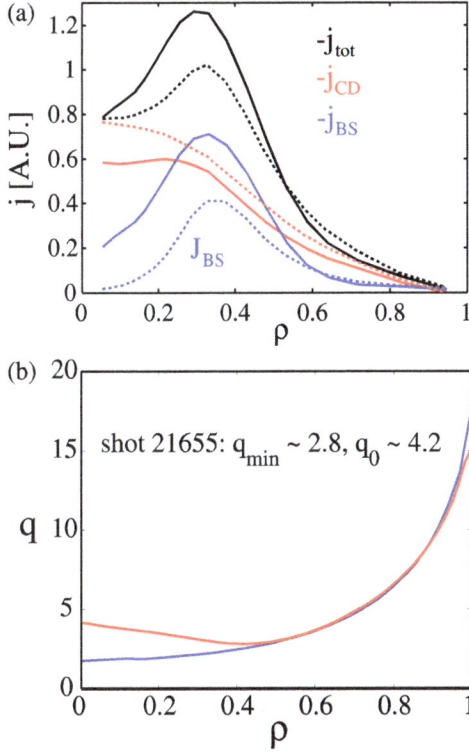

Fig. 1.5 Composition of the toroidal plasma current density and its comparison with simulation data. (a) CQL3D calculated j_{CD}, j_{BS} calculated from the Thomson scattering n_e, T_e profile measurements, and j_{tot} before (——) and after (——) additional central counter-ECCD. (b) The monotonic q-profile calculated by the usual TCV equilibrium reconstruction code LIUQE, and the CHEASE equilibrium code using j_{tot} and the Thomson measurements [Goodman, *et al.* (2003)].

In the example shown in Fig. 1.5, the bootstrap BS current is localized between $\rho = 0.2$ and $\rho = 0.6$, owing to the fact that the mirror ratio R_M that is responsible for electron trapping vanishes as $\rho \to 0$. The fraction of trapped particles vanishes for large ρ owing to the opening of the magnetic field lines. The electron temperature at large ρ, gives the condition of electron orbits be collisionally scattered by ions out of their orbits before making completed mirror orbit. The measure of the number of bounces (mirror reflections) that an electron makes before being collisionally scattered to a different orbit is function of the mirror ratio $R_M = (B_{max} - B_{min})/(B_{max} + B_{min})$ which is approximately given by $2r/R$ in the tokamak. This means that the effective collisionality for the trapped electron is expressed in terms of the number of collisions with the parameter ν_{*e} parameter given by

$$\nu_{*e}(r) = \text{const} \left[\frac{\left(\dfrac{R}{r}\right)^{(3/2)} n_e(r)}{T_e^2(r)} \right] \ln\left[\frac{b_{\max}}{b_{\min}}\right] \qquad (1.2)$$

for a plasma without impurity ions. In the presence of high-ion impurities the collisionality of the plasma is increased and the deuterium-tritium fuel mixture is decreased. In Eq. (1.2) the $b_{\max}$ and $b_{\min}$ are the maximum and minimum impact parameters for the Coulomb scattering of the electrons with the ions [Jackson (1999)].

Calculations for the bootstrap current are given in Sugama, *et al.* (2007) and Helander and Sigmar (2002) and are complicated. Steady-state operation of the tokamak will rely on the parallel plasma currents generated by the bootstrap current with a boost and control feature added by the use of high-powered directional radio frequency waves launched by carefully-designed RF antenna mounted on the chamber walls. This design and how to operate the RF antennas for the lower-hybrid current drive (LHCD) is described in Chapter 7. The current drive uses the absorption of the photon's momentum by the electrons to give a controllable accelerating force on the electrons to maintain the 10-15 MA of plasma toroidal current in the ITER machine and in future DEMO tokamak reactors. The method and theory are described in Chapter 7.

References

Aymar, R., ITER Joint Central Team and ITER Home Teams. (1998). Present status and future prospect of the ITER project, *J. Nucl. Materials* **258-263**, Part 1, pp. 56-64.

Bateman, G., Kritz, A. H., Kinsey, J. E., and Redd, A. J. (1998). Multi-Mode transport modeling of the International Thermonuclear Experimental Reactor (ITER), *Phys. Plasmas* **5**, p. 2355, http://dx.doi.org/10.1063/1.872909.

Bell, M. G., and the TFTR Team. (1989). *Plasma Phys. Control. Nucl. Fusion Res., 1988* (International Atomic Energy Agency, Vienna), Vol. 1, pp. 27-40.

Bickerton, R. J., Connor, J. W., and Taylor, J. B. (1971). Diffusion Driven Plasma Currents and Bootstrap Tokamak, *Nature Phys. Sci.* **229**, 4, pp. 110-112, doi:10.1038/physci229110a0.

Bucalossi, J., Argouarch, A., Basiuk, V., Baulaigue, O., Bayetti, P., Bécoulet, M., Bertrand, B., Brémond, S., Cara, P., Chantant, M., Corre, Y., Courtois, X., Doceul, L., Ekedahl, A., Faisse, F., Firdaouss, M., Garcia, J., Gargiulo, L., Gil, C., Grisolia, C., Gunn, J., Hacquin, S. Hertout, P., Huysmans, G., Imbeaux, F., Jiolat, G., Joanny, M., Jourd'heuil, L., Jouve, M., Kukushkin, A., Lipa, M., Lisgo, S., Loarer, T., Maget, P., Magne, R., Marandet, Y., Martinez, A., Mazon, D., Meyer, O., Missirlian, M., Monier-Garbet, P., Moreau, P., Nardon, E., Panayotis, S., Pégourié, B., Pitts, R. A., Portafaix, C., Richou, M., Sabot, R., Saille, A., Saint-Laurent, F., Samaille, F., Simonin, A., and Tsitrone, E. (2011). Feasibility study of an actively cooled tungsten divertor in Tore Supra for ITER technology testing, *Fusion Eng. Design* **86**, pp. 684-688.

Campbell, D. J. (2001). The physics of the International Thermonuclear Experimental Reactor FEAT, *Phys. Plasmas* **8**, p. 2041, http://dx.doi.org/10.1063/1.1348334.

Cordey, J. G., Balet, B., Bartlett, D. V., Budnya, R. V., Christiansen, J. P., Conway, G. D., Eriksson, L.-G., Fishpool, G. M., Gowers, C. W., de Haasb, J. C. M., Harbour, P. J., Horton, L. D., Howman, A. C., Jacquinot, J., Kerner, W., Lowry, C. G., Monk, R. D., Nielsen, P., Righi, E., Rimini, F. G., Saibene, G., Sartori, R., Schunke, B., Sips, A. C. C., Smith, R. J., Stamp, M. F., Startc, D. F. H., Thomsen, K., Tubbing, B. J. D., and von Hellermann, M. G. (1999). Plasma confinement in JET H-mode plasmas with D, DT and T isotopes, *Nucl. Fusion* **39**, p. 301, http://iopscience.iop.org/0029-5515/39/3/301.

Doyle, E. J., Houlber, W. A., Kamada, Y., Mukhovatov, V., Osborne, Polevoi, A., Bateman, G., Connor, J. W. Cordey, J. G., Fujita, T., Garbet, X., Hahm, T. S., Horton, L. D., Hubbard, A. E., Imbeaux, F., Jenko, F., Kinsey, J. E., Kishimoto, Y., Li, J., Luce, T. C., Martin, Y., Ossipenko, M., Parail, V., Peeters, A., Rhodes, T. L., Rice, J. E., Roach, C. M., Rozhansky, V., Ryter, F., Saibene, G., Sartori, R., Sips, A.

C. C., Snipes, J. A., Sugihara, M., Synakowski, E. J., Takenaga, H., Takizuka, T., Thomsen, K., Wade, M. R., Wilson, H. R., ITPA Transport Physics Topical Group, ITPA Confinement Database and Modeling Topical Grou, and ITPA Pedestal and Edge Topical Group. (2007). Chapter 2: Plasma confinement and transport, *Nucl. Fusion* **47**, S18-S127, doi:10.1088/0029-5515/47/6/S02.

Fowler, T. K. (1997). *The Fusion Quest*, The Johns Hopkins University Press.

Friedberg, R. (2007). *Plasma Physics and Fusion Energy* (Cambridge University Press) ISBN:13-978-0-521-85107-7.

Fujita, T. (2001). Plasma Equilibrium and Confinement in a Tokamak with Nearly Zero Central Current Density in JT-60U, *Phys. Rev. Lett.* **87**, 245001 and 08500.1, `http://dx.doi.org/10.1103/PhysRevLett.87.245001`.

Fujita, T., Kamada, Y., Ishida, S., Neyatani, Y., Oikawa, T., Ide, S., Takeji, S., Koide, Y., Isayama, A., Fukuda, T., Hatae, T., Ishii, Y., Ozeki, T., Shirai, H., and JT-60 Team. (1999). High-performance experiments in JT-60U reversed shear discharges, *Nucl. Fusion* **39**, p. 1627, `http://iopscience.iop.org/0029-5515/39/11Y/302`.

Fujita, T., Hatae, T., Oikawa, T., Takeji, S., Shirai, H., Koide, Y., Ishida, S., Ide, S., Ishii, Y., Ozeki, T., Higashijima, S., Yoshino, R., Kamada, Y. and Neyatani, Y. (1998). High-performance reversed-shear plasmas with a large radius transport barrier in JT-60U, *Nucl. Fusion* **38**, p. 207, doi:10.1088/0029-5515/38/2/305.

Gibson, A., and JET Team (1998). Deuterium-tritium plasmas in the Joint European Torus (JET): behavior and implications, *Phys. Plasmas* **5**, p. 1839, doi:10.1063/1.872854, `http://dx.doi.org/10.1063/1.872854`.

Goodman, T. P., Ahmed, S. M., Alberti, S., Andr'ebe, Y., Angioni, C., Appert, K., Arnoux, G., Behn, R., Blanchard, P., Bosshard, P., Camenen, Y., Chavan, R., Coda, S., Condrea, I., Degeling, A., Duval, B. P., Etienne, P., Fasel, D., Fasoli, A., Favez, J.-Y., Furno, I., Henderson, M., Hofmann, F., Hogge, J.-P., Horacek, J., Isoz, P., Joye, B., Karpushov, A., Klimanov, I., Lavanchy, P., Lister, J. B., Llobet, X., Magnin, J.-C., Manini, A., Marlétaz, B., Marmillod, P., Martin, Y., Martynov, An., Mayor, J.-M., Mlynar, J., Moret, J.-M., Nelson-Melby, E., Nikkola, P., Paris, P. J., Perez, A., Peysson, Y., Pitts, R. A., Pochelon, A., Porte, L., Raju, D., Reimerdes, H., Sauter, O., Scarabosio, A., Scavino, E., Seo, S. H., Siravo, U., Sushkov, A., Tonetti, G., Tran, M. Q., Weisen, H., Wischmeier, M., Zabolotsky, A., and Zhuang, G. (2003). An overview of results from the TCV tokamak, *Nucl. Fusion* **43**, pp. 1619-1631, doi:10.1088/0029-5515/43/12/008.

Hawryluk, R. J. (1998). Results from deuterium-tritium tokamak confinement experiments, *Rev. Mod. Phys.* **70**, 2, pp. 537-587, doi:10.1103/RevModPhys.70.537.

Helander, P., and Sigmar, D. J. (2002). *Collisional Transport in Magnetized Plasmas* (Cambridge Monographs on Plasma Physics).

Horton, W., and Ichikawa, Y. (1996). *Chaos and Structures in Nonlinear Plasmas* (World Scientific) ISBN:81-7764-234-0.

Horton, W. (2012). Turbulent Transport in Magnetized Plasmas (World Scientific) ISBN:978-981-4383-53-0.

Ishida, S., Fujita, T., and the JT-60 Team (1997). Achievement of high-fusion performance in JT-60U reversed shear discharges, *Phys. Rev. Lett.* **79**, pp. 3917-3921, `http://link.aps.org/doi/10.1103/PhysRevLett.79.3917`.

Jackson, J. D. (1999). *Classical Electrodynamics* (Wiley, New York) 3rd Ed., pp. 708-724.

Jacquinot, J., Bhatnagar, V. P., and Gormezano, C. (1993). JET recent results on wave heating and current drive consequences for future devices, *Plasma Phys. Control. Fusion* **35**, A35, doi:10.1088/0741-3335/35/SA/003.

JET Team (1992). Fusion energy production from a deuterium-tritium plasma in the JET tokamak, *Nucl. Fusion* **32**, p. 187, doi:10.1088/0029-5515/32/2/I01.

Kamada, Y. (2001). *Plasma Phys. Control. Fusion* **42**, p. A247.

Kamada, Y., *et al.* (1999). *Nucl. Fusion* **39**, p. 18415.

Kaye, S. M., Greenwald, M., Stroth, U., Kardaun, O., Kus, A., Schissel, D., DeBoo, J., Bracco, G., Thomsen, K., Cordey, J., Miura, Y., Matsuda, T., Tamai, H., Takizuda, T., Hirayama, T., Kikuchi, H., Naito, O., Chudnovskij, A., Ongena, J., and Hoang, G. T. (1997). ITER L-mode confinement database, *Nucl. Fusion* **37**, pp. 1303-1328, doi:10.1088/0029-5515/37/9/I10.

Keilhacker, M., Gibson, A., Gormezano, C., Lomas, P. J., Thomas, P. R., Watkins, M. L., Andrew, P., Balet, B., Borba, D., Challis, C. D., Coffey, I., Cottrell, G. A., De Esch, H. P. L., Deliyanakis, N., Fasoli, A., Gowers, C. W., Guo, H. Y., Huysmans, G. T. A., Jones, T. T. C., Kerner, W., König, R. W. T., Loughlin, M. J., Maas, A., Marcus, F. B., Nave, M. F. F., Rimini, F. G., Sadler, G. J., Sharapov, S. E., Sips, G., Smeulders, P., Söldner, F. X., Taroni, A., Tubbing, B. J. D., von Hellermann, M. G., Ward, D. J., and JET Team. (1999). High-fusion performance from deuterium-tritium plasmas in JET, *Nucl. Fusion* **39** p. 209, doi:10.1088/0029-5515/39/2/306.

Koide, Y., Mori, M., Fujita, T., Shirai, H., Hatae, T., Takizuka, T., Kimura, H., Oikawa, T., Isei, N., Isayama, A., Takeji, S., Kawano, Y., Sakasai, A., Kamada, Y., Fukuda, T., and Ishida, S. (1998). Study of internal transport barriers by comparison of reversed shear and high discharges in JT-60U, *Plasma Phys. Control. Fusion* **40** p. 641, doi:10.1088/0741-3335/40/5/014.

Koide, Y., and JT-60 Team. (1997). *Progress in confinement and stability with plasma shape and profile control for steady-state operation in the Japan Atomic Energy Research Institute Tokamak-60 Upgrade, Phys. Plasmas* **4**, p. 1623, http://link.aip.org/link/?PHPAEN/4/1623/1.

L-H Mode Database Working Group. (1994). *Nucl. Fusion* **34**, pp. 131-167.

McCray, Patrick W. (2010). Globalization with hardware: ITER's Fusion of Technology, Policy, and Politics, *History and Technology* **26**, 4, pp. 283-312, http://dx.doi.org/10.1080/07341512.2010.523171.

Mealer, M. (2013). Y'all smell that? That's the smell of money, *Texas Monthly*.

Nazikian, R., *et al.* (1997). *Fusion Energy*, International Atomic Energy Agency, p. 281.

Shimada, M., Campbell, Mukhovatov, V., Fujiwara, M., Kirneva, N., Lackner, K., Nagami, M., Pustovitov, V. D., Uckan, N., Wesley, J., Asakura, N., Costley, A. E., Donné, A. J. H., Doyle, E. J., Fasoli, A., Gormezano, C., Gribov, Y., Gruber, O., Hender, T. C., Houlberg, W., Ide, S., Kamada, Y., Leonard, A., Lipschultz, B., Loarte, A., Miyamoto, K., Mukhovatov, V., Osborne, T. H., Polevoi, A., and Sips, A. C. C. (2007). Chapter 1: Overview and summary, *Nucl. Fusion* **47**, p. S1, doi:10.1088/0029-5515/47/6/S01.

Stacey, W. M. (2010). *The Quest for a Fusion Energy Reactor* (Oxford University Press), ISBN:978-0-19-973384-2.

Sugama, H., Watanabe, T.-H., and Horton, W. (2007). Collisionless kinetic-fluid model of zonal flows in toroidal plasmas, *Phys. Plasmas* **14**, p. 022502, doi:10.1063/1.2435329, http://link.aip.org/link/?PHPAEN/14/022502/1.

Troyon, F., Gruber, R., Saurenmann, H., Semenzato, S., and Succi, S. (1984). MHD limits to plasma confinement, *Plasma Phys. Control. Fusion* **26**, pp. 209-215, doi:10.1088/0741-3335/26/1A/319.

Yamada, H., Harris, J. H., Dinklage, A., Ascasibar, E., Sano, F., Okamura, S., Talmadge, J., Stroth, U., Kus, A., Murakami, S., Yokoyama, M., Beidler, C. D., Tribaldos, V., Watanabe, K. Y., and Suzuki, Y. (2005). Characterization of energy confinement in net-current free plasmas using the extended International Stellarator Database, *Nucl. Fusion* **45**, 12, pp. 1684-1693, doi:1088/0029-5515/45/12/024.

Chapter 2

Magnetohydrodynamic Description of the Equilibrium and Heating of the Thermal Plasma

Tokamak plasma confinement experiments in 1982 [Wagner, *et al.* (1982)] in the newly designed ASDEX upgrade machine showed that higher quality plasmas are produced by changing from the circular limiter used in previous tokamaks to a poloidal magnetic field configuration with an X-point. The X-point arises from magnetic separatrix in the magnetic surfaces creating an open set of outer magnetic field lines ending (terminating) in a special divertor chamber producing higher quality plasmas. One may think of the analogy of the separatrix in the motion of the pendulum that divides, or separates, phase space between oscillatory and rotating motions. This new magnetic configuration with a magnetic separatrix between the hot core plasma and the outer lower-temperature plasma, in contact with the chamber walls, has now become the standard configuration for fusion tokamaks. For example, in the Alcator upgrade called Alcator C-Mod the builders replaced the earlier circular field-line configuration with the new poloidal field lines that contained two null or X-points with variable control of their positions with respect to the vacuum vessel. The new Alcator C-Mod experiments confirmed the advantage of the variable poloidal magnetic field configuration, showing that the plasma purity was significantly increased.

2.1   Equilibrium with Single Lower-Null X-point Divertor

Earlier plasmas with circular limiters, wall and diaphragms, suffered from an influx of impurity ions such as carbon from the plasma-limiter and plasma-wall interactions. For example, hot plasma ions hitting a carbon-coated metallic wall produce a steady release of carbon ions and a few metal ions from the first wall of the toroidal chamber. In Alcator highly-charged molybdenum ions were found from the molybdenum walls. The walls were then coated with boron, and while this reduced the level of the molybdenum ions, there was a significant level of boron ions in the core plasma. Then, owing to the successful results from the ASDEX divertor machine, the poloidal field coil in Alcator was redesigned to give the configuration shown in Fig. 2.1 [Greenwald, *et al.* (1988)]. There was an immediate lowering of the effective charge $Z_{\text{eff}} = \Sigma n_i Z_i^2 / n_e$, which is the standard measure of the impurity level in a plasma. This impurity measure arises from the formula for the electrical resistivity of plasma containing a mixture of ions with densities n_i and charge $Z_i e$. A comprehensive overview of the confinement and turbulence in plasma physics from the Alcator C-Mod is now available in the *20 Years of Research on the Alcator C-Mod Tokamak* [Greenwald, *et al.* (2014)].

In both the circular and the divertor configurations, data showed that there was a maximum plasma density n_{max} that the machine could achieve for steady-state operation. This density limit, while poorly understood, appeared to be associated with the level of impurities in the plasma and the associated radiative cooling of the plasma edge as the core plasma density increased. The empirical formula for the core plasma density limit was derived in Alcator and applies today to all tokamaks. The density limit, named for the work of the Greenwald team on the Alcator tokamak on exploring the combination of plasma physics and atomic physics associated with impurity ions, is known as the Greenwald density limit.

An early search for the mechanism for the Greenwald density limit that included (1) low-Z impurity density in the edge plasma, (2) low- and high-Z impurity effects in the core plasma and (3) insufficient fueling rates as the cause of the density limit and its scaling with plasma current I_p and machine minor radius a using data from the mid-1980s from ORMAK, DITE and the circular limiter Alcator machine is described in an article titled *A New Look at the Density Limits in Tokamaks* [Greenwald, *et al.* (1988)]. The study was inconclusive as to the physics behind the density limit, but put forth the empirical scaling law for the density limit widely verified and thus used thereafter in the design of fusion reactors. These rules are known as Greenwald density limit and the Hugill limit for confinement and disruptions in the parameter plane of plasma toroidal current I_p versus the normalized electron density $n_e R / B_T$. These limits for the maximum electron density were used in the analysis and design teams for INTOR and ITER.

Figure 2.2 shows a set of simple, classical profiles from the early predictions for the Alcator C-Mod plasma as taken from Greenwald in Fusion Plasmas. The experiments showed a wider range of plasma states which now are known by their acronyms such as L-mode, H-mode, I-mode, and states with edge transport barriers and internal transport barriers. The word "barrier" is to be understood as a region of reduced transport rather than a barrier. These confinement regimes are explained in Chapters 5, 6 and 7.

Experiments with the configuration called the single lower-null divertor configuration proved to be the optimal configuration. So in addition to Alcator C-Mod, the JT-60 machine was converted from its original outside wall divertor at the R_{max} to a new vacuum chamber and poloidal field configuration with a lower single-null divertor configuration. The new lower-divertor configuration is called JT-60U. This configuration of a single null has proven to have better confinement properties than a symmetric double X-point with upper and lower null points. The power required to make the transition from the L-mode to the H-mode confinement is lower in the single-null geometry. It is in the single null-point geometry that gives the lowest

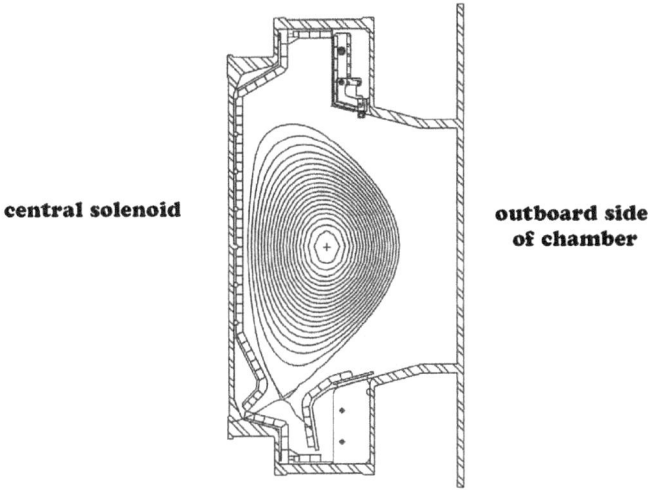

Fig. 2.1 The single lower null divertor configuration for a tokamak with separatrix between open and closed magnetic surfaces. The example is taken from the Alcator C-Mod report of Greenwald, *et al.* (1988). The inner chamber wall on the left butts up against an air core central solenoid that forms the transformer primary for driving the magnetic flux $\Psi(t)$ in the plasma chamber. The maximum transformer flux Ψ_{sol} defines the maximum time interval for the Ohmic driving of the toroidal plasma current $I_p(t)$.

power threshold P_H for the transition from the standard tokamak confinement called the L-mode to the H-mode regime with roughly twice the energy confinement time τ_E.

Clearly, from Fig. 2.1, the single-null geometry breaks the symmetry of the magnetic surfaces. So, in this configuration the reversal of the toroidal magnetic field direction B_T changes the plasma completely. Thus there are two types of plasmas in the single null divertor configuration. One has the grad-B and curvature drifts of ions toward the X-point and one has the grad-B and curvature drift of the ions away from the X-point, that is to the top of the chamber for the lower null point shown in Fig. 2.1. In Fig. 2.1 the central solenoid is not shown but is on the left side of the figure where it consists of an air core coil of windings that produces the primary magnetic flux that is reversed in time to drive the toroidal electric field in the plasma. The strength of the central solenoid is specified equivalently in the units of volt-sec or Webers.

The system is basically a single turn electric transformer with the plasma current being the secondary winding of the transformer. In ITER this magnetic flux of the central solenoid is approximate $\Phi_{sol} = 300$ V-s, which then limits the time of the high toroidal current phase where $I_{tor} = 15$ MA to about 500 s. For reference, a 100 V-s transformer may provide, by definition, a 1 V toroidal electric voltage for 100 s. The toroidal voltage, and thus the toroidal electric filed required to drive the 15 MA current in plasma with a typical Ohmic plasma resistivity is typically less

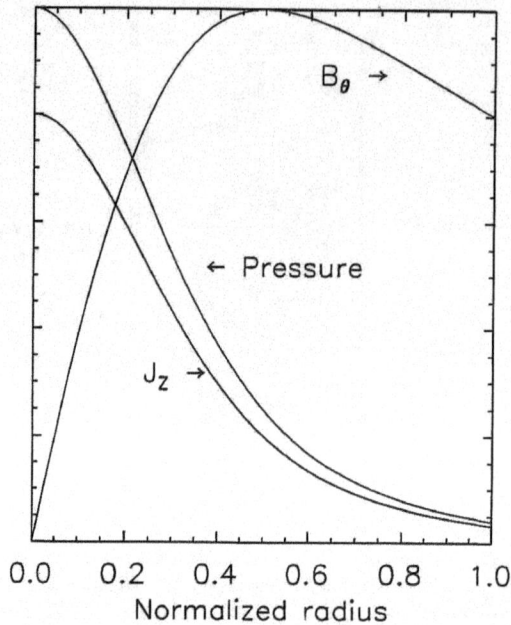

Fig. 2.2 Typical profiles for Alcator C-Mod from for the toroidal current density j_ϕ and the poloidal magnetic field component drive by the plasma current. This magnetic field combines with the vacuum components shown in Fig. 2.1 to give the total poloidal magnetic field in the plasma. The vacuum poloidal magnetic field is created and controlled by the external currents in a large diameter superconducting toroidal coils outside the plasma chamber.

than 100 mV. For a plasma in this regime then the transformer flux would drive the current for 1000 s. In reality a larger toroidal electric field is required in the early phases of the formation and heating of the plasma. The result is the prediction that the central solenoid should be able to drive the 15 MA of current for order 500 s in ITER. Recent experiments on FTU and DIII-D show that by using directed electron cyclotron RF beams for current drive and early heating of the plasma one can conserve, or reduce, the volt-seconds of the transformer required to achieve the final high temperature-high plasma current I_p states. The top or bottom placement of the divertor cavity is not important for the plasma, but, for the architecture of the building placing the plasma exhaust chamber and the vacuum pumps for the plasma at the bottom of the machine, as under the ground floor of the machine in a large basement, is simpler. The basement contains the electric power leads, neutral beam lines and a host of supporting equipment.

Now, the experiments in the single lower-null configuration show that the threshold for the H-mode defined by the net heating power threshold P_H is lower when the toroidal magnetic field direction is such that the ion guiding-center drifts in the vacuum field are toward the divertor. The reason for this is because the ion guiding-center orbits are faster in the weaker poloidal magnetic field region, and thus a stronger radial electric field builds up in this configuration. More precisely, a normal inward-pointing electric field $\boldsymbol{E}_\psi$ is enhanced owing to the ion orbit losses in the weak poloidal field region. This configuration enhances the inward-pointing electric field, which produces better plasma confinement. Numerous reports from the Alcator C-Mod experiments explored the different confinement times of separatrix configurations with upper, lower and balanced X-points and showed that the configuration with the single null with the toroidal field direction so as to have the ion guiding-center drifts in the direction $\boldsymbol{B} \times \nabla \boldsymbol{B}$ toward the separatrix produces the best performance. This X-point geometry is now the standard configuration used in the large machines of JET [Keilhacker, *et al.* (1999)], JT-60U, KSTAR and EAST.

A simple explanation for the better confinement follows from noting the electrons easily follow the ions. Thus, a configuration with inward-pointing electric field $\boldsymbol{E}_r$ is optimal since this pushes the ions in and the electrons rapidly follow the ions.

The C-Mod facility was closed in 2014 owing to the US budget freeze stemming from the trillion dollar national deficit. The tokamak experiments at Alcator and Alcator C lead the way in understanding the Ohmic heating regimes, the Linear Ohmic regime called LOC, and the saturated Ohmic regime called (SOC) and then in the upgrade to Alcator C-Mod with single-null divertor configuration to understanding of the spontaneous onset of toroidal rotation of the plasma, owing to the broken symmetry in that geometry. The power threshold for the onset of spontaneous toroidal rotation is the subject of Chapters 6 and 7. With the broken symmetry by the single null position, it is evident that toroidal rotation has a preferred direction. The studies of plasma rotation without neutral beams injection are particularly important for the future steady state fusion reactors.

For ITER, the toroidal superconducting coils that create the vacuum poloidal magnetic with a single lower null divertor configuration, as shown in Fig. 2.1 for the Alcator C-Mod machine, are from 6 m to 12 m in diameter. These giant coils are wound on location at the ITER laboratory located on a plateau north of Aix-en Provence. The ITER laboratory is adjacent to the CEA laboratory with classified nuclear physics research where fast fission reactors are built and tested. The ITER laboratory is an open international laboratory separated from the high-security CEA Cadarache laboratories (https://www.euro-fusion.org/iter-2/).

2.2   The Single-Fluid Magnetohydrodynamic Model

To obtain the MHD equations from the kinetic equations, or even the two-component finite Larmor radius (FLR) fluid equations, requires a long list of approximations. For those interested in such an effort, see the discussions in Krall and Trivelpiece (1968) and Braginskii (1965). Recognizing that the MHD description is an idealized model, we write down the equations and then discuss some of the key physical processes missing from this description with respect to the subject of drift waves and tearing mode instabilities.

In simple terms, the MHD model describes the plasma as a single fluid with velocity $\mathbf{v}(\mathbf{x},t)$ which recognizes the dominance of the $\mathbf{j}(\mathbf{x},t) \times \mathbf{B}(\mathbf{x},t)$ force in accelerating the highly-conducting fluid. One recognizes that the electrons with a mass of $1/1836$ to $1/10^5$ of the ion mass, depending on the working gas atomic number, are like "mosquitos" or "flies" strongly attracted to the ions and following their motion owing to the very strong Coulomb force between the ions and electrons. In this way the fluid, described by mass density $\rho(\mathbf{x},t)$, is charge neutral and the electric field in the laboratory frame $\mathbf{E}(\mathbf{x},t)$ is such as to vanish in the frame of reference moving with fluid velocity $\mathbf{v}(\mathbf{x},t)$. That gives the electrical dynamics $\mathbf{E}(\mathbf{x},t) + \mathbf{v}(\mathbf{x},t) \times \mathbf{B}(\mathbf{x},t) = 0$. In the case of low-frequency motions, one must add to this constraint the small electric field contribution from the plasma current $\mathbf{j}(\mathbf{x},t)$ flowing in the slightly resistive plasma from the electrical resistivity $\eta(\mathbf{x},t)$ times the current density vector.

The conservation laws for a single ideal, perfectly conducting, dissipationless fluid or gas with mass density ρ and center of mass flow velocity $\boldsymbol{u}$ are

$$\frac{\partial \rho}{\partial t} + \nabla \cdot (\rho \boldsymbol{u}) = 0 \tag{2.1}$$

$$\rho \left(\frac{\partial \boldsymbol{u}}{\partial t} + \boldsymbol{u} \cdot \nabla \boldsymbol{u} \right) = \boldsymbol{j} \times \boldsymbol{B} - \nabla p + \rho \boldsymbol{g} \tag{2.2}$$

$$\frac{\partial \boldsymbol{B}}{\partial t} = \nabla \times (\boldsymbol{u} \times \boldsymbol{B}) \tag{2.3}$$

$$\frac{\partial p}{\partial t} + \boldsymbol{u} \cdot \nabla p + \Gamma p \nabla \cdot \boldsymbol{u} = 0 \tag{2.4}$$

where the current density follows from $\mu_0 \boldsymbol{j} = \nabla \times \boldsymbol{B}$ and the electric field is determined by idealization of a plasma gas with infinite electrical conductivity

$$\boldsymbol{E} + \boldsymbol{u} \times \boldsymbol{B} = 0. \tag{2.5}$$

Equations (2.1) and (2.4) are combined to show that the specific entropy of the plasma gas $S = p/\rho^\Gamma$ is constant in the frame $\boldsymbol{u}$ of the moving plasma. The equation of state (EOS) of the ionized gas is $S = $ const. in the rest frame of the gas. Combining this statement with the mass conservation law, one defines the entropy density, $s(\boldsymbol{x},t) = nk_B \ln(p/\rho^\Gamma)$, is a conserved density satisfying

$$\frac{ds}{dt} = \frac{\partial s}{\partial t} + \nabla \cdot (\boldsymbol{u}s) = 0. \tag{2.6}$$

Here n is the number density, k_B the Boltzmann constant and Γ is the adiabatic gas constant. The MHD system does not describe the thermal transport since the adiabatic gas assumption is used to truncate the chain of coupled fluid equations. This approximation limits the validity of the model to motions that are sufficiently fast and violent that the thermal transport is not significant during that motion. For shocks and fast interchange instabilities this adiabatic model is sufficient.

Equation (2.5) is the condition that the electric field E' vanishes in the local rest frame of the moving plasma. Equation (2.6) is the statement that in a fluid element following the fluid flow, there is no change in the entropy. These two equations are the strong statements that fail in most of the plasma dynamics studied in this book. They become reasonable approximations only for very fast, and thus strongly out-of-force balance plasma dynamics. The fusion reactor plasma must be slowly evolving through a sequence of force-balanced states.

Equation (2.5) then produces the motional EMF in the fixed laboratory frame that is the inertial frame in which the velocity u is measured. For example, in the frame attached to Earth, the solar wind plasma has the velocity u_{sw} that is directed radially outward from the Sun with speeds that range from 200 km/s to 700 km/s. This plasma contains the interplanetary magnetic field B_{IMF} and then combines in the magnetosheath surrounding the Earth's magnetosphere where there is an additional ΔB component. In the Earth's frame of reference this solar wind flowing with velocity u_{sw} across the ambient magnetic field $B(x, t)$ drives the electric field

$$E_{sw} = -u_{sw} \times B. \qquad (2.7)$$

All components of the fields in Eq. (2.7) are measured by spacecraft instruments validating the model. The electric field maps down the magnetic field lines, since $B \cdot E = 0$, to impose the electric field generated by the plasma flow velocity at the magnetopause and into the inner magnetosphere and on to the ionosphere at the polar caps. For this discussion, the polar cap is defined as the region where a field line has one end attached to Earth's magnetic field and the other end goes to the interplanetary magnetic field B_{IMF}. (Note the similarity to the field lines outside the X-point that terminate on the conductive divertor chamber plates). These solar wind-generated potentials range from 40 kV in quiet times to hundreds of kilovolts during magnetic storms. The electric fields are measured by spacecraft surrounding the Earth show good agreement with Eq. (2.7).

In the laboratory plasma the most important example of Eq. (2.5) is of the radial electric field $E_r(r, t)$ that is produced by the high-speed toroidal rotational velocities (10 to 100 km/s) driven by the neutral beam injection (NBI) in some tokamak plasmas. The driven toroidal plasma flow velocity u_ϕ, creates a rotation-driven radial electric field given by

$$E_r^{(0)} = u_\phi B_\theta. \qquad (2.8)$$

The total radial electric field E_r with pressure-gradient driven contributions is known to be very important in influencing the plasma confinement. In the absence of toroidal rotation, the radial electric field in the core plasma points inward, as measured with the Heavy Ion Beam (HIB) diagnostic system [Hallock, *et al.* (1994)].

In using Eq. (2.8) one must be careful to use the same reference frame for $\boldsymbol{E}, \boldsymbol{B}$ and $\boldsymbol{u}_\phi$. A toroidal plasma velocity of $u_\phi = 50\,\text{km/s}$ across the magnetic field $B_\theta = 0.2\,\text{T}$ gives an electric field of $E_r = 10\,\text{kV/m}$ in the lab frame. Often there is also a significant contribution from $-u_\theta B_\phi$, which adds or subtracts from the field created by the toroidal rotation. With respect to the laboratory reference frame, typical values in Eq. (2.8) are $E_r = (50\,\text{km/s})\,(0.2\,\text{T}) = 10\,\text{kV/m}$.

The laboratory measurements of E_r are complex and indirect. Thus, the errors in the data for the radial profile of E_r are rather large. The zero superscript indicates the dominant component of E_r for large toroidal flows $\boldsymbol{u} = \boldsymbol{u}_\phi = R\Omega\hat{e}_\phi$. However, what is known is that, even at the highest velocities, the MHD approximation $E_r = E_r^{(0)}$ given by Eq. (2.8) is incomplete in rotating tokamaks. There are carefully-documented measurements that show that E_r is controlled not only by u_ϕ but by the forces driven by the ion pressure $\nabla p_i / e n_i$ and the impurity pressure gradients as well as the parallel force $F_{\|ab}$ between the working gas ions and the impurity ions. Grierson, *et al.* (2010) develop the measurements of E_r and correlate the profiles of E_r with the predictions of neoclassical transport theory [Hirshman and Sigmar (1977)]. The adjective "neoclassical" means the collisional transport theory taking into account the complex and fast guiding center drifts produced by the toroidal magnetic field geometry.

2.3   Grad-Shafranov MHD Equilibrium for Axisymmetric Systems

Here we review the basis for the axisymmetric plasma equilibrium in toroidal systems. Plasma equilibria are established by transport processes and the sources of particles and power from the external systems. The plasma pressure responds in microseconds from the large $\boldsymbol{j} \times \boldsymbol{B}$ forces driving the pressure profiles to a nearby equilibrium with $\boldsymbol{F} = \boldsymbol{j} \times \boldsymbol{B} - \nabla p = 0$. Thus, early in the history of fusion research H. Grad at NYU, and independently, V. D. Shafranov at Kurchatov in Moscow, developed the basic method for calculating the equilibrium distributions of the pressure and toroidal plasma currents for given boundary constraints assuming the system has a symmetry direction in the toroidal angle ϕ about the major axis of toroidal symmetry.

There are a large class of equilibrium solutions called $p(\psi)$ and $I(\psi) = R j_\phi$ when only the MHD equations and the boundary conditions are specified. Here ψ is the magnetic flux function for the poloidal magnetic field with $\boldsymbol{B}_p = \nabla\phi \times \nabla\psi$. For laboratory plasmas the sources of particles and thermal energy and the transport rates then choose on the transport time scale, typically tens of milliseconds, the particular $p(\psi)$ and $I(\psi) = R j_\phi$ profiles that are achieved, or created, in the plasma from the injection of particles, energy and momentum. Without the transport physics, there is an infinite set of possible MHD equilibria. The selection and prediction of the experimentally-realized profiles is done by using transport simulation codes such as Automated System for Transport Analysis developed by G. V. Pereverzev, P. N.

Yushmanov in reports February 2002IPP 5/98, Max-Planck-Institut für Plasnma-physik, Garching, Germany, IPP 5/98, February 2002IPP 5/98, TRANSP (`http://w3.pppl.gov/transp/`), CRONOS (`http://www-url`) and BALDUR with Multi-Mode transport formulas (`http://w3.physics.lehigh.edu/baldur/index.html`) used to derive the evolution of the profiles created from the sources. These predicted profiles are compared with the limited data from the plasma diagnostics. One may note, however, that the heating systems and ports for diagnostics break the toroidal symmetry to some degree.

The TRANSP code developed at the Princeton Plasma Physics Laboratory (PPPL) is widely used to interpret the diagnostics from fusion plasmas. An example for the TFTR fusion experiments is found in Budny (1994). Analysis of the lithium pellet injection experiments using TRANSP is given in Budny (2011).

For ITER the TRANSP predictions for the high-temperature performance from several plasma heating methods are given in Budny (2009), and the TRANSP predictions for the alpha particle plasma heating for ITER are developed in Budny (2012).

The reader may gain access and learn to use TRANSP by joining the TRANSP, or the FusionGrid. International participants using the FusionGrid include DIII-D and MIT. JET maintains their own transport codes but often uses the FusionGrid for verification. TRANSP is becoming more parallelized so most of the runs use multiple processors. Approval is needed to run on the FusionGrid. The way to get approval is to submit a request following lists in `http://w3.pppl.gov/transp/`. For help using the TRANSP code, contact the developer, Robert Budny (`budny@pppl.gov`).

The transport codes combine the MHD equilibrium constraints with the injected sources of energy, momentum and particles and the diffusion matrix to arrive at the time evolving plasma profiles. Many examples for all major fusion machines are given in the literature.

The equilibrium solutions of the MHD equations for a variety of tokamak plasmas with divertor configurations are derived in Goedbloed, *et al.* (2010). In their Chapter 16 the famous solutions of Shafranov and Zakharov and Soloviev are presented in detail. The Soloviev solutions are shown to give divertor configurations analytically with various values of the elliptical elongation parameter and the triangularity parameters. Chapter 17 of Goedbloed, *et al.* (2010) gives the details of how to derive and compute the numerical equilibria of the Grad-Shafranov equation with problems added for the generalization to fast toroidal rotation $\Omega = v_\phi/R$ and the "Shafranov shift" from the toroidal rotation. Goedbloed, *et al.* (2010) (pp. 34-35, Chapter 17) work out the details of the solutions of the instabilities in the ITER-like geometry. Figures are given for the peeling-tearing instability during the nonlinear states as computed from the JOREK code [Huysmans and Czarny (2007); Huysmans, *et al.* (2009)]. Simple models of the divertor geometry are constructed by adding to the Grad-Shafranov equilibrium the additional magnetic fields generated by the external toroidal current coils I_j, R_j, Z_j (with index j for the j-th toroidal coil) configuration so as to construct the desired poloidal magnetic field topology or configuration. Sometimes these constructions are called "wire models" since the coils added mathematical are equivalent to toroidal current-carrying wires at the positions R_j, Z_j. Examples of these equilibrium are given in Kroetz, *et al.* (2013) and Abdullaev, *et al.* (2006). The single-null divertor is studied in detail with choices of the triangularity and elongation and position of the X-point with respect to magnetic axis of the plasma for modeling JET, ASDEX and ITER. Magnetic field lines are integrated with generic magnetic perturbations created by plasma modes or from the externally applied magnetic coils. The perturbations generate magnetic islands and stochastic magnetic field lines resulting in a complex plasma transport problem. The free streaming of the electrons is investigated for the loss of the electrons, and the electron-thermal energy, in these stochastic magnetic fields [Martins, *et al.* (2014)].

2.4   Particle Energy Distributions

The fusion plasma is raised to and maintained at core solar temperatures $T > 1\,\text{KeV} \gtrsim 10^7\,\text{K}$ and higher by intense external heating mechanisms that distort the particle velocity distributions away from the Maxwell-Boltzmann distributions. Some degree of distortion away from thermal equilibrium is intrinsic to the plasma confinement and heating. Examples include (1) the diamagnetic cross-field currents $j_{\nabla\theta} = dp/Bdr$, (2) the toroidal plasma current j_ϕ used to create the plasma contribution to the poloidal magnetic field component $B_p = rB_T/qR$ and (3) the velocity-space anisotropies described by $A = 1 - T_\perp/T_\parallel$ created by the RF and NBI heating mechanisms. The rate of twisting of the total magnetic field is given by the function q defined here and called the "safety factor" owing to its key role in MHD stability. The distortions in velocity space are used to increase the plasma temperature. The like-particle collisions reduce the anisotropies on the slow collisional time scale and the electron-ion collisions reduce the electrons currents and the electron-to-ion temperature differences. Generally, in the steady-state fusion plasma the velocity-space distribution functions are significantly distorted away from the Maxwell-Boltzmann distribution function.

In Chapter 3 we develop several of the forms of these ion and electron distribution functions in the tokamak. The emphasis on the distortions of the electron distribution is essential in tokamaks to maintain the plasma current for plasma confinement. The tokamak plasma is confined by the poloidal magnetic field maintained by the toroidal plasma current. In ITER this current is required to reach $15\,\text{MA}$ of plasma current to maintain the confinement required by theory and validated by the tokamak confinement database. This is in contrast to the helical confinement system where the plasma confinement is maintained by the magnetic field created by the external helical currents within superconductors. The tokamak database also shows that to raise the central ion temperatures to that required to meet the designed fusion power production, requires the high toroidal-plasma current. Recall that the energy confinement time of the tokamak is determined principally by the strength of the poloidal magnetic field created by the toroidal plasma current. The empirical scaling laws derived from large tokamak confinement databases show that the plasma confinement time is proportional to the toroidal plasma current I_p.

2.5 Magnetic Reconnection from Resistive MHD Dynamics

The magnetic field B_{pl} produced by the toroidal plasma current winds helically around the torus with the winding rate given by the $q(r, \theta, t) = q[\psi(r, \theta, t)]$ function defined in the previous subsection. For rational values of the q-function the magnetic field lines close on themselves. From nonlinear dynamics [White (2014)] one finds that small perturbations added to the system, such as the discrete placement of the heating systems and vacuum ports, produce magnetic islands with a 3D magnetic structure centered on the rational values of the safety function $q = m/\ell$. The magnetic field lines at this rational surface return on themselves after ℓ-revolutions around the torus. Thus, there is a set of nested rational surfaces in the axisymmetric tokamak where small radial magnetic field perturbations produce magnetic "islands".

An example measured in the laboratory is given in Horton and Ichikawa (pp. 10-20) from a 1970 era tokamak that had substantial magnetic perturbations from the wires carrying the current to the poloidal field coils of wire that produce the toroidal magnetic field. In that case the magnetic islands were mapped out in detail by the injection of weak electron beams and recording the "surface of section plots" made by the fluorescence from the electrons striking the semi-transparent screen. The electron fluorescence showed the magnetic island was rather large and led to a solution of the problem.

More intrinsic to the tokamak system, are self-generated magnetic islands on the rational surfaces where $q = m/\ell$. Plasma modes that create such structures are called "tearing modes" since the nested sequence of magnetic surfaces is broken at the rational surface $r_{m/\ell}$ by chains of magnetic islands. The original derivation of the magnetic islands created spontaneously in the plasma were from resistive tearing modes in which the MHD ohm's law is generalized to include an electric field contribution from the plasma resistively $\delta\boldsymbol{E} = \eta\boldsymbol{j}(\boldsymbol{x}, t)$. This small resistive electric field δE has a small effect expected for the MHD motions at the mode rational surface where the effect is to produce a slowly-growing magnetic island. The sensitivity of the rational surfaces to small perturbations extends to all the plasma dynamical models as it is a topological effect. Thus, even without resistivity, the plasma dynamics from temperature gradient driven drift waves develops small magnetic islands about the rational surfaces. A clear example of this role of the finite electron mass producing magnetic reconnection and magnetic islands is found in the Stenzel, *et al.* (2011) laboratory experiments with Argon plasmas and in the LAPD experiments. A truly collisionless environment with well-documented magnetic reconnection is given by Nakamura, *et al.* (2006) from the plasma data from the four closely coordinated spacecraft CLUSTER measuring the magnetic islands created in a plasma current sheet in the collisionless geomagnetic tail plasma. In this case the plasma profiles and all components of the magnetic field are measured accurately with minimal perturbation to the observed system.

Computer simulations by Muraglia (2009), Muraglia, *et al.* (2011), and Ishizawa and Nakajima (2007) with two fluid models show that in the saturated state of the fast growing pressure gradient driven interchange described in Chapter 1, the magnetic perturbations evolve to produce magnetic islands and similar to those modeled with resistive MHD models that are set up at $t = 0$ to have no resistive MHD instabilities. Thus, the formation of magnetic islands around magnetic surfaces that have rational winding numbers is an ubiquitous event when the systems are subjected to significant perturbations.

In tokamaks, the low-frequency core plasma oscillations, under certain conditions, are sufficient to set the formation of magnetic islands. When the toroidal plasma current is pushed to high values, or formed with steep radial gradients, the tokamak plasma undergoes a series of low frequency "sawtooth" relaxation oscillations. These pulsations propagate outward from the core and often are sufficient to set off the formation of small magnetic islands at the outer $q = 3$ rational magnetic surface. An example of this onset of a magnetic island in quiescent H-mode plasma is described in Chapter 6. In the example in Chapter 6, the plasma remains in the H-mode but with a shorter energy global confinement time τ_E produced by the magnetic island. In tokamaks driven to their highest toroidal plasma current levels the onset of the magnetic islands is typical and sets the limit on the energy confinement time and the maximum toroidal plasma current for stable-steady state plasma confinement.

2.6   Vertical Displacement Instabilities and Halo Currents

From Figs. 1.1 and 1.2 the plasma shape in ITER is elliptical with the vertical plasma extension b greater than the horizontal a extension. This ratio of b/a is called the ellipticity parameter κ. Another key parameter for the shape of the plasma is the triangularity δ parameter that creates the "D" shape that is origin of the name of the DIII-D tokamak. The D-shape allows more of the plasma pressure to be inside the major radius where the horizontal component of the magnetic curvature vector produces the inward effective gravity and less of the plasma pressure in the outer region $R > R_{\text{axis}}$ where the outward effective gravity is destabilizing.

These shaping parameters have been explored extensively in the history of tokamak experiments and are contained in the empirical databases used to finalize the ITER design [Campbell (2001)]. The vertical elongation of the plasma shape, however, allows the plasma the possibility of moving vertically and changing its shape as it moves up or down parallel to the vertical axis of symmetry. Experimental studies of the vertical displacement are carried out in the Tokamak Configuration Variable (TCV) giving an extensive data base and comparison with MHD and resistive MHD simulations for these undesirable types of plasma motions. The vertical motions create halo plasma currents that flow in the regions between the confining vessels walls and the core of the hot plasma. The displacement instabilities limit the degree of the shape parameters vertical elongation and ellipticity to domains well-documented in the configuration plasma tokamak databases.

2.7   Electron Temperature Gradient-Driven Turbulence

The electron energy confinement time in tokamaks is determined by the drift wave turbulent transport. Historically, the advantage of the tokamak over other plasma confinement geometries has been that the electron temperature is consistently higher that those found in other types of magnetic confinement systems including the reversed field pinch (RFP), the helical systems or stellarators, and the various types of open traps generically called mirror machines. Tokamaks have routinely reached electron temperatures of a few kilovolts (KeV). The data for the electron thermal confinement is best explained by drift wave turbulence extended to small space scales of the collisionless electrons skin depth where the turbulence is due to the electron temperature-gradient-driven drift wave turbulence abbreviated as ETG turbulence. Under certain conditions a larger turbulent electron thermal diffusivity χ_e can be generated by the trapped electron turbulence. This stronger trapped electron turbulent diffusivity depends on more details of the system and is not universal as is the smaller scale ETG turbulence.

Machines such as the TCV tokamak [Asp, *et al.* (2007)] and NSTX [Kaye, *et al.* (2007)], Tore Supra [Horton, *et al.* (2000)] and JET are well described by the combination of ETG and trapped electron turbulent transport. The details of the turbulent transport modeling are rather involved, so the presentation is deferred to Chapter 4.4. Detailed accounts of the ETG and trapped electron turbulence are in Chapters 12 and 13 of the reference work *Turbulent Transport in Magnetized Plasmas* [Horton (2012)], and in the Reviews of Modern Physics (1999).

2.8   High-Energy Electron Distributions from RF Heating and Toroidal Plasma Currents

The electron distribution function at high-electron temperature T_e in tokamaks is intrinsically skewed in its parallel velocity dependence so as to carry the parallel electron flow that maintains the plasma current producing the confinement magnetic field. The drift velocity in the electron distribution function is $u_\parallel = -j_\parallel/en_e$.

There is an associated parallel thermal flux $q_\| = m_e \int dv_\| f_e v_\| [v^2 - 5T_e/3m_e]$ that also has a strong radial gradient and plays a significant role in the transport of electron thermal energy. There is an intrinsic level of transport from these fluxes from electron-electron and electron-ion collisions. However, the long-range collective modes, particularly those associated with the radial gradients of the phase space distribution functions that are called drift waves, have a much larger contribution to the electron cross-field transport than the electron collisional transport. The simplest collective mode to understand in detail is that driven by the electron temperature gradient $dT_e/dr = -T_e/L_{T_e}$. By examining the Vlasov-Poisson dispersion relation for the low-frequency self-consistent electric field oscillations, one sees immediately that the plasma waves become unstable when the temperature gradient defined by $\eta_e = d \ln T_e/d \ln n_e$ becomes larger than a critical value of order unity. The simplest case is for hot ion plasma, where the critical value for ETG instability reduces to $\eta_e = d \ln T_e/d \ln n_e > 2/3$. In general, the threshold condition is a function of several system parameters that enter the dispersion relation for the drift waves. But, for $T_i \gg T_e$, one can reduce the stability analysis using either fluid models or the Vlasov-Poisson dispersion equation to find that the radial gradient of the dF_e/dr reduces to $d \ln T_e^{3/2}/d \ln n_e > 0$ for reversing the sign of the wave-electron resonance producing unstable growth of the fluctuations with a growth rate $\gamma \cong \text{const}\, v_e/L_{T_e}$. From the fluid perspective the entropy gradient is inverted so that the Carnot cycle running from $T_e(r_1) > T_e(r_2)$ with $r_1 < r_2$ gives energy to the drift wave causing the wave to grow exponentially. Detailed simulations with both fluid models and gyrokinetic models are now widely available with detailed geometric configurations to specify the turbulence and the anomalous thermal flux in detail.

A second source of free energy stored in the electron distribution that adds to or drives the plasma turbulence is from the plateau formed in the parallel velocity part of the electron distribution function carrying the current $j_{\|,e}(r,t)$ and the parallel thermal flux $q_{\|,e}(r,t)$. The plateau in $F(v_\|, v_\perp, r)$ is described as being in the parallel velocity interval (v_1, v_2) and the rate of diffusion of the electrons in parallel velocity $v_\|$ in this interval is given as $D(v_\|, r)$.

In the limit that $D_\|/\nu_e v_e^2 \gg 1$ in the interval (v_1, v_2) solutions for $f(v_\|)$ are derived by geometrical constructions. For the plateau interval one may map the values of $f(v_\| = v_1, v_\perp) \to f(v_\| = v_2, v_\perp)$ since $\partial f/\partial v_\| = 0$ in this interval. The coordinates of the locus of points $(v_2, v_\perp)$ parameterized by the pitch-angle α is $v_2 = v \cos \alpha$ and $v_\perp = v \sin \alpha = v_2 \tan \alpha$. Thus, the value of f along the high velocity resonant plane with $v_\| = v_2$ is given by $f(v_2, v_\perp) = f(v_2, v_2 \tan \alpha)$ with $0 \le \alpha \le \pi/2$ parameterizing the position along the $v_\| = v_2$ surface.

Now the value of $f(v_2, v_\perp)$ is approximately the same as the value of $f(v_1, v_\perp)$ in this high plateau where velocity $\partial f/\partial v_\| = 0$. So, on the surface $v_\| = v_1$ one has $v_\perp = v \sin \alpha$ and $v_\| = v \cos \alpha = v_1$. For parallel velocities below $v_\| = v_1$ the distribution is Maxwellian since this is the solution of $Cf = 0$ in the low-velocity region up to v_1 and for all $\pi/2 < \alpha \leq \pi$ (co-current moving electrons). Thus, f is known (approximately) through the region $[v_1, v_2]$ and on the surface $v_\parallel = v_2$. Pitch-angle collisions are the dominant collisions for $v_\parallel > v_1$ determining the width $T_\perp$ of the local velocity distribution function.

Typical values of the plateau temperatures derived from the X-ray radiation spectrum in the core plasma are $T_\parallel \sim 750\,\text{KeV}$ to $1\,\text{MeV}$, and $T_\perp \sim 150\text{-}300\,\text{KeV}$ with lower energy Maxwellian component with a temperature of 50-100 KeV. A simple analytic model [Stevens, *et al.* (1985)] used in the literature is the three temperature distribution function f_{3T} given by

$$
f_{3T}\left(p_\perp, p_\parallel\right) = \begin{bmatrix} C_n \exp\left(-\dfrac{p_\perp^2}{2T_\perp} - \dfrac{p_\parallel^2}{2T_{\parallel F}}\right) & \text{for} & p_\parallel > 0 \\[3ex] C_n \exp\left(-\dfrac{p_\perp^2}{2T_\perp} - \dfrac{p_\parallel^2}{2T_{\parallel B}}\right) & \text{for} & p_\parallel < 0 \end{bmatrix}
$$

where $3\,\text{T}$ denotes the three-temperature values $T_{\parallel F}, T_\perp$ and $T_{\parallel B}$ derived from the X-ray spectra or from simulations as in those from the LUKE code [Peysson, *et al.* (2012)] for the combination of the RF waves and collisional scattering of the electrons. A sufficient model for the following discussion is to take the fast electrons distribution to be F = forward and B = backward with respect to the direction of the force accelerating the electrons. The phase space distribution with different forward and backward temperature produces the RF driven current. The complex architecture of the RF current drive system is described further in Ekedahl, *et al.* (2009) and in Chapter 7.

The Fokker-Planck equation is solved with the code LUKE (Lower Upper matrix inversion for the Kinetic Equation) with the RF wave spectra used in the phase space transport function in the Fokker-Planck equation. The PSFC (MIT) code developed for RF heating is DKE, the trajectories and resonant wave-particle interactions of the RF waves are the codes R2D2.f for electron cyclotron waves works with a subroutine linked to LUKE related to the DKE code written by A. Ram. The code called R2D2.f calculates the dispersion relation for the cyclotron waves and the C3PO code from Peysson and Decker calculates the dispersion relation for lower hybrid waves and cyclotron waves. The codes use high-order stencils for following the ITER-shaped flux surfaces and LU matrix decompositionwith iterations for inversions [Peysson and Decker (2007); Peysson and Decker (2008); Peysson, *et al.* (2011); Peysson, *et al.* (2012); Peysson, *et al.* (2014); Decker, *et al.* (2011); Decker, *et al.* (2012)] to solve the complex RF heating equations.

The phase space distribution function produces a plasma current modeled by

$$
j_\parallel = \frac{-en(T_{\parallel F} - T_{\parallel B})}{(2\pi m_e T_{\text{avg}})^{1/2}}
$$

using the measured forward (F) and backward (B) electron temperatures derived from the X-ray spectra derived from the plasma X-ray emission and magnetic field data.

In the ITER geometry with flux surfaces $\Psi(R, Z)$ the applied RF fields $\boldsymbol{E}^{\mathrm{RF}}$ creates the force $\boldsymbol{F}$ on the electrons. The electron phase space density $f(\Psi, \epsilon, \xi, t)$ with the dimensionless energy ϵ and pitch-angle variable ξ, then evolves from the 3-D phase space conservation of the number of electrons being accelerated described by

$$\frac{\partial f}{\partial t} + \nabla \cdot \widehat{\boldsymbol{G}} f = S^+ - S_-, \tag{2.9}$$

where the divergence of the electron phase space flux $\widehat{\boldsymbol{G}} f$ for the evolution of the phase space density f is

$$\widehat{\boldsymbol{G}} f = -\underline{\underline{\boldsymbol{D}}} \cdot \frac{\partial f}{\partial \boldsymbol{p}} + \boldsymbol{F} f. \tag{2.10}$$

The divergence of the flux $\widehat{\boldsymbol{G}} f$ is given by

$$\nabla \cdot \widehat{\boldsymbol{G}} = \frac{B}{q\lambda} \frac{\partial}{\partial \psi} \left(\frac{q\lambda |\nabla \Psi| G_\phi}{B} \right) - \frac{1}{p^2} \frac{\partial}{\partial p} \left(p^2 G_p \right) - \frac{1}{\lambda p} \frac{\partial}{\partial \xi} \left[\lambda \left(1 - \xi^2 \right)^{1/2} G_\xi \right], \tag{2.11}$$

as described further in Eq. (5.36) in Chapter 5.

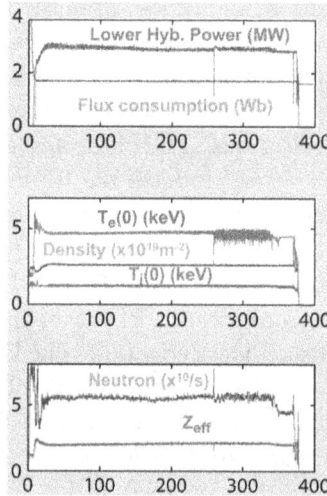

Fig. 2.3   The propagation and penetration characteristics of the LH rays during a current drive experiment on Tore Supra. [Peysson, *et al.* (2012); Peysson, *et al.* (2014); van Houtte, *et al.* (2004)].

Figure 2.3 shows the steady state plasma driven in Tore Supra with 3 MW of lower hybrid current drive with the electron temperature of approximately 5 keV and the ion temperature about 1 keV for a period of 6 minutes. The top panel shows the magnetic flux consumed, the middle panel the electron and ion temperatures and the plasma density and the bottom panel the neutron flux produced in this deuterium plasma. The new WEST upgrade of the Tore Supra machine will test the models of the current drive with two high-powered lower hybrid (LH) antennas at 3.7 GHz operating simultaneously in the divertor configuration that closely models ITER.

The comprehensive RF simulation code named LUKE is available from the CEA developers as described in Decker, *et al.* (2011). The codes for solving these Fokker-Planck Eqs. (2.9), (2.10), and (2.11) for the electron distribution function driven by the RF antennas. The code LUKE works with the CP30 for the RF rays to give the full dynamics of the RF wave propagation and absorption producing the formation of the current carrying electron phase space distribution functions. These codes are being validated on the upgrade of Tore Supra where the poloidal field configuration is changed to be a reduced scale model of the ITER flux surfaces. The presence of the magnetic separatrix and SOL changes the propagation and penetration characteristics of the LH rays and current drive [Peysson, *et al.* (2012); Peysson, *et al.* (2014)] from that measured in the earlier circular cross-section Tore Supra. The new machine is called WEST for Tungsten Experimental Steady-state Tokamak. WEST will have two high-power LH antennas testing the current drive with two high-powered lower hybrid (LH) waves at 3.7 GHz operating in the divertor configuration that closely models the ITER configuration. The plasma-wall interactions are described in Chapter 9.

ITER will have an extensive array of diagnostics that will be used to drive the actuators of the RF systems for fast real-time control of the electron distribution functions driven by the both the LH and the ECRH antennas. Many other components including the poloidal field coils will have real-time controls, but their response time is intrinsically slower than the RF controls. The suite of diagnostics discussed in Chapter 7 will provide input to the predictive code, CRONOS, which contains ultra-fast low-order codes for sending input signals to the complex array of actuators or control systems. This control code was tested for the RF controls and shown capable of maintaining a steady toroidal current for five to six minutes in Tore Supra.

CRONOS is supported by a large team of physicists and developers and documented in Artaud, *et al.* (2010). The code is an open source software code available to the fusion community. The code is installed on computers in all the ITER partner nations and is set up to have reference parameters and runs for many tokamaks. The electron physics modules include the lower-hybrid heating and current drive (LHCD), the electron-cyclotron resonance-current drive (ECCD) and the ion-cyclotron resonance heating for the fast ions. The code includes Orbit Following Monte Carlo dynamics for the neutral beam injection (NBI). There are numerous transport modules with an example of a predictive simulation for fast-wave electron heating (FWEH) with a step up in the heating power from 3 MW to 6 MW over a three-second period in Tore Supra discharge analyzed in Horton, *et al.* (2000) and Hoang, *et al.* (2003).

The RF waves are created with high power klystrons. In Tore Supra [Ekedahl, *et al.* (2009)] 16 klystrons producing up to 0.5 MW per Klyston operating for 1000 seconds driving 32 waveguides that are about 20 m long. The waveguides are designed so as to deliver power with phase shifts through an array of TE_{10} and TE_{30}

waveguides into a forward (toroidal) spectrum with the launched refractive index $n_\parallel = ck_\parallel/\omega_{\mathrm{RF}} = 1.7$. The frequency $\omega_{\mathrm{RF}} = 3.7\,\mathrm{GHz}$ delivers waves (or LH plasmons) to accelerate electrons in the direction opposite to the tokamak current. The success of accelerating the electron is easily measured from the burst of the directed X-ray emission as first reported in PLT and developed in more detail at Alcator C-Mod, Tore Supra, EAST and the FTU machine in Frascati.

The Tore Supra phase-arrayed PAM antenna has launched waves at about 3 MW continuously (cw) with power density of $25\,\mathrm{MW/m^2}$ setting the record for high electron temperature in steady-state tokamak operation. The results of these experiments lead to the upgrade design of the new chamber that has the shape of ITER separatrix and the tungsten (W) walls of ITER for testing the plasma-wall interactions with specially-designed water cooled tungsten blocks described in Chapter 9. The steady-state high-current-high electron operation with the lower hybrid waves maintaining and controlling the toroidal plasma current for periods of 5 to 6 minutes are described in Bucallosi, *et al.* (2011).

At the time of writing (2015), the lower hybrid wave drive is the only proven efficient method to maintain steady state operation in tokamaks. Other methods under development use the injection of currents created by additional external coils that inject magnetic helicity into the toroidal chamber. These helicity injection methods are under development by the University of Washington in Seattle as described by Jarboe, *et al.* (2014) [Wrobel, *et al.* (2013)] and by McCollam, *et al.* (2006) along with another approach under development at the University of Wisconsin [McCollam, *et al.* (2010)]. While the helicity injection method have high current driving efficiency and seem promising for the future, the results achieved to date are well below those achieved with the lower hybrid current drive method.

References

Abdullaev, S. S., Finken, K. H., Jakubowshi, M., and Lehnen, M. (2006). Mappings of stochastic field lines in poloidal divertor tokamaks, *Nucl. Fusion* **46**, p. S113, doi:10.1088/0029-5515/46/4/S02.

Artaud, J. F., Basiuk, V., Imbeaux, F., Schneider, M., Garcia, J., Giruzzi, G., Huynh, P., Aniel, T., Albajar, F., Ané, J. M., Bécoulet, A., Bourdelle, C., Casati, A., Colas, L., Decker, J., Dumont, R., Eriksson, L. G., Garbet, X., Guirlet, R., Hertout, P., Hoang, G. T., Houlberg, W., Huysmans, G., Joffrin, E., Kim, S. H., Köhl, F., Lister, J., Litaudon, X., Maget, P., Masset, R., Pégourié, B., Peysson, Y., Thomas, P., Tsitrone, E., and Turco, F. (2010). The CRONOS suite of codes for integrated tokamak modeling, *Nucl. Fusion* **50**, p. 043001, doi:10.1088/0029-5515/50/4/043001.

Asp, E., Weiland, J., Garbet, X., Parail, X., Strand, P., and the JET EFDA contributors. (2007). JET energy confinement in the hot-electron mode, *Plasma Phys. Control. Fusion* **49**, p. 1221.

Braginskii, S. I. (1965). Transport Processes in a Plasma, *Rev. Plasma Phys.*, **1**, p.205 (Authorized translation from the Russian by Herbert Lashinsky, University of Maryland, USA), Edited by M. A. Leontovich (Consultants Bureau, New York, 1965).

Bucalossi, J., Argouarch, A., Basiuk, V., Baulaigue, O., Bayetti, P., Bécoulet, M., Bertrand, B., Brémond, S., Cara, P., Chantant, M., Corre, Y., Courtois, X., Doceul, L., Ekedahl, A., Faisse, F., Firdaouss, M., Garcia, J., Gargiulo, L., Gil, C., Grisolia, C., Gunn, J., Hacquin, S. Hertout, P., Huysmans, G., Imbeaux, F., Jiolat, G., Joanny, M., Jourd'heuil, L., Jouve, M., Kukushkin, A., Lipa, M., Lisgo, S., Loarer, T., Maget, P., Magne, R., Marandet, Y., Martinez, A., Mazon, D., Meyer, O., Missirlian, M., Monier-Garbet, P., Moreau, P., Nardon, E., Panayotis, S., Pégourié, B., Pitts, R. A., Portafaix, C., Richou, M., Sabot, R., Saille, A., Saint-Laurent, F., Samaille, F., Simonin, A., and Tsitrone, E. (2011). Feasibility study of an actively cooled tungsten divertor in Tore Supra for ITER technology testing, *Fusion Eng. Design* **86**, pp. 684-688, doi:10.1016/j.fusengdes.2011.01.114.

Budny, R. V. (2012). Alpha heating in ITER L-mode and H-mode plasmas, *Nucl. Fusion* **52**, p. 013001, doi:10.1088/0029-5515/52/1/013001.

Budny, R. V. (2011). Comment on Li pellet conditioning in TFTR. *Phys. Plasmas* **18**, p. 092506, http://dx.doi.org/10.2172/1014574.

Budny, R. V. (2009). Comparisons of predicted plasma performance in ITER H-mode plasmas with various mixes of external heating, *Nucl. Fusion* **49**, p. 085008, doi:10.1088/0029-5515/49/8/085008.

Budny, R. V. (1994). A standard DT supershot simulation, *Nucl. Fusion*, **34**, pp. 1247-1262, doi:10.1088/0029-5515/34/9/I06.

Campbell, D. J. (2001). The physics of the International Thermonuclear Experimental Reactor FEAT *Phys. Plasmas* **8**, p. 2041, http://dx.doi.org/10.1063/1.1348334.

Decker, J., Peysson, Y., and Coda, S. (2012). Effect of density fluctuations on ECCD in ITER and TCV, European Physical Journal Web of Conferences, vol. 32, pp. 01016.

Decker, J., Peysson, Y., and Hillairet, J., Artaud, J.-F., Basiuk, V., Becoulet, A., Ekedahl, A., Goniche, M., Hoang, G. T., Imbeaux, F., Ram, A. K., and Schneider, M. (2011). Calculations of lower hybrid current drive in ITER, *Nucl. Fusion* **51**, p. 073025, doi:10.1088/0029-5515/51/7/073025.

Ekedahl, A., Goniche, M., Guilhem, D., Kazarian, F., and Peysson, Y., and the Tore Supra Team. (2009). Lower hybrid current drive in Tore Supra, *Fusion Sci. Tech* **56**, p. 1150.

Goedbloed, J. P., Keppens, R., and Poedts, S. (2010). *Advanced Magnetohydrodynamics* pp. 269-304 (Cambridge Univ. Press) ISBN:978-0-521-87957-5.

Greenwald, M., Terry, J. L., Wolfe, S. M., Ejima, S., Bell, M. G., Kaye, S. M., and Neilson, G. H. (1988). A new look at density limits in tokamaks, *Nucl. Fusion* **28**, 2199, doi:10.1088/0029-5515/28/12/009.

Greenwald, M. (2014). 20 Years of Research on the Alcator C-Mod Tokamak, *Phys. Plasmas.*

Hallock, G. A., Hickok, R. L., and Hornady, R. S. (1994). The TMX heavy ion beam probe, *Plasma Science, IEEE Transactions on Plasma Science* **22**, 4, pp. 341-349, ISSN:0093-3813, doi:10.1109/27.310639.

Hirshman, S. P., and Sigmar, D. J. (1977). Neoclassical transport of a multispecies toroidal plasma in various collisionality regimes, *Phys. Fluids* **20**, p. 418, http://dx.doi.org/10.1063/1.861877.

Hoang, G. T., Horton, W. Bourdelle, C., Hu, B., Garbet, X., and Ottaviani, M. (2003). Analysis of the critical electron temperature gradient in Tore Supra, *Phys. Plasmas* **10**, pp. 405-412, doi:10.1063/1.1534113.

Horton, W. (2012). *Turbulent Transport in Magnetized Plasmas* (World Scientific) ISBN:978-981-4383-53-0.

Horton, W., Zhu, P., Hoang, G. T., Aniel, T., Ottaviani, M., and Garbet, X. (2000). Electron transport in Tore Supra with fast wave electron heating, *Phys. Plasmas* **7**, pp. 1489-1510, doi:10.1063/1.873969.

Horton, W. (1997). Chaos and structures in the magnetosphere, *Physics Reports* **283**, 1-4, pp. 265-302, doi:10.1016/S0370-1573(96)00063-4.

Horton, W. (1999). Drift waves and transport, *Rev. Mod. Phys.* **71**, pp. 735–778.

Huysmans, G. T. A., Pamela, S., van der Plas, E., and Ramet, P. (2009). Nonlinear MHD simulations of edge localized modes (ELMs), *Plasma Phys. Control. Fusion* **51** p. 124012, doi:10.1088/0741-3335/51/12/124012.

Huysmans, G. T. A., and Czarny, O. (2007). MHD stability in X-point geometry: simulations of ELMs, *Nucl. Fusion* **47**, 659-666, doi:10.1088/0029-5515/47/7/016.

Ishizawa, A. and Nakajima, N. (2007). Excitation of macro-magnetohydrodynamic mode due to multi-scale interaction in a quasi-steady equilibrium formed by a balance between micro-turbulence and zonal flow, *Phys. Plasmas* **14**, p. 040702, doi:10.163/1.271669, http://link.aip.org/link/doi/10.1063/1.2716669.

Jarboe, T. R., Hansen, C. J., Hossack, A. C., Marklin, G. J., Morgan, K. D., Nelson, B. A., Sutherland, D. A., Victor, B. S. (2014). A Proof of Principle of Imposed Dynamo Current Drive: Demonstration of Sufficient Confinement, *Fusion Sci. Tech.* **66**, 3, p. 369-384, dx.doi.org/10.13182/FST14-782.

Keilhacker, M., Gibson, A., Gormezano, C., Lomas, P.J., Thomas, P. R., Watkins, M. L., Andrew, P., Balet, B., Borba, D., Challis, C. D., Coffey, I., Cottrell, G. A., De Esch,

H. P. L., Deliyanakis, N., Fasoli, A., Gowers, C. W., Guo, H. Y., Huysmans, G. T. A., Jones, T. T. C., Kerner, W., König, R. W. T., Loughlin, M. J., Maas, A., Marcus, F. B., Nave, M. F. F., Rimini, F. G., Sadler, G. J., Sharapov, S. E., Sips, G., Smeulders, P., Söldner, F. X., Taroni, A., Tubbing, B. J. D., von Hellermann, M. G., Ward, D. J., and JET Team. (1999). High-fusion performance from deuterium-tritium plasmas in JET, *Nucl. Fusion* **39** p. 209, doi:10.1088/0029-5515/39/2/306.

Kroetz, T., Martins, Caroline G. L., Roberto, M., and Caldas, I. L. (2013). Set of wires to simulate tokamaks with poloidal divertor, *Plasma Phys.* **79**, pp. 751-757, http://dx.doi.org/10.1017/S0022377813000391.

Martins, C. G. L., Roberto, M., Caldas, I. L., Vanderlei, and de Sa, P. (2014). Simulation with CRONOS for ITER: Hollow current profile and its relation with the bootstrap current, *Nucl. Fusion* **54**.

McCollam, K. J., Blair, A. P., Prager, S. C., and Sarff, J. S. (2006). Oscillating-Field Current-Drive Experiments in a Reversed Field Pinch, *Phys. Rev. Lett.* **96**, p. 035003, doi:10.1103/PhysRevLett.96.035003.

McCollam, Anderson, J. K., Blair, A. P., Craig, D., Hartog, D. J. Den, Ebrahimi, F., O'Connell, R., Reusch, J. A., Sarff, J. S., Stephens, H. D., Stone, D. R., Brower, D. L., Deng, B. H., and Ding, W. X. (2010). Oscillating-Field Current-Drive Experiments in a Reversed-Field Pinch, *Phys. Plasmas*, **17**, p. 082506.

Muraglia, M., Agullo, O., Yagi, M., Benkadda, S., Beyer, P., Garbet, X., Itoh, S.-I., Itoh, K., and Sen, A. (2009). Effect of the curvature and the β parameter on the nonlinear dynamics of a drift tearing magnetic island, *Nucl. Fusion* **49**, p. 055016 (11pp), http://dx.doi.org/10.1088/0029-5515/49/5/055016.

Muraglia, M., Agullo, O., Benkadda, S., Yagi, M., Garbet, X., and Sen, A. (2011). Generation and Amplification of Magnetic Islands by Drift Interchange Turbulence, *Phys. Rev. Lett.* **107**, p. 095003, doi:http://dx.doi.org/10.1103/PhysRevLett.107.095003.

Nakamura, T. K. M., Fujimoto, M., and Otto, A. (2006). Magnetic reconnection induced by weak Kelvin-Helmholtz instability and the formation of the low-latitude boundary layer, *Geophys. Res. Letts.* **33**, 14, doi:10.1029/2006GL026318.

Peysson, Y., and Decker, J. (2007). Advanced Lower-Hybrid Current Drive Modeling, Radio Frequency Power in Plasmas: 17th Topical Conference on Radio Frequency Power in Plasmas, *AIP Conference Proceedings* **933**, pp. 293-296.

Peysson, Y., and Decker, J. (2008). Fast electron Bremsstrahlung in axisymmetric magnetic configuration, *Phys. Plasmas* **15**, p. 092509, doi:10.1063/1.2981391.

Peysson, Y., Decker, J., Morini, L., and Coda, S. (2011). RF current drive and plasma fluctuations, *Plasma Phys. Control. Fusion* **53** p. 124028, doi:10.1088/0741-3335/53/12/124028.

Peysson, Y., Decker, J., and Morini, L. (2012). A versatile ray-tracing code for studying RF wave propagation in toroidal magnetized plasmas, *Plasma Phys. Control. Fusion* **54**, p. 045003, doi:10.1088/0741-3335/54/4/045003.

Peysson, Y., and Decker, J. (2014). Numerical Simulations of the Radio-Frequency-Driven Toroidal Current In Tokamaks, *Fusion Sci. Tech.* **65**, 1, pp. 22-42, dx.doi.org/10.13182/FST13-643.

Stenzel, O. Wilbrandt, S., Yulin, S., Kaiser, N., Held, M., Tünnermann, A., Biskupek, J., and Kaiser, U. (2011). Plasma ion assisted deposition of hafnium dioxide using argon and xenon as process gases, *Optical Materials Express* **1**, 2, pp. 278-292, http://dx.doi.org/10.1364/OME.1.000278.

Stevens, J., Von Goeler, S., Bernabei, S., Bitter, M., Chu, T. K., Efthimion, P., Fisch, N., Hooke, W., Hosea, J., Jobes, F., Karney, C., Meservey, E., Motley, R., and Taylor, G. (1985). Modeling of the electron distribution based on Bremsstrahlung emission during lower-hybrid current drive on PLT, *Nucl. Fusion* **25**, p. 1529, doi:10.1088/0029-5515/25/11/002.

van Houtte, D., Martin, G., Bécoulet, A., Bucalossi, J., Giruzzi, G., Hoang, G. T., Loarer, Th., and Saoutic, B. (on behalf of the Tore Supra Team) (2004). Recent fully non-inductive operation results in Tore Supra with 6 min, 1 GJ plasma discharges, *Nucl. Fusion* **44**, p. L11, doi:10.1088/0029-5515/44/5/L01.

Wagner, F., Becker, G., Behringer, K., Campbell, D., *et al.* (1982). Regime of improved confinement and high-beta in neutral-beam-heated divertor discharges of the ASDEX tokamak, *Phys. Rev. Lett.* **49**, p. 1408, http://dx.doi.org/10.1103/PhysRevLett.49.1408.

White, Roscoe B. (2014). The Theory of Toroidally Confined Plasmas: 3rd Edition Hardcover, ISBN-13: 978-1783263639 ISBN-10:1783263636, 3rd Edition.

Wrobel, J. S., Hansen, C. J., Jarboe, T. R., Smith, R. J., Hossack, A. C., Nelson. B. A., Marklin, G. J., Ennis, D. A., Akcay, C., and Victor, B. S. (2013). Relaxation-time measurement via a time-dependent helicity balance model, *Phys. Plasmas* **20**, p. 012503 (2013), http://dx.doi.org/10.1063/1.4773401.

Chapter 3

Alfvén Cavity Modes, Fast Ions, Alpha Particles and Diagnostic Neutral Beams

ITER will have a high-energy population of the fusion product alpha particles with a velocity distribution function $f_\alpha(v, r)$ called the slowing-down distribution. At each radius r, or more precisely, on each magnetic flux surface Ψ, the distribution in energy $E = \frac{1}{2} m_\alpha v^2$ of the particles is derived from their birthrate of the 3.5 MeV alpha particles produced by nuclear reaction rate $\langle \sigma v \rangle_{DT}$ times the product of the local deuterium and tritium densities [Bosch and Hale (1992)]. After their birth, the alpha particles slow down and scatter in their direction from the Coulomb collisions with the thermal plasma. In contrast, the 14.1 MeV neutrons born with the alpha particles travel in straight lines through the chamber wall into the outer boron-lithium blanket designed for breeding new tritium atoms for maintaining the tritium fuel supply. The blanket also converts the neutron kinetic energy into thermal power to drive the electric generators for fusion electric power. ITER will test tritium breeding blanket designs, but the conversion of the thermal energy into electric power awaits the next machine called DEMO.

The phase-space distribution of the magnetically-trapped alpha particles is of primary interest since this distribution function is the source of a high-frequency ion-cyclotron wave and the low-frequency Alfvén eigenmodes. The tritium experiments in JET [Keilhacker, *et al.* (1999)] showed the presence of both these instabilities. The high-frequency, ion-cyclotron waves were easily observed in the first DT experiments on TFTR (1996) and then more clearly on JET [Jet Team (1998)]. The power at the fundamental cyclotron frequency of the alpha particle, which is close to the deuteron cyclotron frequency, showed peaks in the wave frequency power spectrum with several cyclotron harmonics excited by the alpha particle distribution function. The amplitude of the fundamental wave at approximately 18 MHz increased linearly with the alpha-particle power as reported in the IAEA publication for TFTR discharges with the progression of 1.8, 4.7 and 8.9 MW of neutral beam heating power.

The alpha particle density was inferred from measurements of the spectrum of the knock-on deuterons produced in the elastic nuclear collisions of the $Z = 2$ alpha particles with the plasma ions D^+ and T^+. The measurements of the spectrum used the neutral particle analyzer that collected the D and T atoms leaving the plasma approximately parallel to the magnetic field.

The analysis of the alpha particle distribution functions shows that the distributions are approximately given by their Coulomb collisional slowing-down process on the thermal plasma. The pitch-angle scattering of the 3.5 MeV isotropically-born alpha particles on the thermal (near Maxwell-Boltzmann) distribution of electrons and hydrogenic ions is more complicated owing to small $\mu = v_{\parallel}/v$ particles being mirror trapped in the minimum $|B|$ on the outboard side of the torus.

A collection of articles on the tokamak data and the modeling of the Alfvén waves driven by alpha particles and the complex trapping structures that evolve from the alpha particle-plasma wave interactions is found in works published from the 12th IAEA Technical Meeting on Energetic Particles in Magnetic Confinement Systems [Berk, *et al.* (2012)]. An article in this same Nuclear Fusion volume, *Alfvén Eigenmodes Stability and Fast Ion Loss in DIII-D and ITER Reversed Magnetic Shear Plasmas* is particularly relevant to the alpha particle-Alfvén interactions to be expected in ITER [Van Zeeland, *et al.* (2012)].

3.1 Plasma Eigenmodes and their Destabilization by High-Energy Ions

This slowing-down distribution of alpha particles at sufficient density $n_\alpha/n_e \gtrsim 10^{-3}$ drives the low-frequency Alfvén waves unstable through the wave-particle resonance $\omega_{k_{\parallel}} - k_{\parallel}v_{\parallel\alpha} = \omega_{D\alpha}$ where $\omega_{D\alpha}$ is the bounce-average of the sum of the gradient B and curvature drift frequency of the alpha particles. This type of instability was simulated extensively [Cheng, *et al.* (1985)] and anticipated from early studies before TFTR was operational. Early NBI injection research on the instabilities driven in the background plasma [Berk, *et al.* (1997); Mikhailovskii (1992)] was carried out in designing the TFTR experiments. The basic instability was found with the NBI injection in the deuterium plasmas in TFTR by Wong, *et al.* (1991) by lowering the magnetic field to so as to achieve the $\omega(k) = k_{\parallel}v_{\text{deuteron}}$ resonance condition with the fast deuterons from the beam injection with the Alfvén waves. Examples of the types of Alfvén modes driven by the NBI beams in TFTR are described in Nazikian, *et al.* (1997, 2003).

These Alfvén wave instabilities arise from the cavity resonant Alfvén waves rather than from the continuum wave spectrum. Thus, the modes are called the TAEs and EAEs for Toroidal Alfvén Eigenmodes and ellipticity-induced Alfvén eigenmodes where we note that the divertor-shaped cross-section of the JET and ITER machines creates the ellipticity and D-shaping of the metallic wall cavity in the eigenvalue problem. Both types (TAE and EAE) of eigenmodes were observed to be destabilized in JET under various magnetic field configurations and NBI injection conditions [Jet Team (1998)]. Figure 3.1 shows a typical spectrum of the TAEs and EAEs for the JET machine [Jet Team (1998)].

The DT experiments in JET were ground-breaking and still, after two decades,

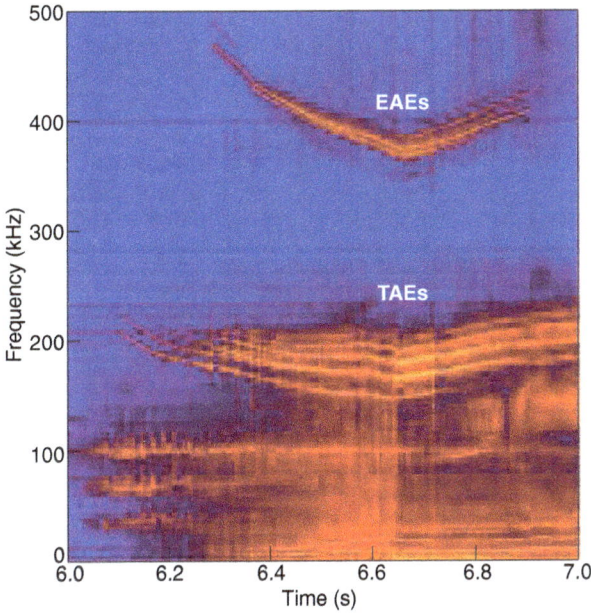

Fig. 3.1 The frequency spectrum of the two types of Alfvén eigenmodes measure from $\delta B(t)$ magnetic probes near the edge of the plasma during the NBI for a 3.8 MA/3.4 T deuterium-tritium discharge created with 10 W of NBI power. The plasma state is that of ELM free period during a hot H-mode confinement state. The label EAE is for elliptic Alfvén eigenmode and TAE is for toroidal Alfvén eigenmode in JET [Jet Team (1998)].

are the only magnetic fusion power experiments clearly demonstrating the possibility of achieving a fusion power reactor. The ITER design draws strongly on the DT experiments in JET and another round of DT experiments are being planned at the time of writing in 2015.

In the 1997-98 DT campaign JET achieved alpha particle power $P_\alpha \lesssim 10.5$ MW from a net injected power of $P_{aux} = 24$ MW for 1-2 seconds. Other longer DT pulsed experiments produced $P_\alpha \lesssim 4$ MW from $P_{aux} = 20$ MW for 4 seconds, as shown in Fig. 3.2. Both ion and electron temperatures increased by the amount forecast from Coulomb drag and angle scattering of the alpha particles on the thermal DT plasma. The ion temperature reached 15 KeV and the electron temperature $T_e \sim 12$ KeV in pulse 42847 in Fig. 3.2 with equal densities (50%-50%) of deuterium and tritium in the thermal plasma. In this discharge the NBI injected power was 10 MW for approximately 3 seconds. The highest gain power amplification in the JET DT plasmas was $P_\alpha/P_{heat} \sim 0.13$ corresponding to net fusion power amplification factor $Q_{fus} = 2/3 \sim 0.7$ where the power is computed with 17.6 MeV per fusion reaction. For reference this corresponds to a neutron production rate of 2×10^{19} n/s (neutrons per second). This DT experiment retains the world record for controlled fusion power at the time of writing in 2015.

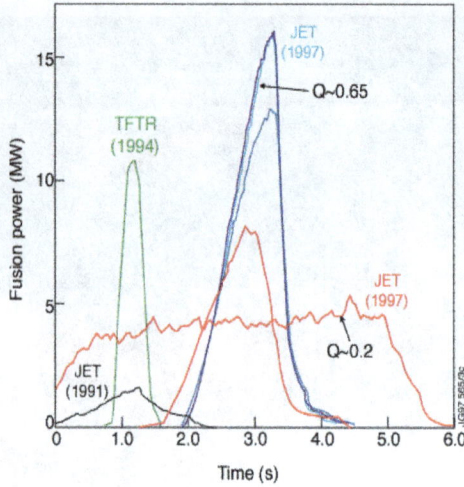

Fig. 3.2   The fusion power produced in JET with the two record shots in the 1997 DT campaign and the early trace shot from 1991. The earlier 1994 short pulse of 10 MW from the end of the TFTR fusion power campaign is shown for comparison. The two high power JET shots remain as the record DT fusion power experiments in 2015 [Keilhacker, *et al.* (1999)].

3.2   Loss Process for High-Energy Ions and Electrons

Fast neutral beam injection (NBI) power that exceeds the stability limit of the internal $m = 1$ modes gives rise to nonlinear relaxation oscillations called "fishbones"[Gunter, *et al.* (1999)] named for the character of the pulsating time-series recorded on the poloidal magnetic field diagnostic probes. Each burst of the $m = 1$ oscillations ejects a fraction of the injected energetic ions and limits their density and pressure. A simple description of the dynamics consistent with the data from Heidbrink, *et al.* (1986) and McGuire, *et al.* (1983), is given by Coppi and Porcelli (1986).

Alpha particle driven toroidal Alfvén eigenmodes (TAEs) were first observed in deuterium-tritium plasmas in TFTR [Nazikian, *et al.* (2003)]. The modes were in the core of the plasma with amplitudes $\delta B/B \sim 10^{-5}$ with toroidal mode numbers $n = 2 - 4$ in a low magnetic-shear regime. The dimensionless alpha particle pressure $\beta_\alpha \sim 2 \times 10^{-4}$ was below the valued expected for onset of the modes from theory. A new interpretation of the alpha particle driven instabilities consistent with the data from the TFTR experiments is given in Nazikian, *et al.* (2003).

The Alfvén waves in the tokamak are coupled in their poloidal angle-θ variation due to the $B_\phi(\theta) = B_0/(1 + r\cos\theta/R)$ variation of the toroidal magnetic field. The coupling results in the formation of discrete eigenmodes where the frequencies $\omega_\pm(m, n)$ of the two uncoupled modes derived in the $\epsilon = r/R \to 0$ limit intersect. The frequency of the gap eigenmode is derived by setting the frequencies to Alfvén

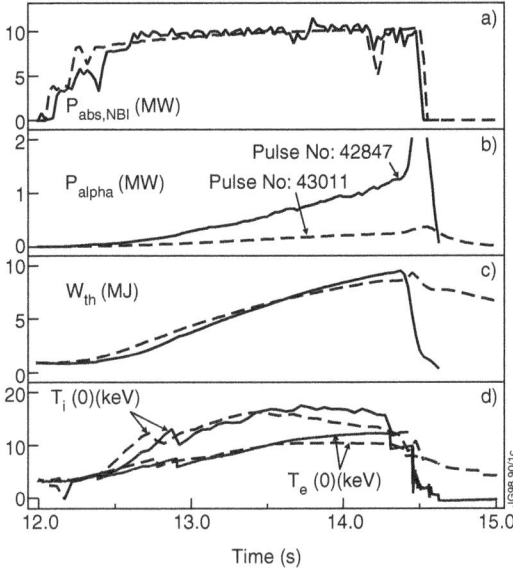

Fig. 3.3   The time profiles for the record JET shot 42847 from 1998. The curves are (a) the absorbed NBI heating power, (b) the alpha article heating power produced by the fusion reactions in the deuterium-tritium plasma, (c) the net thermal energy in the fully ionized DT plasma, (d) the evolution of the core [$r = 0$] central ion temperature for the two pulses: the record shot 42847 and the later pulse 43011. The first pulse has 50:50 mixture of DT and the second shot (43011) was 10:90 mixture of DT.

wave continuum modes $\omega_1^2 = k_\parallel^2(m, n)v_A^2$ and $\omega_2^2 = k_\parallel^2(m + 1, n)v_A^2$ equal to each other in the limit $\epsilon \to 0$. The finite ϵ-coupling of the two continuum modes ω_1, ω_2 at the radial position $r = r_{\text{TAE}}$ generates the gap eigenmode.

The presence of the toroidal Alfvén eigenmodes, or TAE modes, is analogous to the discrete bound electron-hole modes in the semi-conductors energy band-gaps of solid state physics. Whereas the Alfvén wave continuum modes have strong electron Landau damping, the discrete Alfvén gap eigenmodes are weakly damped and easily destabilized by energetic electrons or by ion Landau wave-particle resonances. The Alfvén gap eigenmodes have been extensively developed since the fusion plasma has a high-energy population of alpha (He^4 helium ions) particles that destabilize the Alfvén gap eigenmodes. Laboratory simulation experiments by Wong, *et al.* (1991) with high-energy neutral-beam-produced hydrogen nuclei that created the nonlinear TAE modes with a wide variety of frequency spectra [ITER Physics Expert Group (1999)] of Alfvén waves. These driven low frequency electromagnetic waves produce a fast loss of the resonant ions through turbulent transport processes [Berk, *et al.* (1997); Todo, *et al.* (2010); Gorelenkov, *et al.* (2007); Fu and Van Dam (1989); Breizman and Sharapov (1995)].

In a homogeneous magnetized plasma, linear ideal-MHD arguments show the existence of a shear-Alfvén wave of frequency ω_A with the dispersion relation

$$\omega_A^2 = k_\parallel^2 v_A^2 \tag{3.1}$$

where v_A is the local Alfvén velocity and $k_\parallel$ is the component of the wavenumber vector in the direction of the equilibrium magnetic field $\boldsymbol{B}_0$.

Let us consider this wave in axisymmetric toroidal plasmas. In the cylindrical limit, the periodicities of the system require that there exists two integers, a toroidal mode number n and a poloidal mode number m, such that

$$k_\parallel = \frac{n - m/q(r)}{R_0}, \tag{3.2}$$

where R_0 is the distance from the symmetry axis of the tokamak to the magnetic axis. In an inhomogeneous plasma in a sheared magnetic field, both $k_\parallel$ and v_A are functions of r, where r is the local radial flux coordinate. The simple dispersion relation Eq. (3.1) still applies in this configuration and is called the Alfvén continuum. Since phase velocity is a function of radius, a wave packet with finite radial extent would suffer from phase-mixing, giving rise to the so-called continuum damping. Except for energetic particle modes, resonant drive-by fast particles are not enough to overcome this damping of a single continuum wave. However, the toroidal coupling of two successive poloidal modes m and $m+1$ breaks up the continuous spectrum. This is illustrated in Fig. 3.4, which shows the Alfvén continuum for $n = 1$, $m = 2$, and $n = 1$ and $m = 3$, in cylindrical geometry, where the two poloidal continuum modes are decoupled, and in toroidal geometry, with a two-mode coupling creating the gap eigenmodes. The latter is obtained with equilibrium plasma parameters corresponding to JT-60U shot E32359 at $t = 4.2\,\text{s}$ assuming concentric circular magnetic flux surfaces, while retaining toroidicity effects in the first order in inverse aspect ratio. Though we show only the $\omega > 0$ half-plane, the continuum spectrum is symmetric with respect to $\omega \to -\omega$. Coupled modes are (n, m) and $(-n, -m-1)$ for $\omega > 0$, and $(n, m+1)$ and $(-n, -m)$ for $\omega < 0$. The frequency gap is centered at a radius r_A such that $q(r_A) = (m+1/2)/n$, where the two continuous spectra would cross in the absence of the $\cos\theta$ coupling and where $|k_\parallel| = 1/2qR_0$. The resulting discrete eigenmode is a TAE mode at the frequency $\omega_A = v_A/2qR_0$.

For a deuterium plasma with typical magnetic field $B_0 \sim 1\,\text{T}$ and density $n_i \sim 10^{20}\,\text{m}^{-3}$, the Alfvénic energy is $E_A \equiv m_i v_A^2/2 \sim 10\,\text{KeV}$, which is in the range of passing particles induced by neutral beam injection (NBI). For ITER parameters, $E_A \sim 1\,\text{MeV}$, which is in the range of passing α-particles born from the fusion reactions. In both cases, TAEs can be driven unstable by resonance with energetic particles. For fast-passing particles, the resonance condition is $\Omega = \omega_A$, where

$$\Omega = n\omega_\zeta + l\omega_\theta, \tag{3.3}$$

where $\omega_\zeta = v_\parallel/R_0$ and $\omega_\theta = v_\parallel/qR_0$ are frequencies of toroidal motion and poloidal motion, respectively, and $l = -m$ for co-passing particles, $l = m$ for counter-passing particles. For TAEs driven by co-injected ions, we can simplify the following analysis by considering only co-passing particles. Then, the resonance condition is

$$\omega_A - n\frac{v_\parallel}{R_0} + m\frac{v_\parallel}{qR_0} = 0. \tag{3.4}$$

In JET the TAE modes are routinely observed and used to interpret the $q(r,t)$ profiles in the evolving plasma. An early example is given by Fasoli, *et al.* (1995) for the measurement of the damping of toroidicity-induced Alfvén eigenmodes.

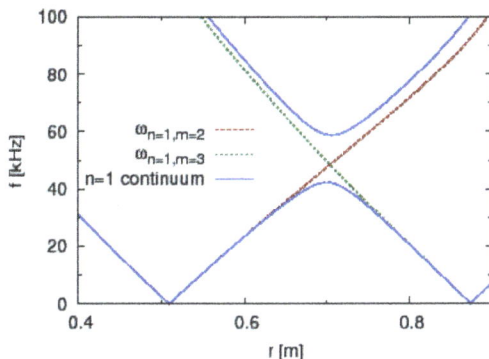

Fig. 3.4   Alfvén continuum for $n = 1$ with (solid line) and without (dashed line) coupling between $m = 2$ and $m = 3$ poloidal modes. Note that the discrepancy, relatively far from resonance, between the upper branch of coupled continuum and the uncoupled $m = 2$ branch, is accounted by terms of second-order in the inverse aspect ratio.

An important low-order TAE mode example, as shown in Fig. 3.4, occurs for the coupling of the $m = 1$ and $m = 2$ forward $\omega_+(1, n, r)$ and backward $\omega_-(m+1, n, r)$ Alfvén wave continuum modes. Writing the condition for the two waves

$$\omega_+ = k_\parallel(m, n, r)v_A \quad \text{and} \quad \omega_- = -k_\parallel(m, n, r)v_A \tag{3.5}$$

to be degenerate $\omega_+ = \omega_-$ gives

$$\left(\frac{n}{R} - \frac{m}{qR}\right)v_A = -\left(\frac{n}{R} - \frac{m+1}{qR}\right)v_A, \tag{3.6}$$

which reduces to

$$\frac{2n}{R} = (m+1+m)/qR \tag{3.7}$$

giving the resonant radius r_0 for the gap mode defined by

$$q(r_0, m, n) = \frac{m+1/2}{n}. \tag{3.8}$$

Extended radial TAE-modes occur for $m = 1, n = 1$ coupled to $m = 2, n = 1$ modes at $q(r_0, m, n) = (m + 1/2)/n$, as shown in Figs. 3.1 and 3.4.

Nonlinear simulations of the energetic particle destabilization and evolution of these TAE modes are described in Todo, *et al.* (2010). The saturation level is shown

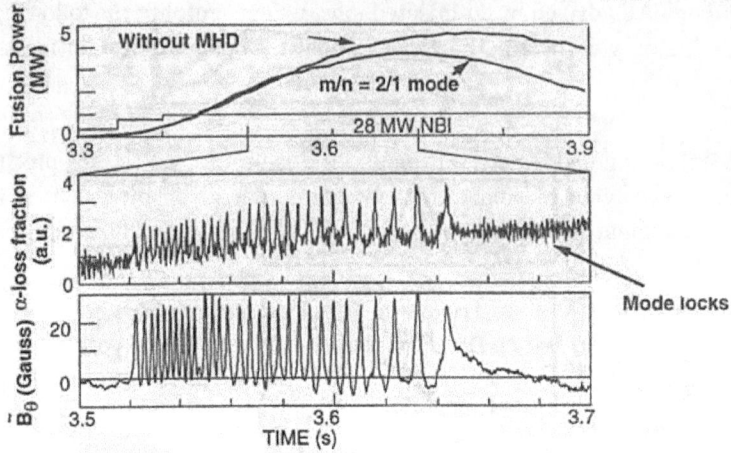

Fig. 3.5   With DT operations, it is found that the alpha particle loss can be significant because of the presence of the neoclassical MHD modes.

to require both nonlinear particle dynamics and nonlinear MHD dynamics when the amplitudes reach high levels with $\delta B/B \sim 10^{-2}$. There is a zonal flow and nonlinear coupling to high n-modes up to $n \sim 10$.

Other low-frequency MHD modes where the high energy-injected ions create instabilities are called energetic particles modes, or EP-modes, for short. One of the more interesting and still incompletely understood of the energetic particle modes is the fish-bone mode. How "fishbone" modes might appear in the burning plasma driven by the fusion products of alpha particles is a troublesome question. The name is derived from the plot of the fluctuating magnetic field versus time which has the structure resembling the skeleton with backbone of a large-flat Bonita type of fish.

Recent progress in the research of fishbones is given in Okabayashi, *et al.* (2011). The waves are driven by a population of high-energy ions resonating with the low-frequency modes through the bounce averaged gradient-B drift frequency with the transit motion frequency. The polarization of the magnetic field fluctuations indicate the mode has a compressional magnetic component in addition to the line-bending shear Alfvén wave component. The radial gradient of the fast ion distribution function is important. The mode ejects some fraction of the fast ions from the plasma and repeats in rapid succession, creating the "fishbone" signature on the magnetic field probes. New results on DIII-D show that when the mode occurs sufficiently off-axis the magnetic activity destabilizes or drives the resistive-wall mode. Experiments showing the importance of the toroidal Alfvén eigenmodes were carried out on TFTR by Wong, *et al.* (1991). A model of the pulsations observed from energetic particle modes is given in Lilley and Nyquist (2014).

3.3 Diagnostic Neutral Beam Injection

Real-time diagnostics are used to determine the state of the plasma. The diagnostics planned for ITER are extensive (`http://www.iter.org/mach/diagnostics`). Real-time diagnostics and controls will be installed on ITER to provide the measurements necessary to optimize plasma performance over the run-time of a discharge. ITER diagnostics will open a new regime for understanding the complex large-scale long-time dynamics of plasmas. The diagnostics include measurements of temperature, density, impurity concentration, and particle and energy confinement times.

The diagnostic systems will have about 50 individual measuring systems drawn from the full range of modern plasma diagnostic techniques, including lasers, X-rays, neutron cameras, impurity monitors, particle spectrometers, radiation bolometers, pressure and gas analysis, and optical fibers.

Because of the harsh environment in the vacuum vessel, these systems will have to cope with a range of phenomena not encountered in earlier diagnostic implementations in tokamaks. Higher levels of neutral particle fluxes and neutron fluxes will be respectively about 10 and 10,000 times higher than the harshest fluxes experienced in previous fusion machines. The pulse length of the fusion reaction, or the amount of time the reaction is sustained, will be about 100 times longer than in previous machines.

The intensity of the atomic line emission allows an estimate of the density of the impurity ions provided their density is above a threshold level. The charge exchange recombination spectroscopy (CERS) method is to inject a beam of neutral hydrogen atoms and measure the line radiation from the atomic recombination lines of the electrons in excited states of the atoms created by the charge-exchange atomic collision. CERS is used for measuring light impurity ions, such as C, B Be, He, or Li, which are fully ionized in a high temperature plasma and have no line emission without the neutral beam collisions. CERS can directly measure impurity transport and provides important parameters for turbulence models: ion temperature, impurity density, and plasma rotation. Figure 3.5 shows a schematic of the charge exchange collision of injected neutral hydrogen atom H^0_{beam} or neutral deuterium atom D^0_{halo} in a charge exchange collision with a helium ion. The collision results in transferring an electron from the neutral atom to the helium ion He^{+2} and leaves the fast-injected neutral beam atom as a singly charged fast ion. The charge exchange reaction has a high probability of putting the newly formed He^{+1} ion in an excited state, which promptly decays, emitting a photon measured spectroscopically.

Thus, the collection optics shown measures these emitted photons and uses the width and shift of the frequency of the 656 nm emission line to determine the mean speed or velocity component along the line of sight of the fully ionized plasma ions. $\varepsilon(r, \Omega, \lambda)$ is the light of wavelength $[\lambda, \lambda + d\lambda]$ emitted in the direction Ω. Table 3.1 gives the atomic reactions for the neutral hydrogen beam injected in a

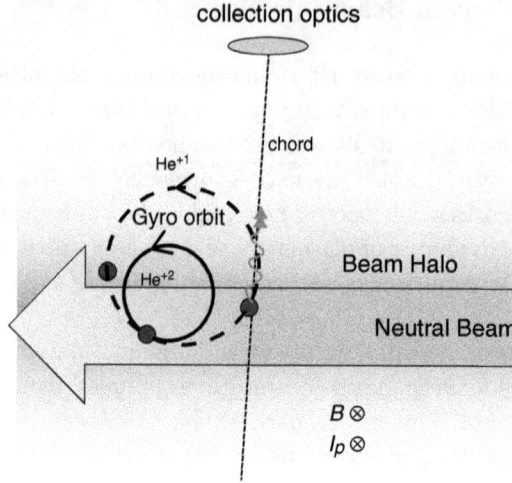

Fig. 3.6   Schematic of the Charge Exchange Recombination Spectroscopy system (CERS) used in fusion research. The injector accelerates hydrogen ions and then neutralizes most of the ions as they pass through a second chamber. Finally, a magnetic field is used to bend and remove any remaining hydrogen ions. The CERS diagnostic is used to find the level, the velocity, and the temperature of the impurity species, and to infer the corresponding data about the working gas ions [Liao (2014)].

helium plasma.

Table 3.1   Charge Exchange Recombination Spectroscopy

---

1. Diagnostic neutral beam charge exchange
   $$He^{+2} + H^0_{beam} \rightarrow He^{+1*} + H^+_{fast}$$
2. Halocharge exchange
   $$He^{+2} + D^0_{halo} \rightarrow He^{+1*} + D^+$$
3. Spontaneous emission
   $$He^{+1*} \rightarrow He^{+1} + \gamma$$
4. Ionization and return to initial ionized Helium atom (alpha particle)
   $$He^{+1} + e^- \rightarrow He^{+2} + 2e^-$$

The spectroscopic emission formulas used for the CERS analysis is of the form:

$$\varepsilon(\mathbf{r}, \Omega, \lambda)dV\, d\Omega\, d\lambda = \frac{1}{4\pi} \int \int f_n(\mathbf{r}, \mathbf{v}_n) f(\mathbf{r}, \mathbf{v}) q_{em}(|\mathbf{v}_r|) \epsilon(\mathbf{v}, \widehat{\mathbf{s}}, \Omega, \lambda) d^3\mathbf{v} d^3\mathbf{v}_n \quad (3.9)$$

where $\varepsilon(r, \Omega, \lambda)$ is the light of wavelength $[\lambda, \lambda + d\lambda]$ emitted in direction Ω. In Eq. (3.9) the kernel function ϵ is the atomic line emission model for the collision and f_n is the injected neutral atom phase space density. The line emission formula for $\epsilon(\boldsymbol{v}, \widehat{\boldsymbol{s}}, \boldsymbol{B}, \lambda)$ is a summation over each Zeeman fine structure component a_n.

$$\epsilon(\mathbf{v}, \widehat{\mathbf{s}}, \mathbf{B}, \lambda) = \sum_n a_n(\Omega)\delta\left[\lambda - \lambda_n\left(1 + \frac{\mathbf{v}\cdot\widehat{\mathbf{s}}}{c}\right)\right]. \quad (3.10)$$

An atomic spectral-diagnostic code called STRAHL [Dux (2006)] is used to analyze the CERS data. STRAHL uses a 1D transport code to solve for the density profile of species $\langle n_s \rangle$

$$\frac{\partial \langle n_s \rangle}{\partial t} = -\left(\frac{\partial V}{\partial \rho}\right)^{-1} \frac{\partial}{\partial \rho}\left(\frac{\partial V}{\partial \rho} \langle \Gamma_s^\rho \rangle\right) + \langle g_s \rangle \tag{3.11}$$

where the surface average flux for each ion species is given by

$$\langle \Gamma_s^r \rangle = -D_s^r(r)\frac{\partial n(r)}{\partial r} + V_s^r(r)n(r) \tag{3.12}$$

for a specified flux geometry.

The $D(r)$ and $V(r)$ transport coefficients of the transport flux $\langle \Gamma_s^r \rangle$ in Eq. (4.2) are determined by minimizing the deviation measured by the variance G functional given by

$$G[D(r), V(r)] = \sum_i^{N_r} \sum_j^{N_t} \left(n_{\text{meas}}(r_i, t_j) - \frac{\langle n_{\text{meas}} \rangle}{\langle n_{\text{pred}} \rangle} n_{\text{pred}}(r_i, t_j)\right)^2 \tag{3.13}$$

over an ensemble of $N_r \times N_t$ data points.

An atomic spectral emission code for plasmas named STRAHL generates a new fit by adding random value to current best fit and repeating the calculation of the error measure G defined in Eq. (3.13). Neural network software would be used to derive the best parameter set efficiently from large samples with $N_r N_t$ data points.

Extensive analysis for the helium density, temperature, velocity profiles for ^{3}He and ^{4}He on Alcator C-Mod shows that the transport from $D(r, t)$ and $V(r, t)$ is consistent with the turbulent transport of these ions from drift waves discussed in the next chapter.

Beam emission spectroscopy (BES) has been used to measure density fluctuations during impurity gas injection experiments on DIII-D and JET in discharges with negative central magnetic shear or reversed shear (RS) discharges. The results verify that the RS plasma have reduced the density fluctuation and reduced outward turbulent transport. The BES data also gives information on the radial electric field through the radial force balance on the impurity ions. The data shows that there is a positive feedback mechanism between the fluctuations and momentum transport such that increased rotation velocity increases the radial electric field enchanting the transport barrier [Doyle, *et al.* (2007)].

A diagnostic beam system was developed for the RFX reversed field pinch machine in Padova, Italy. Currently the system is loaned to Alcator C-mod, MIT, Boston. The diagnostic beam system is primarily used for measurement of the ion temperature by charge-exchange recombination spectroscopy and for internal magnetic field measurements emission frequency shifts arising from the motional Stark effect. The system comprises an ion source, beam duct equipped with vacuum pumps and with an array of various diagnostics of the beam [Ivanov, *et al.* 2000); Ivanov, *et al.* 2008); Korepanov, *et al.* (2004); Deichuli, *et al.* (2006)].

3.4   Resistive Wall Modes

The finite electrical resistivity η of the metal walls of the vacuum chamber along with its thickness, Δ, gives rise to the resistive wall diffusion time given by $\tau_{\text{wall}} = \mu_0\, d\Delta/\eta$, which is the time for the currents induced by edge-electric field from $\delta B/\delta t$ at the inner chamber wall to soak through the wall. For times longer than the resistive wall time τ_{wall}, the chamber wall is not forcing the boundary condition of vanishing δB_r at the wall as used in MDH stability analysis. Instead, new slow-growing modes become unstable that are known as resistive wall modes (RWM). These resistive wall modes will become more limiting as the tokamak machines operate for long-time pulses and particularly in the steady state treated in Chapter 7.

Under certain conditions toroidal plasma rotation can stabilize the resistive wall modes while under other conditions the plasma rotation can destabilize one of the two branches of the stabilized kink-ballooning modes if the plasma pressure is not far below the critical plasma β_{MHD} for onset of the kink-ballooning instability. Under these conditions, close to the critical β_{MHD} one of the two branches of the MHD mode is Doppler shifted to zero frequency by the plasma rotation and then becomes unstable from the electromagnetic interaction with the resistive wall. The growth rate $\gamma_{\text{res}-\text{wall}}$ mode is slow but nevertheless, the mode possesses a serious threat to the plasma confinement in a quasi-steady state plasma [Bondeson and Ward (1994); Berkery, *et al.* (2010); Hirota and Aiba (2014)]. The long pulse ITER machine is needed to investigate the ultimate role of the resistive wall modes on fusion plasmas.

An extended magnetohydrodynamics (MHD) code M3D [Park, *et al.* (1999)] includes separate density and temperature, over the entire region toroidal simulation [Sugiyama and Park (2000); Sugiyama (2008); Sugiyama (2013); Sugiyama (2014)]. The simulations show that Edge Localized Modes (ELMs) in high temperature, toroidal fusion plasmas a rise from a nonlinear plasma instability. This instability is from steep temperature gradients near the magnetic separatrix. The work builds on work from Strauss (2009) on the complex nature of resistive modes taking into account the complex shape of the machine chamber walls and the currents induced in the walls as the plasma moves towards the walls.

References

Berk, H. L, Breizman, B. N., and Petviashvili, N. V. (1997). Spontaneous hole-clump pair creation in weakly unstable plasmas, *Phys. Lett. A* **234**, 3, pp. 213-218, doi:10.1016/S0375-9601(97)00523-9.

Berk, H. L. (2012). 12th IAEA Technical Meeting on Energetic Particles in Magnetic Confinement Systems, ed. H. L. Berk, *Nucl. Fusion* **52**, p. 9.

Berkery, J. W., Sabbagh, S. A., Betti, R., Hu, B., Bell, R. E., Gerhardt, S. P., Manickam, J., and Tritz, K. (2010). Resistive Wall Mode Instability at Intermediate Plasma Rotation, *Phys Rev. Lett.* **104**, p. 035003, http://dx.doi.org/10.1103/PhysRevLett.104.035003.

Bondeson, A., and Ward, D. J. (1994). Stabilization of external modes in tokamaks by resistive walls and plasma rotation, *Phys. Rev. Lett.* **72**, p. 2709, http://dx.doi.org/10.1103/PhysRevLett.72.2709.

Bosch, H.-S., and Hale, G. M. (1992). Improved formulas for fusion cross-sections and thermal reactivities, *Nucl. Fusion* **32**, p. 611, doi:10.1088/0029-5515/32/4/107.

Breizman, B. N., and Sharapov, S. E. (1995). Energetic particle drive for toroidicity-induced Alfvén eigenmodes and kinetic toroidicity-induced Alfvén eigenmodes in a low-shear tokamak, *Plasma Phys. Control. Fusion* **37**, p. 1057, doi:10.1088/0741-3335/37/10/001.

Cheng, C. Z., Chen, L., and Chance, M. S. (1985). High-n ideal and resistive shear Alfvén waves in tokamaks, *Ann. Phys.* **16**, pp. 21-47 doi:10.1016/0003-4916(85)90335-5.

Connor, J. W., Hastie, R. J., and Taylor, J. B. (1991a). Resonant magnetohydrodynamic modes with toroidal coupling. Part I: Tearing modes, *Phys. Fluids B* **3**, p. 1532, http://dx.doi.org/10.1063/1.859724.

Connor, J. W., Hastie, R. J., and Taylor, J. B. (1991b). Resonant magnetohydrodynamic modes with toroidal coupling. Part II: Ballooning-twisting modes, *Phys. Fluids B* **3**, p. 1539, http://dx.doi.org/10.1063/1.859993.

Coppi, B., and Porcelli, F. (1986). Theoretical model of fishbone oscillations in magnetically-confined plasmas, *Phys. Rev. Lett.* **57**, p. 2272, doi:http://dx.doi.org/10.1103/PhysRevLett.57.2272.

Deichuli, P. P., Abdrashitov, G. F., Ivanov, A. A., Kolmogorov, V. V., Mishagin, V. V., Shul'zhenko, G. I., Stupishin, N. V., Beals, D. and Granetz, R. (2006). Ion source with LaB6 hollow cathode for a diagnostic neutral beam injector, *Rev. Sci. Instrum.* **77**, p. 03B514, doi:10.1063/1.2171754.

Doyle, E. J., Houlberg, W. A., Kamada, Y., Mukhovatov, V., Osborne, T. H., Polevoi, A., Bateman, G., Connor, J. W., Cordey, J. G., Fujita, T., Garbet, X., Hahm, T. S., Horton, L. D., Hubbard, A. E., Imbeaux, F., Jenko, F., Kinsey, J. E., Kishimoto, Y.,

Li, J., Luce, T. C., Martin, Y., Ossipenko, M., Parail, V., Peeters, A., Rhodes, T. L., Rice, J. E., Roach, C. M., Rozhansky, V., Ryter, F., Saibene, G., Sartori, R., Sips, A. C. C., Snipes, J. A., Sugihara, M., Synakowski, E. J., Takenaga, H., Takizuka, T., Thomsen, K., Wade, M. R., Wilson, H. R., ITPA Transport Physics Topical Group, ITPA Confinement Database and Modeling Topical Group and ITPA Pedestal and Edge Topical Group (2007). Chapter 2: Plasma confinement and transport *Nucl. Fusion* **47**, pp. S18-S127, doi:10.1088/0029-5515/47/6/S02.

Duane, B. H. (1972). Fusion Cross-Section Theory, Rept. BNWL-1685 (Brookhaven National Laboratory).

Dux, R. (2006). STRAHL User Manual, Max-Planck-Institut für Plasmaphysik, IPP 10/30.

Fasoli, A., Borba, D., Bosia, G., Campbell, D. J., Dobbing, J. A., Gormezano, C., Jacquinot, J., Lavanchy, P., Lister, J. B., Marmillod, P., Moret, J.-M., Santagiustina, A., and Sharapov, S. E. (1995). Direct measurement of the damping of toroidicity-induced Alfvén eigenmodes, *Phys. Rev. Lett.* **75**, 4, pp. 645-648, doi:http://dx.doi.org/10.1103/PhysRevLett.75.645.

Fu, G. Y., and Van Dam, J. W. (1989). Excitation of the toroidicity-induced shear Alfvén eigenmode by fusion alpha particles in an ignited tokamak, *Phys. Fluids B* **1**, p. 1949, http://dx.doi.org/10.1063/1.859057.

Gorelenkov, N. N., Berk, H. L., Fredrickson, E., Sharapov, S. E., and JET EFDA Contributors. (2007). Predictions and observations of low-shear beta-induced shear Alfvén-acoustic eigenmodes in toroidal plasmas, *Phys. Letts. A* **370**, 1, pp. 70-77, doi:10.1016/j.physleta.2007.05.113.

Günter, S., Gude, A., Lackner, K., Maraschek, M., Pinches, S., Sesnic, S., Wolf, R., and ASDEX Upgrade Team (1999). The influence of fishbones on the background plasma, *Nucl. Fusion* **39**, p. 1535, doi:10.1088/0029-5515/39/11/304.

Heidbrink, W. W., Bol, K., Buchenauer, D., Fonck, R., Gammel, G., Ida, K., Kaita, R., Kaye, S., Kugel, H., LeBlanc, B., Morris, W., Okabayashi, M., Powell, E., Sesnic, S., and Takahashi, H. (1986). Tangential neutral-beam-driven instabilities in the Princeton Beta Experiment, *Phys. Rev. Lett.* **57**, p. 835, doi:http://dx.doi.org/10.1103/PhysRevLett.57.835.

Hirota, M., and Aiba, N. (2014) preprint.

Hirota, M., and Aiba, N. (2009). *Plasma Sci. Tech.* **11**, p. 409.

ITER Physics Expert Group on Confinement and Transport (1999). *Nucl. Fusion* **39**, p. 2175, doi:10.1088/0029-5515/39/12/302.

Ivanov, A. A., Davydenko, V. I., Deichuli, P. O.,

Ivanov, A. A., Davydenko, V. I., Deichuli, P. P., Kreter, A., Shulzhenko, G. I., and Stupishin, N. V. (2008). Ion sources with arc-discharge plasma box driven by directly heated LaB$_6$ electron emitter or cold cathode (invited), *Rev. Sci. Instrum.* **79**, p. 02C103, doi:10.1063/1.2798503.

Jet Team. (1998). International Atomic Energy Agency conference proceedings.

Keilhacker, M., Gibson, A., Gormezano, C., Lomas, P.J., Thomas, P. R., Watkins, M. L., Andrew, P., Balet, B., Borba, D., Challis, C. D., Coffey, I., Cottrell, G. A., De Esch, H. P. L., Deliyanakis, N., Fasoli, A., Gowers, C. W., Guo, H. Y., Huysmans, G. T. A., Jones, T. T. C., Kerner, W., König, R. W. T., Loughlin, M. J., Maas, A., Marcus, F. B., Nave, M. F. F., Rimini, F. G., Sadler, G. J., Sharapov, S. E., Sips, G., Smeulders, P., Söldner, F. X., Taroni, A., Tubbing, B. J. D., von Hellermann, M. G., Ward, D. J., and JET Team. (1999). High-fusion performance from deuterium-tritium plasmas in JET, *Nucl. Fusion* **39** p. 209, doi:10.1088/0029-5515/39/2/306.

Korepanov, S. A., Abdrashitov, G. F., Beals, D., Davydenko, V. I., Deichuli, P. P., Granetz, R., Ivanov, A. A., Kolmogorov, V. V., Mishagin, V. V., Puiatti, M.,

Rowan, B., Stupishin, N. V., Shulzhenko, G. I., and Valisa, M. (2004). Neutral beam injector for active plasma spectroscopy, *Rev. Sci. Instrum.* **75**, 5, pp. 1829-1831, doi:10.1063/1.1699513.

Liao, K. T. (2014). Helium Charge Exchange Spectroscopy on Alcator C-Mod Tokamak. PhD Thesis. The University of Texas at Austin.

Lilley, M. K., and Nyquist, R. M. (2014). Formation of Phase Space Holes and Clumps, *Phys. Rev. Lett.* **112**, p. 155002, doi:10.1103/PhysRevLett.112.155002.

McGuire, K., Goldston, R., Bell, M., Bitter, M., Bol, K., *et al.* (1983). Study of high-beta magnetohydrodynamic modes and fast-ion losses in PDX, *Phys Rev. Lett.* **50**, p. 891, doi:http://dx.doi.org/10.1103/PhysRevLett.50.891.

Mikhailovskii, A. B. (1992). *Electromagnetic Instabilities in an Inhomogeneous Plasma* (Institute of Physics Publishing). ISBN:0-7503-0182-1.

Nazikian, R., Kramer, G. J., Cheng, C. Z., Gorelenkov, N. N., Berk, H. L., and Sharapov, S. E. (2003). New interpretation of alpha-particle-driven instabilities in deuterium-tritium experiments on the Tokamak Fusion Test Reactor, *Phys Rev. Lett.* **91**, pp. 125003-1, doi:http://dx.doi.org/10.1103/PhysRevLett.91.125003.

Nazikian, R., Fu, G. Y., Batha, S. H., Bell, M. G., Bell, R. E., Budny R. V., Bush, C. E., Chang, Z., Chen, Y., Cheng, C. Z., Darrow, D. S., Efthimion, P. C., Fredrickson, E. D., Gorelenkov, N. N., Leblanc, B., Levinton, F. M., Majeski, R., Mazzucato, E., Medley, S. S., Park,, H. K., Petrov, M. P., Spong, D. A., Strachan, J. D., Synakowski, E. J., Taylor, G., Von Goeler, S., White, R. B., Wong, K. L., and Zweben, S. J. (1997). Alpha particle driven toroidal Alfvén eigenmodes in the tokamak fusion test reactor, *Phys. Rev. Lett.* **78**, p. 2976, doi:http://dx.doi.org/10.1103/PhysRevLett.78.2976.

Okabayashi, M., Matsunaga, G., deGrassie, J. S., Heidbrink, W. W., In, Y., Liu, Y. Q., Reimerdes, H., Solomon, W. M., Strait, E. J., Takechi, M., Asakura, N., Budny, R. V., Jackson, G. L., Hanson, J. M., La Haye, R. J., Lanctot, M. J., Manickam, J., Shinohara, K., and Zhu, Y. B. (2011). Off-axis fishbone-like instability and excitation of resistive wall modes in JT-60U and DIII-D, *Phys. Plasmas* **18**, p. 056112, doi:10.1063/1.3575159, http://dx.doi.org/10.1063/1.3575159.

Park, W., Belova, E. V., Fu, G. Y., Tang, X. Z., Strauss, H. R., and Sugiyama, L. E. (1999). Plasma simulation studies using multilevel physics models, *Phys. Plasmas* **6**, p. 1796, http://dx.doi.org/10.1063/1.873437.

Strauss, H. R., Sugiyama, L., Park, G. Y., Chang, C. S., Ku, S., and Josheph, I. (2009). Extended MHD simulation of resonant magnetic perturbations, *Nucl. Fusion* **49**, 5, p. 055025, doi:10.1088/0029-5515/49/5/055025.

Sugiyama, L. E. (2014). Compressible magnetohydrodynamic sawtooth crash, *Phys. Plasmas* **21**, p. 022510, http://dx.doi.org/10.1063/1.4865571.

Sugiyama, L. E. (2013). On the formation of $m = 1, n = 1$ density snakes, *Phys. Plasmas* **20**, p. 032504, http://dx.doi.org/10.1063/1.4793450.

Sugiyama, L. E. (2008). Guiding center plasma models in three dimensions, *Phys. Plasmas* **15**, p. 092112, http://dx.doi.org/10.1063/1.2977981.

Sugiyama, L. E., and Park, W. (2000). A nonlinear two-fluid model for toroidal plasmas, *Phys. Plasmas* **7**, p. 4644, http://dx.doi.org/10.1063/1.1308083.

Todo, Y., Berk, H. L., and Breizman, B. N. (2010). Nonlinear magnetohydrodynamic effects on Alfvén eigenmode evolution and zonal flow generation, *Nucl. Fusion* **50**, pp. 084016-084025, doi 10.1088/0029-5515/50/8/084016.

Van Zeeland, M. A., Gorelenkov, N. N., Heidbrink, W. W., Kramer, G. J., Spong, D. A., Austin, M. E., Fisher, R. K., Garcia Muñoz, M., Gorelenkova, M., Luhmann, N., Murakami, M., Nazikian, R., Pace, D. C., Park, J. M., Tobias, B. J., and White,

R. B. (2012). Alfvén eigenmode stability and fast ion loss in DIII-D and ITER reversed magnetic shear plasmas, *Nucl. Fusion* **52** p. 094023, doi:10.1088/0029-5515/52/9/094023.

Wong, K. L., Fonck, R. J., Paul, S. F., Roberts, D. R., Fredrickson, E. D., Nazikian, R., Park, H. K., Bell, M., Bretz, N. L., Budny, R., Cohen, S., Hammett, G. W., Jobes, F. C., Meade, D. M., Medley, S. S., Mueller, D., Nagayama, Y., Owens, D. K., and Synakowski, E. J. (1991). Excitation of toroidal Alfvén eigenmodes in TFTR, *Phys. Rev. Lett.* **66**, pp. 1874-1877, doi:10.1103/PhysRevLett.66.1874.

Chapter 4

Turbulent Transport from the Temperature Gradients

This chapter derives and surveys the instabilities producing the turbulence that limits the energy confinement in the tokamaks. The turbulence is driven by the strong radial gradients of temperature $T_e(r), T_i(r)$ profiles and the gradients at scrape-off layer (SOL) of plasma density $n_e(r)$. These density and temperature gradients combine to determine the pressure gradients of $p(r,t) = n(r)T(r)$. The profiles and their gradients are driven by the intense injected power and relax by the turbulent transport of particles and thermal energy.

As discussed in Chapters 1 and 2, the pressure gradient produces a basic MHD instability that arises from the unfavorable magnetic curvature, as described in Fig. 1.4 of Chapter 1. This is the plasma analog of the Rayleigh-Taylor instability of neutral fluids in a gravitational field, with the rising columns of lower density fluid in the atmosphere and the oceans. The temperature gradient instabilities require thermal transport equations so they are beyond the adiabatic gas description of MHD. The dynamics of the local density fluctuation and temperature fluctuations quickly become nonlinear and create vortex structures that have their own dynamics as coherent structures [Waelbroeck, *et al.* (2004)]. The larger structures have a significant local polarization electric field that accelerates them out of the regions of steep radial gradients at the magnetic separatrix where they grow large to become vortices or blobs and bubbles near the magnetic separatrix (SX) into the scrape-off layer outside the magnetic separatrix. (The name for these coherent nonlinear structures varies depending on the type and origin.) Most of the hot plasma from the large vortex fluxes flows into the divertor chamber.

When the plasma is driven hard to reach the maximum fusion power performance, the pressures and temperatures increase and the plasma turbulence grows stronger attempting to lower the gradients by these expulsion events also known by a variety names in the literature depending on their size and location. The largest type is seen to occur repeatedly at the hot plasma edge and is known as Edge Localized Modes (ELMs). The ELMs become large and rapid as the injected heating power pushes the core temperatures and pressure to their highest levels. The ELMs are transient events in which hot plasma "blobs" or vortex structures propagate out across the magnetic separatrix were into the scrape-off layer (SOL) beyond the magnetic separatrix. These Edge Localized Modes, now universally referenced as ELMs, occur as quasi-periodic releases of about 5 to 10% of the stored plasma energy when the plasma is driven hard. In Chapter 5, Fig. 5.3 shows an example of an ELM-ing discharge. During the expulsion of the hot plasma blob by a local vortex structure the plasma mass is conserved with a high-pressure blob propagating outward and a low-pressure "bubble" propagating inward. Both structures produce drops in the thermal energy of the confined hot plasma. When the plasma heating is lowered, these structures change character into less severe and more frequent pulsations named Type III ELMs. While Type III ELMs and another regime called the I-mode for improved confinement hold promise for the steady-state fusion power regimes. The search for record-high confinement temperatures and neutron production levels has led experimentalists to push the plasmas into the ELMy H-mode regime to achieve new fusion power records of neutron production rates and the record values of the fusion measure Q_{DT}. In the future, the large volume ITER machine may have improved confinement sufficiently to operate in the I-mode regime or in the Type III ELM regime with a fusion Q_{DT} below the record level, but still greater than unity.

Figures 5.3 and 5.4 show the data from a relatively long discharge that makes a transition into a state where there is a series of these ELM events, as the plasma adjusts to the intense-injected power absorbed inside the separatrix. These periods called ELM-ing states are of intense interest and concern, and thus are treated in detail in Chapter 5. Currently, there are plans to install – outside the first chamber wall – an auxiliary set of current-carrying coils with particular mode numbers (m, n) to fight (or restrain) the growth of large ELMs. As described in some detail in Chapter 5, these external current-carrying coils are basically antennas that produce Magnetic Resonance Perturbations (RMPs). Extensive simulation modeling is being performed to design RMP coils for ITER using tests on several tokamaks with fusion grade plasma for validation. A final decision on adding RMP coils has not been made.

An example of both energy and particle losses induced in an ELMy H-mode plasma in JT-60U is presented in Asakura, *et al.* (1997). Data from a 15-channel photomultiplier (PMT) is used to reconstruct the both the ELMs propagating across the outboard separatrix and MARFE structures in the X-point region connecting the hot plasma to the low-temperature divertor chamber plasma. Deuterium gas is injected into the hot plasma and the resulting spectral profiles of the line radiation at 656 nm is measured for the reconstruction of the plasma dynamics. The ELM frequency is from 110 to 160 Hz. The drops of the plasma stored energy are measured with the diamagnetic loop antennas giving $W_{dia}(t)$ and from infrared TV camera images. For this machine, the ELMs set in when the injected neutral beam power P_{NBI} exceed 6 MW and the plasma density was high approaching the Greenwald density limit $n_e/n^{Gr} \lesssim 1$. Within a constant, the Greenwald density limit is given by $n^{Gr} = I_p/\pi a^2$ where a is the minor radius and I_p the plasma current.

First we begin the turbulent transport discussion with the classic density and temperature gradient driven instabilities for which many basic plasma physics experiments have been performed around the world, verifying the universality of the drift wave instabilities and their turbulent transport [Horton (2012)]. Then we turn to the ELMs and other more complex turbulent transport events and structures.

4.1 Drift Wave Instabilities from Density and Pressure Gradients

The uniform plasma has acoustic wave with the dispersion relation for waves with electric fields and density waves $\cos[(\boldsymbol{k} \cdot \boldsymbol{x}) - \omega t]$ is determined by $\delta n_e = \delta n_i$ for low frequencies ($\lesssim 500\,\mathrm{kHz}$) and wavelengths, long compared to the Debye length $\lambda_{De} = v_e/\omega_{pe} = (\epsilon_0 T_e/e^2 n_e)^{1/2}$, which is on the scale of millimeters. The Debye length is a very small space scale compared with the wavelengths of the drift waves and MHD modes of principal concern in large tokamaks. Only pure electron waves exist on the millimeter scale.

The drift waves have a vector wavenumber $\boldsymbol{k}$ that has both a component $k_\parallel$ parallel to the local magnetic field and a perpendicular vector component $\boldsymbol{k}_\perp$ to the local magnetic field vector. The electric field is largely electrostatic given by the fluctuating electric potential $\delta\phi_{\boldsymbol{k},\omega}$ and the fluctuating plasma density $\delta n_{\boldsymbol{k},\omega}$. Owing to the long wavelength $2\pi/k$ compared to the Debye length, the ion and electron density fluctuations are equal in the order of $k^2\lambda_{De}^2$. The electrons and ions $\boldsymbol{E} \times \boldsymbol{B}$ drift across the magnetic field with ions moving somewhat slower than the electrons due to their averaging the wave over their larger gyroradius. This effect is called the Finite Larmor Radius (FLR) effect and introduces a form factor given by $J_0(k_\perp v_\perp/\Omega)$ in the effective electric potential acting on the ions. The effective electric fields of the response functions of particles to fluctuations $\boldsymbol{E}(\boldsymbol{k})J_0(k_\perp v_\perp/\Omega)\exp[i(\boldsymbol{k}_\perp \cdot \boldsymbol{x})]$ is the electric field in the plasma which reduces the drift velocity of the particles with larger gyroradius $\rho = v_\perp/\Omega$ orbits.

The electron response to these low-frequency waves is called adiabatic since the electrons move so fast over the wavelengths that the wave is almost stationary from the point of view of the electrons. The dispersion relation then becomes that of ion acoustic waves $[1 + k^2\rho_s^2]\omega^2 - \omega_{*e}\omega - k_\parallel^2 c_s^2 = 0$ where the important new frequency ω_{*e} results from the $\boldsymbol{E} \times \boldsymbol{B}$ convection of the plasma over the nonuniform plasma density $n = n(r)$ profile. This drift frequency ω_{*e} measures the effect of the local density gradient on the ion acoustic wave.

One branch of this dispersion relation moves faster and the other branch slower owing to the density gradient. The faster branch becomes unstable in the drift wave instability. The literature for the drift waves introduces the local density and temperature gradient scale lengths defined by L_n

$$\frac{1}{L_n} = -\frac{d}{dr}\ln n(r) = -\frac{1}{n}\frac{dn}{dr} \tag{4.1}$$

and by L_T with the corresponding definitions such as for the ion and electron temperature gradients $1/L_{T_i}$ and $1/L_{T_e}$ scale lengths. Basic plasma experiments measuring the properties of the drift waves are described in historical perspective in Horton (2012).

Often it is sufficient to take L_n as a local constant parameter characteristic of a particular region of the density profile. The same type of local gradient scales lengths L_{T_e}, L_{T_i} and L_p are used for the gradients of the electron and ion temperatures and the plasma pressure, which depend on all three scale lengths $L_n, L_{T_e}, L_{T_i}, T_e$ from the measured temperature profiles $T_e(r)$ and $T_i(r)$. In the following chapters and in the literature for turbulence, the various stable and unstable plasma waves are defined by the eigenmodes from the vanishing of the determinant $D(\boldsymbol{k}, \omega) = 0$ for response matrix expressed in terms of the usual Stix matrix structure. The matrix elements are the complex dielectric response functions $\epsilon_{i,j}(\omega, \boldsymbol{k})$ being functions of the characteristic drift wave frequencies ω_{*j} and the FLR Bessel functions $J_0^2(k_\perp v_\perp/\Omega)$.

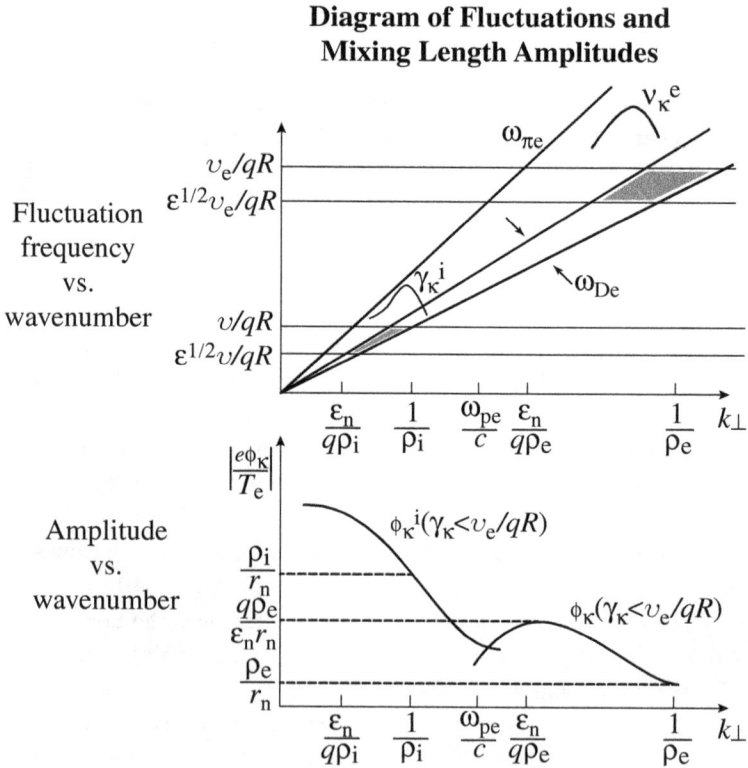

Diagram of Fluctuations and Mixing Length Amplitudes

Fig. 4.1　A diagram for the distribution of the fluctuations and wave amplitudes ranging from the ion transit frequency through the ion temperature gradient-trapped electron mode (ITG-TEM) regime. The high-frequency, small-scale regime of the ETG turbulence is above that of the ITG turbulence. The upper part (a) gives the frequency versus wavenumber region of the turbulence and particle transit and bounce motions. The lower part (b) gives the amplitudes for the mixing length level of saturation, where the estimate is for the regime with $L_T \sim L_n < a \ll R$ for the typical well-confined tokamak plasma [Horton, *et al.* (1988)].

Figure 4.1 gives a broad view of the spectrum of frequency versus wavenumber for the drift waves in a tokamak. The following sections of Chapter 4 describe these turbulence regimes in some detail. The frequency spectrum extends from a few kHz to MHz. The general scheme of the electron temperature gradient driven turbulence is shown in Fig. 4.1.

The general form of the diamagnetic gradient drift wave frequencies for plasma species $s = i, e, \alpha$ is

$$\omega_{*,s} = k_y \frac{T_s}{e_s B n_s} \frac{dn_s}{dr}. \tag{4.2}$$

4.1.1 *Drift wave frequencies and instabilities from density and temperature gradients*

One sees that for magnetically confined plasma the drift wave frequency for the electrons is in the positive $d\theta/dt$ direction as is the $\boldsymbol{E} \times \boldsymbol{B}$ rotation for an inward-pointing radial electric field $E_r < 0$, which is typically the direction for the radial electric field for well-confined plasmas. The exception occurs when the plasma has a high bulk flow velocity in the laboratory reference frame as occurs with strong unbalanced neutral beam injection that produce fast toroidal rotation. The origin of the sign $E_r < 0$ is that the electrons follow the ions, owing to their very small mass, so the ions must be pushed inward by the electric field to keep the plasma well confined and then the electrons will follow the positive ions.

Of course, this is a rather simplified explanation of a complex equilibrium problem and some plasmas are produced with outward radial electric fields. In general these plasmas are called as having the "electron root" of the ambipolar plasma neutrality condition. Plasmas with the inward-pointing electric field are called ion root ambipolar plasmas. The roots can be controlled by external biasing of the chamber walls as in the Large Plasma Device (LAPD) and the Helimak or by strongly non-ambipolar particle drifts of the guiding centers as in the complex helical system plasmas. Generally, plasmas with the inward-pointing radial electric fields show better confinement and lower fluctuations levels. We return to this point in Chapter 7 where the transitions between the low (L-mode) confinement state and the high (H-mode) confinement states are described. The transition to the H-mode state requires less auxiliary plasma heating power when the initial plasma has the inward pointing electric field.

4.1.2 *Instabilities from magnetic curvature and toroidal plasma currents*

The plasma literature uses η_s to define the ratio of the temperature gradient to the density gradient as

$$\eta_s = \frac{d}{dr} \ln T_s(r) / \frac{d}{dr} \ln n_s(r).$$

As in neutral fluids, the ratio of the temperature gradient to the density gradient is important for determining the stability of the fluid. When either the ion or the electron cross-field plasma pressure gradients

$$\frac{dp_e}{dr} = \frac{-p_e(1 + \eta_e)}{L_{n_e}}$$

and/or

$$\frac{dp_i}{dr} = \frac{-p_i(1 + \eta_i)}{L_{n_i}}$$

become sufficiently high from the auxiliary electron and ion heating power, the plasma dynamics becomes fast and the very light electrons move rapidly so as to neutralize the electric field in the plasma rest frame, giving the idealized state with $E + v \times B = 0$. This equation defines the highly-idealized infinity conductive MHD model.

By Faraday's law the magnetic field changes from the curl of the electric field and thus the $v \times B$ motion produces a changing magnetic field given by

$$\frac{\partial B}{\partial t} = \text{curl}(v \times B). \tag{4.3}$$

In this regime we have, or the plasma has, the Alfvén waves with the dispersion relation

$$\omega_k^2 = k_{\parallel}^2 \frac{B^2}{[\mu_0 \sum_s m_s n_s]} \equiv \omega_A^2 \tag{4.4}$$

where the total mass density is $\rho = \sum_s m_s n_s$. The restoring force in these oscillations arises from $j \times B$ force, where in the linear regime the $B = B_0(r)$ of the unperturbed magnetic field structure defines the eigenmodes $\delta B_{k,\omega}(x)$.

This single highly-conducting fluid description has nice mathematical properties but is too idealized to describe the laboratory plasma dynamics for magnetic confinement in tokamaks. Nevertheless there are many simulation codes based on the mathematical properties of the single-fluid ideal MHD model dynamics. The model gives the user a guide to the real plasma by several comparison theorems. For example, this single-pressure $p(x, t)$ plasma has infinite or idealized infinite electrical conductivity and thus the magnetic field is frozen into the plasma, like a superconductor, and moves with the plasma's motion $v(x, t)$.

Waves in the local plasma pressure gradient in the confinement regions follow from a generalization of Eq. (4.4) that includes currents from local pressure gradient and the associated magnetic field gradients. The resulting dispersion relations shows that there is a maximum local pressure gradient $dp/dr = -p/L_p$ for stable MHD motions. The balance of the energy released by convection across the pressure gradient with the wave's magnetic energy density $(\delta B)^2/2\mu_0$ gives the stability limit on the plasma pressure. The formula for the critical plasma pressure is usefully written as

$$\beta_{\text{mhd}}^{\text{crit}} = \frac{L_p}{q^2(r)R} \leq \frac{\epsilon_p}{q_{\text{min}}^2}, \tag{4.5}$$

where the pressure gradient is expressed locally as $dp/dr = -p/L_p$ and a local value of the parallel wavenumber for the Alfvén wave $k_\parallel = 1/q(r)R$ is used. Since the pressure gradient length L_p is connected to the mirror radius a of the machine, Eq. (4.5) is used to determine the maximum plasma pressure for the machine. This pressure gradient limit from MHD is a fundamental rule of plasma confinement. The local beta limit $\beta_{\text{mhd}}^{\text{crit}}$ is rather restrictive when evaluated with $q = 2$ to 3 owing to the low value of the stabilizing $k_\parallel v_A$ restoring force. The tokamak database shows this formula to be very predictive for the maximum achievable plasma pressure. A form of this limit is called the Troyon beta limit, as discussed in Chapter 1. The confinement time formula in Eq. (1.1) in Chapter 1 shows the strong dependence on the strength of the poloidal magnetic field B_p produced by the toroidal plasma current I_p through the factor $1/q^2$ in the Troyon beta limit formula.

As one approaches the rational magnetic surfaces defined in Chapter 2 in Eq. (2.2), where locally $k_\parallel(r) = 0$ and the magnetic field winding number $1/q = n/m$ is rational, the wave function is now of the form $f[m\theta - n\phi]$ and the stabilization from Eq. (4.4) vanishes altogether. One then must solve a radial eigenvalue problem called the Syudam problem when performed in the cylindrical limit and called the ballooning mode eigenvalue stability problem when carried out in the toroidal geometry.

From the discussion in Chapter 1.4 we see that the effective outward force from bending a cylindrical chamber into a torus gives the pressure gradient driving force up the magnetic field gradient ∇B on the inside $\theta = +\pi$ and down the pressure gradient on the outboard $\theta = 0$ side of the torus. The analysis gives the local formula for the frequency (squared) in this region as

$$\omega^2(\theta) = \frac{-k_\theta^2}{(k_r^2 + k_\theta^2)} \frac{(T_i + T_e)}{\langle m_i \rangle RL_p} \times [\cos(\theta) + s\theta \sin(\theta)].\tag{4.6}$$

Formula (4.6) shows quantitatively how the driving force for the ballooning mode varies with radius through the magnetic shear $s(r)$ and through poloidal position with θ. Formula 4.6 clearly shows the strongly unstable ($\omega^2 < 0$) region on the outboard side of the torus where $\theta = 0$ with growth rate of order $v_T/(RL_p)^{1/2}$ of order $10^6/\text{s}$.

The variation in θ clearly depends on the sign and magnitude of the magnetic field shearing given by

$$s(r) = \left[\frac{r}{q(r)}\right]\frac{dq}{dr}.\tag{4.7}$$

Since $q(r)$ defines the twisting of the helical magnetic field, the shearing rate $s(r)$ has the radial derivative of the rotational magnetic field given by $1/q(r)$. It is shown in Chapter 6 that the plasma confinement data improves in discharges with regions of negative magnetic shear $s(r) \leq 0$, with the plasmas having markedly different confinement properties than those with positive definite magnetic shear. These discharges are favored and are called reversed shear (RS) and enhanced reversed shear (ERS) plasmas.

The ITER database shows a weak dependence of the confinement time on B_T and a strong dependence on the poloidal magnetic field B_p proportional to I_p/a or q_{95}. Thus, the use of Eqs. (4.6) must be modified by other functions of q and the ion mass for the isotope effect when used to interpret or predict plasma confinement.

As we discuss the data in Chapter 6 the reversal of the sign of $s(r)$ from positive to negative has a strong effect on the ballooning-interchange mode stability, as given in Eq. (4.6). Plasmas with negative or reversed magnetic shear often develop a transport barrier, as described in Chapter 6, across which the plasma remains stable even for steep pressure gradients. The reversed shear discharges in JET, JT-60U, and TFTR have produced the highest fusion Q-values of all the numerous types of discharges as described in Chapter 6.

There are highly-developed computer simulation codes for finding the eigenvalues of the MHD linearized equations. The corresponding eigenmode structures are complex and vary for all types of shear, pressure, and current profiles. The code most widely used for ITER is the JOREK, code [Huysmans and Czarny (2007); Huysmans, *et al.* (2009)]. The USBPO team has two well-known codes: (1) the TRANSP code developed at PPPL and (2) the NIMROD code [Sovinec, *et al.* (2004)], M3D [Park, *et al.* (1999)], and M3D-C1 [Ferraro, *et al.* (2010)] developed at the University of Wisconsin (https://nimrodteam.org/presentations/index.html). These codes are widely available with support from their developers and can be run to find the fastest-growing instability as an initial value problem. Owing to the large number of MHD time steps (δt of order microseconds) in a transport time scale of (of order 100 milliseconds) combined with high spatial resolution required for accurate wave functions, the MHD simulations required code parallelization with MPI on large parallel computers. In addition, when the toroidal plasma current j_ϕ becomes too large or has a strong radial gradient, the plasma develops what is called the peeling-ballooning mode or instability. With these gradients of the current and pressure at or beyond the stability limit it is not feasible to run these simulation codes with near to real-time for feedback control of the plasma actuators. Simpler monitors of the stability of the code, such as low-order models with systems of ordinary differential equations (ODE), are used for real-time control algorithms.

4.2 Ballooning-Interchange Modes and Resistive-g Modes

Internal ballooning modes are described in Huysmans, *et al.* (p. 347) along with frequency spectra for the various Alfvén Eigenmodes (AE) described in the literature as BAE, TAE, and EAE eigenmodes. The subject of the MHD structure from equilibrium through linear waves to nonlinear dynamics is thoroughly developed in the magnetic fusion literature, allowing detailed simulations in the single-fluid MHD description of the plasma dynamics on supercomputers. As more physics is added to the idealized infinity conductivity MHD system, a rich spectrum of slower instabilities appears. One well-explored example occurs from releasing the frozen-in magnetic field constraint by adding the resistivity ηj term to the Ohm's Law. This allows slow instabilities to growth with growth rates that are hybrids of the fast MHD growth rate and the slow resistive diffusion rate $\gamma_{\text{res}} = \eta/\mu_0 L_p^2$. These are the resistive-interchange modes and the magnetic reconnection modes in the fusion literature that grow before the plasma reaches the Troyon pressure gradient limit for the fast MHD interchange modes.

4.3 Temperature Gradient Instabilities Driving Turbulent Thermal and Density Transport

ITER will have a turbulent transport of electron thermal energy that determines the maximum core electron temperature as a function of the core plasma heating power from collisional slowing-down of the alpha particles and the auxiliary-injected power. Auxiliary electron heating in ITER will be in first phase experiments from the neutral beam injectors, from electron cyclotron resonance heating (ECH). In the second phase experiments lower hybrid (LH) waves will be used to maintain the toroidal plasma current. The ohmic heating, while important for the initial breakdown phase of the plasma formation, is relatively unimportant in heating power compared to the auxiliary heaters driving the ITER plasmas.

The problem of turbulent electron thermal losses has plagued all types of magnetic confinement experiments from the beginning of the fusion research activities. In the early tokamak experiments the plasma confinement was characterized by the number of Bohm confinement times that could be achieved. The definition of the Bohm diffusivity D_B is $T_e/16\,eB$ where the coefficient of $1/16$ has no particular theoretical significance, but instead is historical from the early work of Bohm (1949) in low temperature plasma. Later the early stellarator plasma at PPPL showed the Bohm diffusion rate limiting the plasma temperature to few hundred electron-volts. The tokamak was the breakthrough in toroidal confinement that produced kilovolt electron plasmas with a turbulent type of confinement $\nu_{\text{eff}}\rho_{\text{pol}}^2$ transport for the electron thermal energy discovered by Artsimovitch and the Kurchatov team in 1967-1969.

Now, it is conventional in the turbulent transport literature to define the Bohm D_B and related gyroBohm D_{gB} diffusivities by

$$D_B = \frac{k_B T_e}{eB} \quad \text{and} \quad D_{gB} = \frac{\rho_s k_B T_e}{aeB} \tag{4.8}$$

where the scale length a is physically associated with the scale length of radial plasma gradients driving the turbulence such as L_{T_e} or L_{n_e}. But for comparison between different tokamaks, the gradient scale length in the diffusivity is taken as the minor radius a of the machine which largely determines the gradient scale length of the plasma. For example, discharges with high-core electron temperatures, the gyroBohm scaling is from the ETG turbulence and the radial scale length would be the L_{T_e} from the electron temperature profile. A basic laboratory on the Columbia Linear Machine by Sen, *et al.* (2006) shows this dependence with $1/a \leq 1/L_{T_e}$ in a series of controlled steady state experiments in a 10 eV hydrogen plasma [Sen, *et al.* (2006)].

To have machine-to-machine comparisons, the gradient scale length is given the reference value, independent of the particular profile, of the machine radius a, and is used in D_{gB}. (A few simulations studies use the major radius R of the machines so care must be taken in comparing the results quoted for the coefficient of the gyroBohm confinement data.) In MKS units for the electron charge (e) and magnetic field and electron volts for the temperature, the Boltzmann constant is dropped. For example, a plasma with $T_e = 1\,\text{keV}$ plasma in a $B = 1\,T = 10^4\,\text{G}$ magnetic field and system with minor radius $a = 1\,\text{m}$, the formula for $D_B = 1000\,V/1\,T = 10^3\,\text{m}^2/\text{s}$ and $\rho_s = 3\,\text{mm}$, so the gyroBohm diffusivity is $D_{gB} = 3\,\text{m}^2/\text{s}$. The best confinement experiments typically have thermal diffusivities χ on the order of magnitude given by the gyroBohm formula. The SOL plasma may have much larger radial transport of order a fraction of D_B.

While power balance studies in large tokamaks typically observe confinement times comparable to those implied by D_{gB}, the scaling of the confinement is not as given by $1/B^2$, but rather has a strong scaling with the poloidal magnetic field $B_p \propto I_p/a$ and a weak variation with B_T. A typical example of this poloidal field or plasma current dependence is shown in Eq. (1.1) where one notes the small exponent (0.03) on the toroidal magnetic field and the large exponent (0.96) on the plasma current I_p of the global energy confinement time.

The transport database gives plasma confinement characterized as making a transition from one form to another form. When good confinement is obtained the numerical factor $\chi_e = c_B D_B$ in the Bohm or gyroBohm diffusivity formula is given as a measure of the success of the confinement regime. The toroidal quadrupoles and octupoles were the first toroidal devices that obtained the low values of $c_B \lesssim 1/100$, setting records in magnetic fusion confinement in the 1970s. In these quadrupoles and octupoles the poloidal magnetic field is produced by internally supported toroidal wire currents rather than toroidal plasma currents. Parenthetically, the early toroidal octupole experiments of Kerst at General Atomics and then University of Wisconsin led to the Ohkawa design of the DIII-D machine through a complicated path of experiments.

Scaling with the gyroBohm formula is easy to understand in terms of drift wave turbulence which brings in the scale length ρ_s of the turbulence driving the transport in a natural way. The formula for D_{gB} follows simply from using the drift wave wavelengths for the mixing lengths, and the growth rate time scale for the decorrelation time scale for the turbulence. Thus, the numerical simulations almost exclusively express their diffusivity results in units of D_{gB}. For Bohm turbulent diffusivity the mixing length of the turbulence increases from ρ_s to $(\rho_s L_T)^{1/2}$ and the correlations time remains of order L_T/c_s.

The electron thermal diffusivity, χ_e, is reported in the first high-electron temperature magnetic confinement experiments greatly exceeded the collisional transport value. This situation remains true today. Thus, it seems clear that the electron thermal transport is due to the drift wave turbulence over a wide range of conditions. Kadomtsev, in his 1992 monograph, devotes several chapters to the universality of the large, anomalous value of the electron thermal diffusivity χ_e. Kadomtsev argues that intrinsic electromagnetic drift wave processes are responsible for this universal anomalous thermal transport and emphasizes the role of the δB_r turbulence.

Rebut and Lallia (1988) put forth a formula based on general arguments about electromagnetic fluctuations in magnetic confinement systems. The resulting formula was used extensively during the 1980-1990 period to interpret the electron confinement data from JET [Keilhacker, *et al.* (1999)]. Since the safety factor q enters strongly in tokamak drift wave turbulence, the parameter scaling of the thermal diffusivity is found to be between q to $q^2 \times D_{gB}$. As stated earlier the large ITER confinement database has confinement scaling varying strongly with I_p and weakly with B. To obtain this result from the gyroBohm formula one introduces a factor of q increasing the radial correlation length in the turbulence. Database studies with a mixture of Bohm and gyroBohm formulas have been presented in several works and interpreted physically by Ottaviani, *et al.* (1997).

Since the mid-1990s the theoretical models for the anomalous electron transport based on drift waves are now almost exclusively used to interpret the electron transport data. There remain some disagreements on the particular species of drift wave fluctuations that dominate in determining the turbulent thermal diffusivity. In terms of broad categories, the two types of fluctuations that are found in theory and simulations are (i) the larger scale drift wave fluctuations on the scale of the ion inertial gyroradius scale length ρ_s and (ii) the smaller scale, faster growing skin depth c/ω_{pe} electron scale drift wave turbulence. Both forms of turbulence have strong spectral cascades so that a broad range of scale lengths evolve from the scale of the source at the wavenumber for the maximum growth rate. There are cases where the power balance studies compared with databases favor one or the other types, or species, of drift wave instabilities. Since the turbulence is on two different space-time scales separated by the $(m_i/m_e)^{1/2}$, both the smaller scale electron temperature gradient driven turbulence and the larger scale ion temperature gradient driven turbulence are generally present in high performance plasmas. Both forms of the turbulence simultaneously play a role in the fusion confinement experiments producing the measured anomalous electron and ion thermal losses. This may account for the historical difficulty in determining the χ_e formula when the first reports from Artsimovitch on the original tokamak machine showed a large anomalous thermal diffusivity. Now, however, sophisticated plasma diagnostics along with longer period quasi-steady-state discharges make much clearer the comparison and validation of the drift wave turbulent transport formulas that predict plasma confinement.

Electron energy confinement analysis of the Tore Supra, TCV, NSTX, and

MAST plasmas support the conclusion that the ETG model is able to explain a wide range of anomalous electron transport data. Detailed thermal transport analysis supports the drift waves thermal loss models in ASDEX [Ryter, *et al.* (2001a); Ryter, *et al.* (2001b)], Tore Supra [Hoang, *et al.* (2003)], and the Frascati Tokamak Upgrade [Jacchia, *et al.* (2002)].

Now we concentrate on the dynamics and transport from the small spatial scale-fast electron temperature gradient driven turbulence. The thermodynamic nature of the electron temperature gradient (ETG) physics is similar to that of that of the ion temperature gradient (ITG) modes with the role of the electrons and ions interchanged. Thus, the space scales are smaller and the time scales shorter for the ETG turbulence than the ITG and the trapped electron mode turbulence. The gyrokinetic simulations of the two types of turbulence fully support this difference in the space and time scales of the turbulence. The basic laboratory experiment CLM for the ETG turbulence is used to validate the slab ETG turbulent transport in a SciDAC project [Lin, *et al.* (2002)] that shows the inverse cascade increases the wavelength of the peak of the turbulence to near that of the ion gyroradius scale in both the simulations and laboratory experiments. The data shows a strong turbulent transport of the electron thermal energy in a hydrogen plasma at 10-15 eV.

4.4   Electron Temperature Gradient-Driven Transport Instabilities Producing Anomalously Low-Electron Temperatures and Regions of Ergodic/Stochastic Magnetic Field Lines

In the ETG turbulence the parallel wave phase velocity $\omega/k_\parallel$ is higher than and comparable to the electron thermal velocity in contrast to the lower phase velocity of the ion temperature gradient ITG-TEM turbulence. The smaller correlation lengths make the ETG turbulence less sensitive to the details of geometry than for the lower-frequency longer wavelength ITG-TEM turbulence. The ETG instability saturates with a strong inverse cascade, as shown theoretically and in computer simulations. This inverse cascades extends to the ion inertial length scale, which means that the nonlinear transfers of fluctuation energy extends to the high-k part of the ITG instability scale. Thus, wave energy from nonlinear interactions is transferred from the ETG to the ITG-TEM regime. ETG spectral simulations are typically in boxes of order $3\rho_i \times 3\rho_i$ so that the pile-up of low-k energy can be absorbed into the dynamics of the ITG modes at the box-size limit of the k-space. The formulas are for a tokamak geometry with parameters $\epsilon_n = L_n/R$ and magnetic shear $s \sim 1$. Here $\rho_i = v_i/\omega_{ci}$ is the thermal ion gyroradius and $\rho_e = v_e/\omega_{ce}$ is the electron thermal gyroradius. For ITER the ratio of the machine size to gyroradius scales ρ_i and ρ_e is an order of magnitude larger than the values from the tokamaks used to create the ITER database. The ITG and ETG range of scales and frequencies is typically two orders of magnitude and dictates the use of different diagnostic methods to measure the turbulence.

Owing to the faster parallel phase velocity, the quiver velocity of the electrons along the magnetic field from the parallel electric field and pressure gradient is relatively high for a given wave amplitude. This builds up the wave parallel electric current $\delta j_\parallel$ giving a significant magnetic perturbation δB_r even at relatively low-electron plasma beta or thermal-to-magnetic pressure ratio $\beta_e = 2(\omega_{pe}v_e/\omega_{ce}c)^2 = 8\pi n_e T_e/B^2$. Thus, in the ETG turbulence one introduces the parallel vector potential $A_\parallel$ to describe the electromagnetic parts of the ETG turbulence. The parallel vector potential is solved from Ampere's law with the fluctuating parallel electron current

$$\nabla^2 A_\parallel = -\mu_0 \delta j_\parallel. \tag{4.9}$$

ETG turbulence dominates the thermal transport in numerous tokamaks. Examples are well documented for NSTX, TCV, and Tore Supra.

A typical simulation for NSTX plasma shows contours of isopotentials for the three fluctuating fields $\phi(x,y,t)$, $A_\parallel(x,y,t)$ and $\delta T_e(x,y,t)$ in the saturated state. The figure shows that there are different dominant spatial scales for the three fields. The electron temperature fluctuations have the smallest spatial scale lengths of order ρ_e, the electrostatic potential has the intermediate scale length size of order $\sqrt{\rho_e L_{T_e}}$ to $q\rho_e$ and the vector potential field has the scale length of order c/ω_{pe}. The $\boldsymbol{E} \times \boldsymbol{B}$ flow velocities and the $\boldsymbol{\delta B}_\perp$ vectors are given by the local tangents to the isolines of the electric potential and the vector potential.

The directional derivatives along these vector fields are given by the Poisson bracket operators that arise from the directional derivatives of $\boldsymbol{v} \cdot \nabla$ and $\boldsymbol{\delta B} \cdot \nabla$. With the potentials these directional derivatives of a scalar f field become the Poisson brackets of $[\phi, f]$ and $[A_\parallel, f]$, respectively. These convective nonlinearities create vortex structures in the fields as incompressible fluid dynamics.

The electromagnetic component of the turbulence introduces a magnetic flutter described by Callen (1977) and Horton (1984). The space scale of the magnetic flutter is the electron collisionless skin depth defined by $\delta_e = c/\omega_{pe}$. This scale length is larger than the electron gyroradius but smaller than the ion gyroradius and ion inertial scale length ρ_s. The relationship of the skin depth scale to the electrostatic scale of the inverse cascaded electrostatic potential is important to determine the parameter scaling of the electron thermal diffusivity. In Fig. 4.1 the regime is shown with the skin depth larger than the electrostatic scale length $l_{es} = q\rho_e/\epsilon_n = qR\rho_e/L_n$. First one considers that $L_n \sim L_{T_e}$, typical of L-modes and I-modes plasma regimes. Then one considers H-modes $L_n > L_{T_e}$ where the scale length becomes L_n of order R which must be small compared with the temperature gradient scale length L_{T_e} for a confined hot plasma.

As the plasma pressure-to-magnetic pressure given by β increases to exceed $\beta_{e,\text{crit}}$, then the skin depth $\delta_e = c/\omega_{pe}$ becomes smaller than the electrons' scale length $l_{es} = q\rho_e L_s/L_n$. Now the mixing length is controlled by the electromagnetic skin depth and the theoretical turbulent diffusivity, χ_e, is reduced compared to the extrapolated higher values given by electrostatic turbulence formula which increases as $T_e^{3/2}$. In the electromagnetic regime where

$$\beta_e > \beta_{e,\text{crit}} = \frac{2m_e}{m_i} \qquad (4.10)$$

the mixing length, l_{em} becomes

$$l_{em} = \delta_e = \frac{c}{\omega_{pe}}. \qquad (4.11)$$

In this regime the turbulent thermal diffusivity increases more slowly with increasing electron temperature and is in agreement with the Tore Supra discharge where T_e is stepped between two steady states with the first state at approximately 2 keV from 3 MW of Fast Wave electron heating and the second state at approximately 4 keV core temperature from 6 MW of Fast Wave heating. This discharge is modeled with the CRONOS transport simulation code and the turbulent diffusivity, as shown in Fig. 4.2 taken from Artaud (2010) and Asp (2008). The dotted lines are from the plasma diagnostic systems and the solid lines are from the CRONOS simulation of the discharge with drift wave turbulent transport. The electromagnetic transport

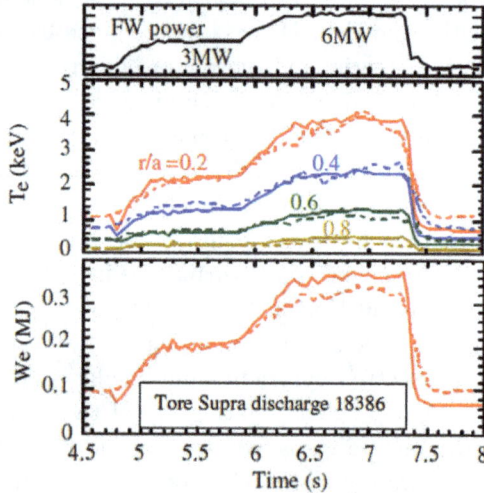

Fig. 4.2 Predictive simulation well documented and researched Tore Supra discharge driven through two stages with fast wave electron heating FWEH defined by discharge number and parameters (#18368, $I_p = 0.6\,\text{MA}$, $B = 2.2\,\text{T}$, $n_e(0) = 4 \times 10^{19}\,\text{m}^{-3}$). The top panel with the fast wave heating power rising from 3 to 6 MW. Middle panel shows the time evolution of electron temperature (T_e) at various radii. The bottom panel shows the rise of the total electron energy context at (W_e). (First published in Horton, *et al.* (2000) and recently in the publication [Artaud, *et al.* (2010)] that documents the suite of codes called CRONOS for integrated modeling with general 2D magnetic equilibrium and full range of plasma heaters.)

formulas for the electron thermal diffusivities with this transition was derived and used in Horton, *et al.* (2000, 2003) and Hoang, *et al.* (2003) for the Tore Supra transport experiments. The turbulent electron diffusivity for $\beta_e > \beta_{\text{crit}}$ is given by

$$\chi_e^{\text{em}} = C_{\text{em}} q^\nu \frac{c^2}{\omega_{pe}^2} \frac{v_e}{(L_{T_e} R)^{1/2}} \left(\frac{R}{L_{T_e}} - \frac{R}{L_c} \right) \tag{4.12}$$

and for $\beta_c < \beta_{\text{crit}}$ by

$$\chi_e^{es} = C_e^{es} q^\nu \left(\frac{R}{L_{T_e}} \right)^{3/2} \left(\frac{\rho_e^2 v_e}{L_{T_e}} \right) \left(\frac{R}{L_{T_e}} - \frac{R}{L_c} \right). \tag{4.13}$$

The magnetic shear s and q dependence of the growth rate are used in choosing the exponents ν in the thermal flux. The q exponent ν varies between 1 and 2. The critical gradient $(\nabla T_e)_c$ is derived from the marginal stability analysis of the dispersion relation. The presence of a critical gradient in the thermal flux expressions gives the flux formula the form of a diffusivity plus a radial transport velocity $V_r(r, t)$ where the relationship with the critical gradient is

$$V_r = \chi_e \frac{(\nabla T_e)_c}{T_e} = -\frac{\chi_e}{L_{T_{e,c}}} \tag{4.14}$$

giving an inward thermal pinch velocity with $L_{T_{e,c}} V_r / \chi_e \sim 1$.

Power balance studies yield $V_r = \chi_e^{pb} / L_{Te,c} \geq 10 \, \text{m/s}$. The gyroBohm estimate for χ_e is $\chi_e / L_{T_e} = c_s (\rho_s / L_{Te})^2$. The transport velocity V_r is inward giving rise to the off-diagonal heat pinch effect. The sheared slab regime has

$$V_r = -1.9 |s| \chi_e / qR \left(1 + Z_{\text{eff}} T_e / T_i \right) \tag{4.15}$$

from Eq. (4.14). In the absence of magnetic shear the compression stabilization from $\nabla \cdot v_e$ rather than electron Landau damping determines the transport velocity V_r with $(\nabla T_e)_c$. Peaked density profiles with small L_{ne} arise from an inward particle pinch velocity V_r. Pinch terms are equivalent to off-diagonal transport terms and are a standard feature of turbulent transport matrices.

The formulas, Eq. (4.12) through Eq. (4.15), are able to interpret the extensive Tore Supra database where the plasma β_e is above the $\beta_{\text{crit}} = 2m_e/m_i$ in the core region and below the $\beta_{\text{crit}} = 2m_e/m_i$ in the outer part of the profile. The electron temperature profile is accurately measured with multiple cords of Thomson scattering data and the RF electron heating is into the thermal distribution in the core of the plasma.

The transport formulas in Eqs. (4.12) and (4.13) were used in the CRONOS transport code along with the WAVE code for the plasma heating profile from the Fast Wave Ion Cyclotron heating used to drive the Tore Supra plasma. The results show a good predictive performance as shown in the example in Fig. 4.2 where the RF heating was first held at 3 MW and then ramped up to 6 MW during a long $\Delta t = 2$ s pulse.

An important difference in the large-scale ITG turbulence and the small scale ETG turbulence is that the electron parallel fluctuating currents $j_\parallel(k, \omega)$ are larger in terms of their parallel acceleration of the electron fluid. This means that the electron inertial is important in determining the phase of the accelerated electrons.

Thus, the scale length $\delta_e = c/\omega_{pe}$ is the important dispersion scale length in the turbulent electron transport. The ion inertial scale length $\rho_s = c_s/\omega_{ci}$ is the scale length in ion gyroscale turbulence. As the plasma pressure increases these dispersion scale lengths become equal when the dimensionless electron pressure $\beta_e = \beta_{e,crit} = 2m_e/m_i$ given in Eq. (4.10). At this $\beta_{e,crit}$ pressure value the electron thermal pressure, the Alfvén wave velocity is equal to the electron thermal velocity, so it is natural that the waves and fluctuations become electromagnetic. When the cross-field correlation length for the electron transport is taken as the electron skin depth $\delta_e = c/\omega_{pe}$ then some of the features of the empirical scaling laws introduced by Rebut, et al. (1991) and Kadomtsev (1992) are present. The first principle of understanding this anomalous electron transport, however, comes from the electron temperature gradient-driven drift waves and their nonlinear dynamics. Experiments in the Tore Supra device over a wide range of electron heating powers (Ohmic heating to 8 MW of RF Fast Wave heating power) confirmed the ability of the ETG transport models to systematically explain the anomalous electron thermal transport rates.

The dynamics of the large-scale structures in the toroidal ETG turbulence are explained with theory and simulations. Gyrokinetic simulations [Li and Kishimoto (2004); Jenko and Dorland (2002)] describe how the radially extended convective streamers emerge from the nonlinear interactions in the ETG turbulence.

The simulations using ETG formulas reproduced the Tore Supra FWEH experiments and CLM experiments [Wei, et al. (2010)] in two different plasma regimes. The most direct evidence of fluctuations in the tokamak follows from the electromagnetic scattering of microwaves off the ETG driven electron density fluctuations. The techniques described in Chapter 2 are extended to smaller space scales by Mazzucato (2009, 2010) on NSTX. Earlier studies of the power balance on a large database of RF heated discharges in Tore Supra by Horton, et al. (2000, 2003), and Hoang, et al. (2001, 2003) showed that theory reproduces the $T_e(r,t)$ profiles measured with Thomson laser scattering for plasma heating powers from Ohmic to 8 MW of radio frequency (RF) heating power where the core temperature increased from below 1 keV to 3 keV. Both the spatial profile and the time dependence with a step-up in heating power are well reproduced.

Similar agreement with the profiles of $T_e(r,t)$ were found in the NSTX experiment with high-resolution Thomson scattering diagnostics in Kaye, et al. (2007). In the NSTX and MAST experiments the ion thermal transport flux remains close to the neoclassical collisional level while the electron transport is larger and explained by the ETG level turbulent transport ETG models prediction of the measured electron temperature profiles from the deposited power. A typical example of the comparison of the ETG thermal flux calculated from simulations compared with the power balance flux for a fast wave heated NSTX discharge [Kaye, et al. (2007); Stutman, et al. (2009)] is shown in Fig. 4.3.

Measurements with coherent scattering of 280 GHz microwaves in NSTX with

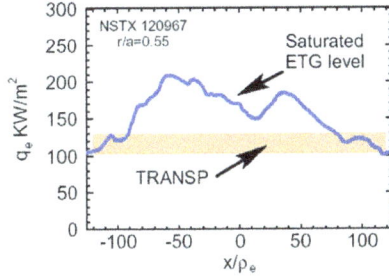

Fig. 4.3 NSTX validation of the turbulent thermal flux computed from a local pseudospectral gyrofluid simulation. The system of partial differential equations used in the turbulence simulation is given in Eqs. (4.12)-(4.15).

a five-channel heterodyne receiver verified the presence of the ETG fluctuations at $k_\perp \rho_e \sim 0.2\text{-}0.4$ with fluctuation amplitudes strongly correlated with the strength of the electron temperature gradient. The plasma beta is in the range of 3% to 6% so that the plasma density fluctuation produce the microwave scattering confirming the turbulence with $k_\perp \delta_e \sim 2$. The electron temperature gradient varied from $R/L_{T_e} = 4$ to 8 in the experiments producing an order of magnitude variation in the spectral intensity of the scattered microwave power.

The validation of the electron temperature gradient driven instability is given by Rhodes, *et al.* (2007). In lower single-null plasma with $B_T = 1.9\,\mathrm{T}, \mathrm{I_p} = 0.7\,\mathrm{MA}$ and cord averaged electron density of $1.5 \times 10^{19}\mathrm{m}^{-3}$ an elaborate set up of RF scattering diagnostics is used to measure simultaneously the low-k wavenumbers of the ion temperature gradient and trapped electron instabilities and with backscattering of millimeter [94 GHz] RF waves the high-k wavenumbers measuring the electron temperature gradient driven turbulence. The DIII-D discharge 120329 experiment used a stair-step series of increased levels ECH plasma heating power starting from the initial Ohmic state with $T_e = 1.5\,\mathrm{keV}$ while monitoring small k, medium-k and high-k density fluctuations and the density and electron temperature radial profiles. The data shows clearly that the high $k\rho_s \gg 1$ turbulence increases and explains the change in the electron temperature profiles correlating with the increase in the turbulent electron thermal flux.

The ion temperature profile and the associated ion thermal flux change little. The electron thermal flux inferred from power balance increases strongly and correlates with increased electron density and electron temperature increases with the step-ups in the ECH power through the sequence $P_{\mathrm{ECH}} = 0(\mathrm{ohmic}), 0.5, 1.0, 1.5,$ and 2.0 MW as predicted from ETG theory and simulations.

4.5   Thermodynamic Properties of Electron Temperature Gradient Driven Transport

In the high-k regime of ETG turbulence, the thermodynamics of the electron gas explains the direct drive of the turbulence from the electron temperature gradient ∇T_e. Thermodynamics shows that there is a critical temperature gradient $\nabla T_e \equiv -T_e/L_{T_e}^{\mathrm{crit}}$ for the ETG turbulence $W_{\mathrm{ETG}}(k, \omega)$ to extract energy from the temperature difference, $\Delta T = T_2 - T_1$, over the correlation length l_c.

The Carnot cycle shown in Fig. 4.4 gives the upper bound on the released turbulence energy $W \leq W_c = \Delta T \Delta S_e$, where ΔS_e is the change in the electron entropy density along the isothermal side T_2 in the core and T_1 in the lower temperature zone [Horton, *et al.* (2004)] of the nonlinear convective heat engine. The simulations with the associated partial differential equations for the electron gas confirm the onset of the turbulence above the critical gradient and the presence of elongated convection cells called streamers, characterized by a correlation length l_c.

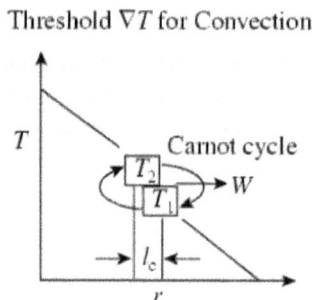

Fig. 4.4   The Carnot engine driven by the temperature difference $T_2 - T_1$ over the cell size ℓ_c gives a bound for the turbulent energy W released per cycle.

As the gradient rises above the critical value, stronger nonlinear features appear in the fluctuations. In addition to thermal energy the fluctuations produce a transport of parallel electron momentum and magnetic flux that also develop into strong coherent structures for the parallel electron current and the magnetic flux function in their variations across the magnetic field. The transport of magnetic flux results in the turbulence developing into a form of the small-scale tearing mode turbulence be a microscale form of magnetic reconnection. This form of magnetic reconnection is also called electron MHD or e-MHD.

Meso-scale tearing modes are shown to evolve for longer times in the simulations with modes changing their parity in the nonlinear state. Driven by both the temperature gradient and the gradient of the toroidal current density profile, complex nonlinear structures are formed from the initial temperature gradient-driven turbulence [Muraglia, *et al.* (2009); Ishizawa and Nakajima (2010)]. Even with the absence of the dj/dr driven tearing modes, magnetic structures evolve from the

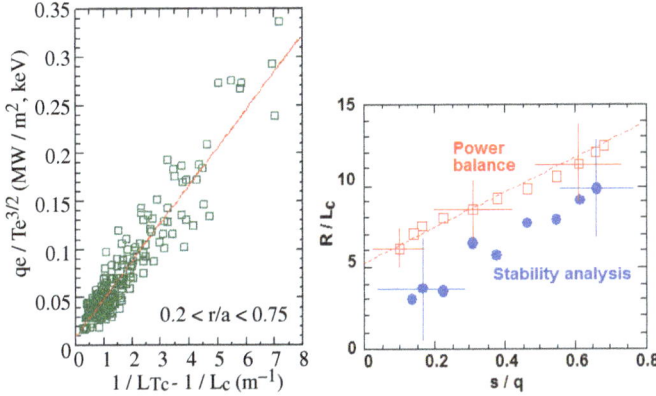

Fig. 4.5   Reduction of the large Tore Supra database for the scaling of the electron thermal flux q_e versus the temperature gradient for discharges steady state, hot electron discharges driven by fast wave electron heating (FWEH). The core deposited heating power ranges from 0.5 MW to 7 MW. The scaling with $T_e^{3/2}$ agrees with Eq. (4.12) and the increase of the critical gradient scale with magnetic shear s/q agrees with Eq. (4.15) [Hoang, *et al.* (2001); Hoang, *et al.* (2003)].

nonlinear ETG modes for longtime simulations. Thus, the finite electron pressures produce a small-scale or microtearing magnetic turbulence when the resistive MHD tearing modes are stable. This ∇T_e driven magnetic turbulence twists and shears the magnetic field lines which then reconnect releasing magnetic energy. This process transforms a fraction of poloidal magnetic energy into the parallel electron parallel acceleration and into electron thermal energy.

The ETG turbulence is characterized by long or extended radial cells giving relatively large correlation scales $\ell_c \gg 1/\Delta k_\theta$. The cells are called streamers. Figure 4.4 shows the extended radial structures that convect the plasma in such a way as to take thermal energy from the high-temperature region T_2 to the lower temperature region T_1 across the extended radial convection cells. One can place an upper bound on the energy per cycle released to do the available work on the turbulence and convective cells from the Carnot cycle theorem of thermodynamics. The cycle periods are of order microseconds.

On the adiabatic sides of the Carnot cycle created by the convection cells or streamers, a gas constant, $\Gamma = (d+2)/d$, describes the electron gas, where $d = 1, 2, 3$ for the number of degrees of freedom active in the dynamics. Kinetic theory guides the choice of d with the slab and toroidal modes having $d = 1$ and $d = 3$, respectively. The drift-wave vortex or streamer gives the convective $\boldsymbol{E} \times \boldsymbol{B}$ motion between (n_1, T_1) and (n_2, T_2) producing the Carnot cycle for the convection period. From the Carnot-cycle calculation, the change of entropy density is given by,

$$\Delta S_e = n_e k_B \left(\frac{3}{2} \frac{\Delta T_e}{T_e} - \frac{\Delta n_e}{n_e} \right) \tag{4.16}$$

for $d = 3$ where k_B is the Boltzmann constant. Using the correlation length ℓ_c and the gradients $T_2 - T_1 = -\ell_c dT_e/dr$ and $n_2 - n_1 = -\ell_c dn_e/dr$ gives the maximum released electron energy density W_e released by each cycle is

$$W_e = \frac{3}{2} n_e T_e \ell_c^2 \frac{dT_e}{dr} \left[\frac{d\ln T_e}{dr} - \frac{2}{3} \frac{d\ln n_e}{dr} \right]. \tag{4.17}$$

In this ideal limit without dissipation in the system, the critical temperature gradient is expressed in terms of the major radius, R, of the torus as

$$\frac{R}{L_{T_e}^{\text{crit}}} = \frac{2}{d} \frac{R}{L_{ne}} \tag{4.18}$$

that follows from condition $W_e > 0$. Here d is the number of degrees of freedom in the electron dynamics. This critical gradient value $R/L_{T_e}^{\text{crit}}$ as expressed in terms of the temperature-to-density gradient ratio $\eta_e = L_n/L_T$ is well known from the Nyquist stability analysis for both ITG and ETG instabilities. Adding wave dissipation induced by magnetic shear, $R/L_s = s/q$, increases the critical temperature gradient [Horton, et al. (2004)] in proportion to $|s|/q$. The corresponding linear contribution is given in Hahm and Tang (1989) from eigenmode calculation for the offset in the critical gradient from magnetic shear given in Eq. (4.18).

4.5.1   Two-space scales for electron transport

The role of the two space scales in the electron thermal heat flux, $q_e(r,t) = -n_e \chi_e dT_e/dr$, is made clear from the kinetic theory formulas for q_e. Kinetic theory gives the total energy Q_e flux in terms of the particle flux Γ_e and the thermal flux q_e as

$$Q_e = \frac{3}{2} T_e \Gamma_e + q_e \tag{4.19}$$

Here Γ_e is the particle flux taking the average electron thermal energy with it and the flux q_e is the thermal conduction flux that flows down the temperature gradient even in the absence of a particle flux $\Gamma_e = 0$. This is important division since the particle flux can be inward but is typically too small by an order of magnitude to explain the outward thermal transport even when the particle flux is outward.

The whole concept of magnetic confinement for thermonuclear fusion of plasmas depends on the thermal insulating principle by the strong magnetic field. From the deuterium-tritium reactivity function shown in Chapter 1, Fig. 1.6, one knows that the core ion plasma temperatures greater than 10^8 K or 10 keV are required for significant fusion nuclear reactivity for producing the required fusion power.

Magnetized plasmas with high temperature gradients across the confining magnetic field are a common owing to the thermal-insulating properties of the magnetic field. Such temperature gradients occur in the magnetosphere, the solar corona and with particularly high gradients in the laboratory experiments aimed at magnetic confinement for nuclear fusion power. In the large tokamak experiments we see that the core ion temperatures have reached 40 keV in both the TFTR device and in the JT-60U machine with $R/a = 3.1\,\text{m}/0.7\,\text{m}$. Thus, we know that the magnetic field can provide strong thermal insulation withstanding gradients greater than 20 keV/m under practical laboratory conditions with closed irrational magnetic flux surfaces, with and without a magnetic separatrix.

Table 4.1 gives the parameters for a well-known JT-60U discharge that has a core ion temperature of 38 keV and a core electron temperature of 12 keV. The ion (deuterium) quantities T_i and toroidal velocity v_ϕ are inferred from spectroscopic measurement of the partially ionized carbon ions as described in Chapter 3.

For the JT-60U discharges, transport profiles were generated using a 1.5-D transport analysis code and an orbit following Monte Carlo code for the neutral beam injection.

The results for the model ion and electron thermal diffusivities were compared with the neoclassical diffusivities and the radial electric field $E_r(r,t)$ for the reversed-shear plasmas with $I_p = 1.5\,\text{MA}, B_t = 3.5\,\text{T}$ with enhanced confinement associated with the presence of an ITB. One of these plasmas produced the record shot that is reported in Ishida *et al.* (1997) as having the Q_{DD} that would convert to a deuterium-tritium equivalent discharge with a break-even $Q_{DT}^{\text{equiv}} = 1$. This assertion has been the subject of some controversy and, in any event, is not a demonstration

Table 4.1   Two high-temperature fusion plasmas from the JT-60U tokamak [8,9]

Shot	E27969	E31872
Mode	Reversed shear	Reversed Shear
	L-mode edge	L-mode edge
Divertor	Open	W shaped
Time (s)	7.315	6.944
I_p (MA)	2.79	2.61
B_t (T)	4.34	4.37
P_{NB}^{abs} (MW)	16.0	11.77
a (m)	0.70	0.69
R_p (m)	3.10	3.08
q_{95}	3.15	3.17
W_{dia} (MJ)	10.9	8.17
$S_n\ (10^{16}/\text{s})$	4.52	3.63
Z_{eff}	3.49	3.19
$n_e(0)\ (10^{19}\,\text{m}^{-3})$	9.7	8.5
$n_D(0)\ (10^{19}\,\text{m}^{-3})$	4.9	4.8
$T_i(0)$(keV)	16.5	16.8
$T_e(0)$(keV)	18.4	7.2
τ_E (s)	0.97	1.07
$H_{ITER89P}$	3.23	3.21
β_N	1.88	1.53
β_p	1.15	0.98
β_t (%)	1.66	1.32
$n_D(0)\tau_E T_i(0)\ (10^{20}\,\text{m}^{-3}\cdot\text{s}\cdot\text{keV})$	7.80	8.59
$Q_{DD}\ (10^{-3})$	4.7	5.6
Q_{DT}^{eq}	1.05	1.25

[8] Ishida, S., *et al.* (1997). *Phys. Rev. Lett.* **79** 3917.
[9] Fujita, T., *et al.* (1999). *Nucl. Fusion* **39**.

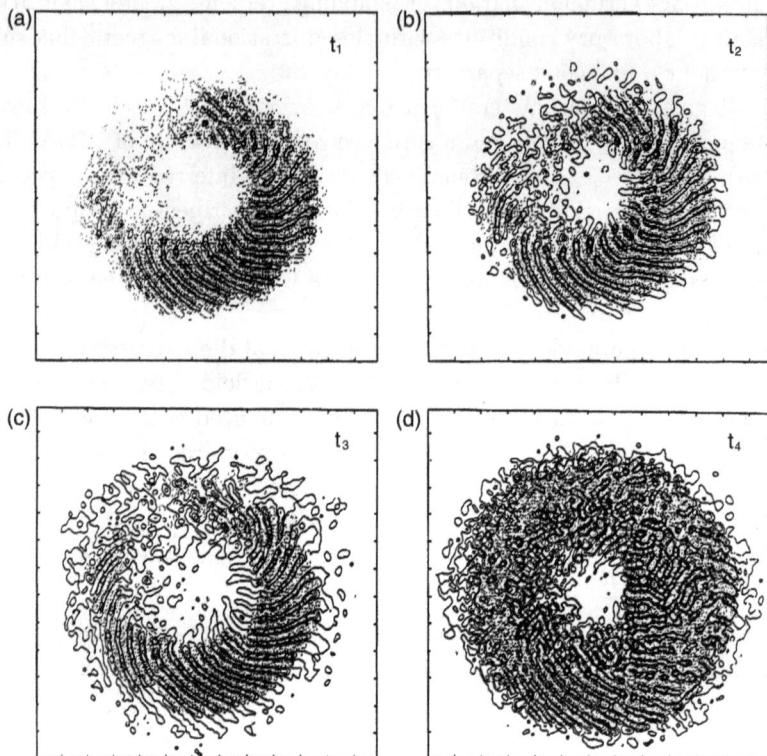

Fig. 4.6   Evolution of the isopotential contours for the ITG turbulence for $a/\rho_i = 28$, $L_T/R = 0.2$, $T_i = T_e$ with hyperbolic tangent profiles for the temperature. The density profile is flat [Sydora (1996)] giving strong ITG turbulence. The figure shows the rapid universal spreading of the drift wave turbulence through the plasma.

of fusion power break-even since the Q_{DD} is three order of magnitude smaller than the Q_{DT} at ion temperatures in the 15 to 20 keV range.

The radial electric field $E_r(r,t)$ in JT-60U was calculated from a force balance equation for carbon impurity density profile and using the toroidal rotation velocities measured from charge exchange recombination spectroscopy along with the poloidal rotation velocity calculated from the neoclassical transport theory. The analysis described in Ishida, *et al.* (1997) shows that there is a thin layer with an ITB (internal transport barrier) in which the inward pointing E_r rapidly increases in strength from few keV/m to -20 to -40 keV/m in the region from $\rho = 0.5$ to 0.6. Inside this layer the inferred ion thermal diffusivities are low a few times the neoclassical value. The electron thermal diffusivity also drops and is comparable in value to ~ 2-$3\,\mathrm{m^2/s}$ which is comparable but larger than the ion thermal diffusivity as expected from turbulence theory. There is an edge transport barrier at

$0.8 < \rho \lesssim 1.0$ just inside the last closed flux surface inside the magnetic separatrix.

The broken up-down symmetry from the lower single null divertor configuration gives rise to nonlinear conversion of the poloidal rotation with toroidal flow helicity vector to a vertical component of flow helicity with the fast toroidal rotation. This effect is common in fluid dynamics as seen in the dust devil and the hurricanes and is driven by the strong plasma heating, as in these two analogous planetary systems. The clearest theoretical example is given by simulation of Morales, *et al.* (2012).

For a background plasma with local Maxwellian electrons velocity distribution parameterized by $n_e(r)$, $T_e(r)$ and the energy density $g(\epsilon)d\epsilon$ for $\epsilon = m_e v^2/2T_e$, the cross-field electron thermal flux is given by the summation over both the lower-frequency, large-scale drift wave fluctuations q, Ω and the small-scale fluctuations k, ω as given by

$$
q_e = n_e T_e \left[\sum_{q,\Omega} \frac{q_y |\overline{\Phi}_q|^2}{BT_e} f_t \int_0^\infty d\epsilon \left(\epsilon - \frac{3}{2} \right) g(\epsilon) \, \mathrm{Im} \left\{ h(\epsilon, q, \Omega) \right\} \tag{4.20}
$$

$$
+ \sum_{k,\omega} \frac{k_y |\Phi_k|^2}{BT_e} \int_0^\infty d\epsilon \left(\epsilon - \frac{3}{2} \right) g(\epsilon) \, \mathrm{Im} \{ h(\epsilon, k, \omega) \} \right] \tag{4.21}
$$

where $\int_0^\infty d\epsilon g(\epsilon) = 1$, $\overline{\Phi}_q$ is the bounce-averaged potential fluctuation, f_t is the fraction of trapped electrons, $h(\epsilon, k, \omega)$ is the non-adiabatic electron phase-space-density response function, and $\mathrm{Im}\{h\}$ is the imaginary part of reduced wave-electron response function $h(\epsilon)$ for each fluctuation component of the turbulence. The notation is used that k and ω are the small-scale, high-frequency part of the fluctuation spectrum while Ω, q are for the large-scale, low-frequency part of the fluctuation spectrum.

In the thermal-flux formula, the low-frequency, large-scale contribution arises only from radial gradients in the trapped electron distributions, since the bounce-averaged potential $\overline{\Phi}_q(r) = 0$ is for passing electrons (except for r exactly on a rational surface). The small-scale response Φ_k involves resonance with both passing and trapped electrons through the response function,

$$
h_k(\epsilon, \omega) = \frac{\left\{ \omega - \omega_{*e} \left[1 + \eta_e \left(\epsilon - \frac{3}{2} \right) \right] \right\} J_o^2 \left(\frac{k_\perp v_\perp}{\omega_{ce}} \right) f_M}{\omega - \epsilon \omega_{De} - k_\parallel v_\parallel + i\delta^+} \tag{4.22}
$$

where the wave frequency, $\omega_k \gg \Omega_q$ is given by

$$
\omega_k = -\frac{\omega_{*e} \left[I_o(b_e) e^{-b_e} + \eta_e b_e \left(I_1(b_e) - I_o(b_e) \right) e^{-b_e} \right]}{Z_{\mathrm{eff}} T_e/T_i + 1 - I_o(b_e) e^{-b_e} + k^2 \lambda_{De}^2}. \tag{4.23}
$$

The low-frequency turbulence is given by

$$
\Omega_q = -\frac{\omega_{*i} \left[I_o(b_i) e^{-b_i} + \eta_i b_i (I_1(b_i) - I_o(b_i)) e^{-b_i} \right]}{(T_i/ZT_e) + [1 - I_o(b_i) e^{-b_i} + q^2 \lambda_{De}^2]} \tag{4.24}
$$

where $b_e = (k_\perp \rho_e)^2$ and $b_i = (q_\perp \rho_i)^2$ in Eqs. (4.22)-(4.24).

The formula, Eq. (4.22), uses the nonadiabatic contribution to the fluctuating electron distribution function computed from $\delta f_{k,\omega} = [1 - h_k(\epsilon, \omega)][e\phi_k/T_e]$. Equation (4.23) extends the electron response up to frequencies that are a fraction of $v_e/L_{T_e} \leq 10^7/\text{s}$. Thus, thermal and supra-thermal electrons with both large and small pitch angles carry the anomalous electron heat flux in the ETG turbulence. The pitch angle of a particle is the angle between the velocity vector and the magnetic field vector. For the ion or electron to be trapped in the low magnetic field on the outboard side, this pitch angle of the velocity vector must be sufficiently close to 90°. The value of the angle for trapping depends on the value of r/R, which determines the depth of magnetic well for magnetic mirror trapping.

For example, Eq. (4.23) yields the megaHertz frequency in the case of TCV 29892 discharge at $r/a = 0.7$ shown in Fig. 4.7. For the η_e values in this discharge, the ETG mode changes direction of rotation at $k_y\rho_e \simeq 0.3\text{-}0.8$ and this is where the linear ETG growth rate reaches a maximum. The direction of rotation is in the plasma rest frame which may be rapidly rotating in the laboratory frame of reference. For both regimes, the transport flux is dominated by longer wavelength turbulence generated by the mode-coupling nonlinearities described in the following subsections. This inverse cascade of the turbulent energy is observed in both the ion and electron drift wave experiments and simulations.

4.5.2   *Nonadiabatic ion response*

The effect of nonadiabatic ions occurs in the transitional wavenumber space-scale region where $k_\perp \rho_i \sim 1\text{-}5$. The wave response is associated with the part of the frequency spectrum shown in Figs. 4.1, 4.6, and 4.7, where the wave propagates in the ion diamagnetic direction. In this low-k part of the ETG spectrum the wave resonates with the guiding-center drift motion of the ions with a dimensionless resonant energy $\epsilon = m_i v^2/2T_i$, determined by resonance condition

$$\omega_{k_y} = \omega_{Di}\epsilon + k_\parallel v_i \epsilon^{1/2} \cos(\alpha)$$

where α is the pitch angle of the ion velocity. At $B = B_{\min}$ the pitch angle is defined by $\cos\alpha = v_\parallel/v$. The guiding center drift ω_{Di} dominates the particle-wave resonance for the conditions present in the TCV data.

In the low-$k_y\rho_e$ part of the spectrum the waves are resonant with ion guiding-center drifts. The resonant ions have energy $\epsilon_r = E_{\text{res}}/T_i = \omega_{k_y}/\omega_{Di} \lesssim 1$ in the low-k_y part of the spectrum and has a weak ion resonance when the direction of rotation changes to that of the electron diamagnetic direction. These resonant ions add the dissipative, phase-shifted response to the ion density (and temperature fluctuations) through the energy-integrated ion response to the wave given by

$$\delta n_i = -\frac{Ze\phi}{T_i} n_{i0} \left\langle 1 - \frac{\omega_k - \omega_{*i}(\epsilon)}{\omega_k - \omega_{Di}\epsilon + i0^+} J_0^2 \right\rangle. \tag{4.25}$$

From Eq. (4.25) one calculates by making an expansion for $k_\perp \rho_i > 1$ in γ_k/ω_k, that

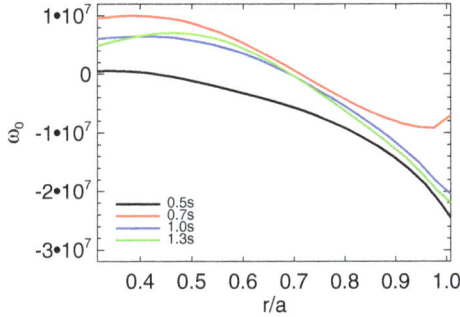

Fig. 4.7 The ETG mode frequencies along the radial positions r/a for $k_y \rho_e = 1.0$ in the discharge TCV 29892. In the core region the frequency increases monotonically in time with $t = 0.5, 0.7, 1.0$ from Eq. (4.23) until $t = 1.3$ where a magnetic island has formed. The plasma is heated with ECH power after the initial $t = 0.5\,\mathrm{s}$ ohmic H-mode.

the contribution of the resonant ions to the growth rate is

$$\gamma_i(k) = \frac{\pi^{1/2}\omega_k(\omega_{*i} - \omega_k)}{|\omega_{Di}|(1 + b_i)^{1/2}}. \qquad (4.26)$$

Equation (4.26) applies for fluctuations rotating in the ion diamagnetic direction (with respect to the plasma rest frame) at low-$k_y\rho_e$. The upper bound on this growth rate is given by v_i/L_{T_e} which is lower than the maximum ETG growth rate from the shorter wavelength part of the spectrum where the fluctuations rotates in the electron direction. In some works on ETG turbulence the ions are taken as adiabatic, which is typically not well satisfied in the experiments. At the same time this resonant ion contribution gives a particle flux due to the difference in the phase between the density fluctuation and the potential fluctuation from Eq. (4.25). Figure 4.7 illustrates for the four regimes of a single discharge the change in direction of rotation with radius of a typical high-frequency ETG wave.

4.5.3   *Electron thermal transport in TCV*

The Tokamak Configuration Variable (TCV) [Tonetti, *et al.* (1991)] is a plasma with high-power density (up to $8\,\mathrm{MW/m^3}$) deposited in the core of the plasma by Electron Cyclotron Resonance Heating (ECRH) at significant plasma densities ($\leq 7 \times 10^{19}\,\mathrm{m^{-3}}$) with the plasma exhibiting a well-defined turbulent electron thermal transport χ_e.

The discharge chosen for analysis is distinguished by having four distinct high-confinement mode (H-mode) phases. First there is an Ohmic H-mode with Type III edge localized modes (ELMs), which turns into a Type I ELMy H-mode when the ECRH is switched on. The ELMs then vanish, which gives rise to a long quasi-stationary ELM-free H-mode. This ELM-free phase can be divided into two parts, one without magnetohydrodynamics (MHD oscillations) and subsequently a steady state with a (one) resonant MHD oscillation. The MHD mode in the last stage causes the plasma energy confinement to drop by $\sim 15\%$. For all four phases both large-scale trapped electron (TEM) mode and the small-scale electron temperature gradient (ETG) modes are found to co-exist. One finds, however, that the TEM formulas have difficulty predicting both the magnitude and the radial profile of the electron thermal flux. The ETG turbulence gives a thermal flux between the profiles that agrees more closely with the power balance data as shown by several measures of the variance of the model with the data. Collisionality governs the drive of the TEM, which for the discharge in question implies the TEM instability can be driven by either (1) the temperature or (2) density gradient. The TEM response function is calculated to be relatively small and to have sharp resonances in its energy dependence defined by the condition $\Omega_q = \overline{\omega}_{De}$. Only the magnetically-trapped electrons contribute to the TEM transport. The ETG turbulence, on the other hand, is driven solely by the electron temperature gradient and both passing and trapped electrons contribute. Both trapped and passing electrons add to the ETG instability and the associated turbulent thermal flux.

4.5.4   *Average relative variance (ARV)*

For a quantitative comparison of the results of the TEM and ETG modes in this discharge, and more generally for comparison of transport data and simulation predictions, one introduces the Average Relative Variance (ARV). The Average Relative Variance is used in weather forecasting models and space physics as a way to reduce the dependence of the model validation on the large versus small variations in the measured data. The method was used in the analysis of the TCV electron transport data with the TCV Equipe for the high electron temperature gradient experiments [Asp, *et al.* (2008)]. Two models were competed, or compared, for the analysis of the electron thermal transport inferred from the Thomson scattering data for the electron temperature profiles. The first transport model was the Weiland model described in Weiland, *et al.* (2005) and the second model was the ETG model described in Horton, *et al.* (1988).

The dimensionless error measure called the average relative variance (ARV) is introduced to take into account the wide swings in the level of the electron temperature and thermal flux calculated from the plasma data. The ARV is a method is designed to compensate for the fact that one requires more accuracy to predict daily temperatures in regions where the temperature variation is mild as in Hawaii, for example, than required for meaning temperature variations in Texas where the daily temperature variations can be $20°$ C. To make this compensation for wide versus narrow swings in the data, the variance of the data from the model prediction is normalized by the variance in the data itself. Thus, the model must predict the temperature more accurately when the daily variation in the temperature is weak or small.

The ARV method was applied to the variance of the electron thermal flux inferred from TCV power balance data derived from a suite of diagnostic signals. The data is then compared with the value given by the ETG turbulence model and the Weiland transport model based on the TEM-ITG turbulence. The ARV metric applied to the ETG model explains 70% of the variation in the electron heat diffusivity whereas the TEM-ITG model gave an ARV less than 50% of the variation in the data is explained. These results for TCV turbulent thermal transport lead to the conclusion that the ETG model is needed to explain the anomalous electron thermal transport when there is a significant electron thermal transport. All tokamak power balance studies have shown significant to large anomalous turbulent electron thermal losses.

A wide range of anomalous electron transport data is available from tokamak data published over decades. The ETG turbulence model gives the best explanation for the electron thermal transport data. The model gives the smallest ARV.

Other evidence for the anomalous electron transport is found in ASDEX [Ryter, *et al.* (2001a); Ryter, *et al.* (2001b)], Tore Supra [Hoang, *et al.* (2003)], and the Frascati Tokamak Upgrade [Jacchia, *et al.* (2002)]. The simulations and power balance studies of the electron transport data is an ongoing research activity at the time of writing.

The Tokamak Configuration Variable (TCV) [Tonetti, *et al.* (1991)] is a compact tokamak with high-power electron cyclotron resonance heating (ECRH), which is ideally suited for studies of turbulent electron thermal transport. In the configuration of TCV analyzed here, the plasma energy confinement is consistent with the empirical formula of $H_{98(2,y)} \geq 1$. The important feature of the particular discharge, TCV 29892, analyzed here, is that the relatively long duration of the quasi-steady-state plasma contains four different regimes all in the high-confinement mode (H-modes). The first stage of the discharge shown is formed with ohmic heating and is in the H-mode. Then the ECRH heating is started. This heating power increases the temperature and its gradient, giving a transition to an ELMy H-mode phase. Then there are two long quasi-stationary phases without edge-localized modes (ELMs) [Porte, *et al.* (2006)]. These different confinement regimes are described further in Chapter 5. The discharge is unique in having a series of steady states with four different confinement regimes. The differences between this quasi-stationary regime and other regimes obtained at DIII-D [Lohr, *et al.* (1988)] and Alcator C-Mod [Fielding, *et al.* (1996)], are that in the TCV the tokamak confinement is achieved with pure core electron heating at low-q_{95} (< 3), with no auxiliary ion heating, no active fueling and no cryo-pumping. Clearly, this makes the discharge much simpler to analyze without the multiple heating mechanisms.

In addition, the TCV discharge is an ITER-relevant plasma, as the plasma has a relatively flat core q-profile with low-magnetic shear in the central profile. This is a desirable high-performance regime for the future ITER experiment fusion reactors.

New DIII-D experiments are being planned with a variable tilting of the neutral beam injectors (NBI) to vary the beam energetic particle (EP) radial profiles. The simulation results by Heidbrink, *et al.* (2013) and Waltz and Bass (2010) provide evidence that there is a critical density profile for the onset of the energetic particle driven Alfvén wave instability that can be controlled by tilting the neutral beam injection angle [Heidbrink, *et al.* (2013)]. The maximum local Alfvén eigenmode (AE) growth rate is larger than the local ITG/TEM rate at the low-toroidal mode numbers of the energetic particle (EP) modes. Large computer simulations with GYRO [Bass and Waltz (2010)] are used to investigate the nonlinear state of these stiff EP mode transport at the AE marginal stability gradient. The on-going work makes projections for ITER. The Alfvén eigenmode (AE) driven alpha particle confinement losses keep the beam-like ion distribution functions closer to the collisional slowing down distribution.

References

Asakura, N., Shimizu, K., Shirai, H., Koide, Y., and Takizuka, T. (1997). *Plasma Phys. Control Fusion* **39**, pp. 1295-1314.

Artaud, J. F., Basiuk, V., Imbeaux, F., Schneider, M., Garcia, J., Giruzzi, G., Huynh, P., Aniel, T., Albajar, F., Ané, J. M., Bécoulet, A., Bourdelle, C., Casati, A., Colas, L., Decker, J., Dumont, R., Eriksson, L. G., Garbet, X., Guirlet, R., Hertout, P., Hoang, G. T., Houlberg, W., Huysmans, G., Joffrin, E., Kim, S. H., Köhl, F., Lister, J., Litaudon, X., Maget, P., Masset, R., Pégourié, B., Peysson, Y., Thomas, P., Tsitrone, E., and Turco, F. (2010). The CRONOS suite of codes for integrated tokamak modeling, *Nucl. Fusion* **50**, p. 043001, doi:10.1088/0029-5515/50/4/043001.

Asp, E., Kim, J.-H., Horton, W., Porte, L., Alberti, S., Karpushov, A., Martin, Y., Sauter, O., Turri, G., and the TCV TEAM. (2008). Electron Thermal transport in Tokamak Configuration Variable, *American Institute of Physics*, doi:10.1063/1.2965828.

Bass, E. M., and Waltz, R. E. (2010). Gyrokinetic simulations of mesoscale energetic particle-driven Alfvénic turbulent transport embedded in microturbulence, *Phys. Plasmas* **17**, p. 112319, http://dx.doi.org/10.1063/1.3509106.

Bohm, D., Burhop, E. H. S., and Massey, H. S. W. (1949). "The use of probes for plasma exploration in strong magnetic fields" in *The characteristics of electrical discharges in magnetic field*, edited by A. Guthrie, and R. K. Wakerling (McGraw-Hill, New York).

Callen, J. D. (1977). Drift-wave turbulence effects on magnetic structure and plasma transport in tokamaks, *Phys. Rev. Lett.* **39**, pp. 1540-1543, http://link.aps.org/doi/10.1103/PhysRevLett.39.1540.

Ferraro, N. M., Jardin, S. C., and Snyder, P. B. (2010). Ideal and resistive edge stability calculations with M3D-C, *Phys. Plasmas* **17**, p. 102508, http://dx.doi.org/10.1063/1.3492727.

Fielding, S. J., Ashall, J. D., Carolan, P. G., Colton, A., Gates, D., Hugill, J., Morriss, A. W., Valovic, M., and the COMPASS-D and ECRH Teams. (1996). *Plasma Phys. Control. Fusion* **38**, p. 1091, doi:10.1088/0741-3335/38/8/002.

Hahm, T. S., and Tang, W. M. (1989). Properties of ion temperature gradient drift instabilities in H-mode plasmas, *Phys. Fluids B* **1**, p. 1185, http://dx.doi.org/10.1063/1.859197.

Heidbrink, W. W., Van Zeeland, M. A., Austin, M. E., Bass, E. M., Ghantous, K., Gorelenkov, N. N., Grierson, B. A., Spong, D. A., and Tobias, B. J. (2013). The effect of the fast-ion profile on Alfvén eigenmode stability, *Nucl. Fusion* **53**, 9, doi:10.1088/0029-5515/53/9/093006.

Hoang, G. T., Horton, W. Bourdelle, C., Hu, B., Garbet, X., and Ottaviani, M. (2003).

Analysis of the critical electron temperature gradient in Tore Supra, *Phys. Plasmas*
10, pp. 405-412, doi:10.1063/1.1534113.

Hoang, G. T., Bourdelle, C., Garbet, X., Giruzzi, G., Aniel, T., and Ottaviani,
M., Horton, W., Zhu, P., and Budny, R. V. (2001). Experimental determination of critical threshold in electron transport on tore supra, *Phys. Rev. Lett.*
87, p. 125001, doi:10.1103/PhysRevLett.87.125001, http://link.aps.org/doi/10.
1103/PhysRevLett.87.125001.

Holland, C., and Diamond, P. H. (2002). Electromagnetic secondary instabilities in electron
temperature gradient turbulence, *Phys. Plasmas* **9**, p. 3857, http://dx.doi.org/
10.1063/1.1496761.

Horton, W. (2012). *Turbulent Transport in Magnetized Plasmas* (World Scientific)
ISBN:978-981-4383-53-0.

Horton, W., Hoang, G. T., Bourdelle, C., Garbet, X., Ottaviani, M., and Colas, L. (2004).
Electron transport and the critical temperature gradient, *Phys. Plasmas* **11**, 5,
p. 2600, http://dx.doi.org/10.1063/1.1690761.

Horton, W., Hong, B.-G., and Tang, W. M. (1988). Toroidal electron temperature gradient
driven drift modes, *Phys. Fluids* **31**, p. 2971-2983, doi:10.1063/1.866954.

Huysmans, G. T. A., and Czarny, O. (2007). MHD stability in X-point geometry: Simulation of ELMs, *Nucl. Fusion* **47**, pp. 659-666, doi:10.1088/0029-5515/47/7/016,
stacks.iop.org/NF/47/659.

Ishizawa, A., and Nakajima, N. (2010). Turbulence driven magnetic reconnection causing long-wavelength magnetic islands, *Phys. Plasmas* **17**, p. 072308,
doi:10.1063/1.1388176, http://link.aip.org/link/doi/10.1063/1.1388176.

Jacchia, A., Luca, F. D., Cirant, S., Sozzi, C., Bracco, G., Bruschi, A., Buratti, P., Podda,
S., and Tudisco, O. (2002). Gradient length driven electron heat transport study
in modulated electron cyclotron heating FTU tokamak, *Nucl. Fusion* **42**, p. 1116,
doi:10.1088/0029-5515/42/9/310.

Jenko, F., and Dorland, W. (2002). Prediction of significant tokamak turbulence at electron
gyroradius scales, *Phys. Rev. Lett.* **89**, p. 225001, http://link.aps.org/doi/10.
1103/PhysRevLett.89.225001.

Kadomtsev, B. B. (1992). *Tokamak Plasma: A Complex Physical System* (Institute of
Physics Pub., Bristol, UK and Philadelphia) ISBN: 0750302348.

Karpushov, A. N., Duval, B. P., Schlatter, C., Afanasyev, V. I., and Chernyshev, F.
V. (2006). Neutral particle analyzer diagnostics on the TCV tokamak, *Rev. Sci.
Instrum.* **77**, p. 033503, doi:10.1063/1.2185151, http://dx.doi.org/10.1063/1.
2185151.

Kaye, S. M., Levinton, F. M., Stutman, D., Tritz, K., Yuh, H., Bell, M. G., Bell, R. E.,
Domier, C. W., Gates, D., Horton, W., Kim, J.-Y. LeBlanc, B. P., Luhman, Jr.,
N. C., Maingi, R., Mazzucato, E., Menard, J. E., Mikkelsen, D., Mueller, D., Park,
H., Rewoldt, G., Sabbagh, S. A., Smith, D. R., and Wang, W. (2007). Confinement
and Transport in the National Spherical Torus Experiment, *Nucl. Fusion* **47**, p. 499,
doi:10.1088/0029-5515/47/7/001.

Keilhacker, M., Gibson, A., Gormezano, C., Lomas, P.J., Thomas, P. R., Watkins, M. L.,
Andrew, P., Balet, B., Borba, D., Challis, C. D., Coffey, I., Cottrell, G. A., De Esch,
H. P. L., Deliyanakis, N., Fasoli, A., Gowers, C. W., Guo, H. Y., Huysmans, G. T. A.,
Jones, T. T. C., Kerner, W., König, R. W. T., Loughlin, M. J., Maas, A., Marcus, F.
B., Nave, M. F. F., Rimini, F. G., Sadler, G. J., Sharapov, S. E., Sips, G., Smeulders,
P., Söldner, F. X., Taroni, A., Tubbing, B. J. D., von Hellermann, M. G., Ward,
D. J., and JET Team. (1999). High Fusion Performance from Deuterium-Tritium
Plasmas in JET, *Nucl. Fusion* **39** p. 209, doi:10.1088/0029-5515/39/2/306.

Li, J., and Kishimoto, Y. (2004). Numerical study of zonal flow dynamics and electron transport in electron temperature gradient driven turbulence, *Phys. Plasmas* **11**, pp. 1493-1511, doi:10.1063/1.1669397, http://dx.doi.org/10.1063/1.1669397.

Lin, Z., Ethier, S., Hahm, T. S., and Tang, W. M. (2002). Size Scaling of Turbulent Transport in Magnetically Confined Plasmas *Phys. Rev. Lett.* **88**, p. 195004, doi:http://dx.doi.org/10.1103/PhysRevLett.88.195004.

Lohr, J., Stallard, B. W., Prater, R., Snider, R. T., Burrell, K. H., Groebner, R. J., Hill, D. N., Matsuda, K., Moeller, C. P., Petrie, T. W., St. John, H., and Taylor, T. S. (1988). Observation of H-mode confinement in the DIII-D Tokamak with electron cyclotron heating, *Phys. Rev. Lett.* **60**, pp. 2630-2633, doi:http://dx.doi.org/10.1103/PhysRevLett.60.2630.

Mattor, N., and Diamond, P. H. (1988). Momentum and thermal transport in neutral-beam-heated tokamaks, *Phys. Fluids* **31**, p. 1180, http://dx.doi.org/10.1063/1.866747.

Mazzucato, E. (2010). Study of turbulent fluctuations driven by the electron temperature gradient in the National Spherical Torus Experiment, *Nucl. Fusion* **50**, p. 029801, http://iopscience.iop.org/0029-5515/50/2/029801.

Mazzucato, E., Bell, R. E., Ethier, S., Hosea, J. C., Kaye, S. M., LeBlanc, B. P., Lee, W. W., Ryan, P. M., Smith, D. R., Wang, W. X., Wilson, J. R., and Yuh, H. (2009). Study of turbulent fluctuations driven by the electron temperature gradient in the National Spherical Torus Experiment, *Nucl. Fusion* **4**, p. 5, doi:10.1088/0029-5515/49/5/055001.

Morales, Jorge A., Bos, Wouter, J. T., Schneider, Kai, and Montgomery, David C. (2012). Intrinsic Rotation of Toroidally Confined Magnetohydrodynamics, Phys. Rev. Lett. **109**, p. 175002, doi:10.1103/PhysRevLett.109.175002.

Muraglia, M., Agullo, O., Yagi, M., Benkadda, S., Beyer, P., Garbet, X., Itoh, S.-I., Itoh, K., and Sen, A. (2009). Effect of the curvature and the β parameter on the nonlinear dynamics of a drift-tearing magnetic island, *Nucl. Fusion* **49** pp. 055016-055027, http://dx.doi.org/10.1088/0029-5515/49/5/055016.

Ottaviani, M., Horton, W., and Erba, M. (1997). The long wavelength behaviour of ion-temperature-gradient-driven turbulence and the radial dependence of the turbulent ion conductivity, *Plasma Phys. Control. Fusion* **39**, 9, pp. 1461-1477, ISSN:0741-3335.

Peeters, A. G., Angioni, C., Bortolon, A., Camenen, Y., Casson, F. J., Duval, B., Fiederspiel, L., Hornsby, W. A., Idomura, Y., Hein, T., Kluy, N., Mantica, P., Parra, F. I., Snodin, A. P., Szepesi, G., Strintzi, D., Tala, T., Tardini, G., de Vries, P., and Weiland, J. (2011). Overview of toroidal momentum transport, *Nucl. Fusion* **51**, p. 094027, doi:10.1088/0029-5515/51/9/094027.

Porte, L., Coda, S., Alberti, S., Arnoux, G., Blanchard, P., Bortolon, A., Fasoli, A., Goodman, T. P., Klimanov, Y., Martin, Y., Maslov, M., Scarabosio, A., and Weisen, H. (2006). Plasma Dynamics with Second and Third Harmonic ECRH on TCV, *21st International Atomic Energy Agency Fusion Energy Conf.* (Vienna) EX/P6-20.

Rebut, P. H., Watkins, M. L., Gambier, D. J., and Boucher, D. (1991). A program toward a fusion reactor, *Phys. Fluids B* **6**, 3, pp. 2209-229, http://dx.doi.org/10.1063/1.859638.

Rebut, P. H., Lallia, P. P., and Watkins, M. L. (1988). The critical temperature gradient model of plasma transport: applications to JET and future tokamaks, *Plasma Phys. Control. Nucl. Fusion*, http://www.iaea.org/inis/collection/NCLCollectionStore/_Public/21/008/21008733.pdf.

Rhodes, T. L., Peebles, W. A., DeBoo, J. C., Prater, R., Kinsey, J. E., Staebler, G. M.,

Candy, J., Austin, M. E., Bravenec, R. V. Burrell, K. H., deGrassie, J. S., Doyle, E. J., Gohil, P., Greenfield, C. M., Groebner, R. J., Lohr, J., Makowski, M. A., Nguyen, X. V., Petty, C. C., Solomon, W. M., St. John, H. E., Van Zeeland, M. A., Wang, G., and Zeng, L. (2007). Broad wavenumber turbulence and transport during Ohmic and electron cyclotron heating in the DIII-D tokamak, *Phys. Plasmas Control. Fusion* **49** p. B183, doi:10.1088/0741-3335/49/12B/S17.

Rice, J. E., Ince-Cushman, A., deGrassie, J. S., Eriksson, L.-G., Sakamoto, Y., Scarabosio, A., Bortolon, A., Burrell, K. H., Duval, B. P., Fenzi-Bonizec, C., Greenwald, M. J., Groebner, R. J., Hoang, G. T., Koide, Y., Marmar, E. S., Pochelon, A., and Podpal, Y. (2007). Inter-machine comparison of intrinsic toroidal rotation in tokamaks, *Nucl. Fusion* **47**, p. 1618, doi:10.1088/0029-5515/47/11/02.

Rice, J. E., Bonoli, P. T., Goetz, J. A., Greenwald, M. J., Hutchinson, I. H., Marmar, E. S., Porkolab, M., Wolfe, S. M., Wukitch, S. J., and Chang, C. S. (1999). Central impurity toroidal rotation in ICRF heated Alcator C-Mod plasmas, *Nucl. Fusion* **39**, p. 1175, doi:10.1088/0029-5515/39/9/310.

Ryter, F., Imbeaux, F., Leuterer, F., Fahrbach, H.-U., Suttrop, W., and ASDEX Upgrade Team. (2001a). Experimental characterization of the electron heat transport in low-density ASDEX Upgrade Plasmas, *Phys. Rev. Lett.* **86**, pp. 5498-5501, doi:http://dx.doi.org/10.1103/PhysRevLett.86.5498.

Ryter, F., Leuterer, F., Pereverzev, G., Fahrbach, H.-U., Stober, J., Suttrop, W., and ASDEX Upgrade Team. (2001b). Experimental evidence for gradient length-driven electron transport in tokamaks, *Phys. Rev. Lett.* **86**, pp. 2325-2328, doi:http://dx.doi.org/10.1103/PhysRevLett.86.2325.

Sen, A. K., Sokolov, V., and Wei, X. (2006). A new paradigm for plasma transport and zonal flows, *Phys. Plasmas* **13**, p. 055905, http://dx.doi.org/10.1063/1.2196267.

Sovinec, C. R., Glasser, A. H., Gianakon, T. A., Barnes, D. C., Nebel, R. A., Kruger, S. E., Schnack, D. D., Plimpton, S. J., Tarditi, A., Chu, M. S., and the NIMROD Team (2004). Nonlinear magnetohydrodynamics simulation using high-order finite elements, *J. Comput. Phys.* **195**, p. 355.

Stutman, D., Delgado-Aparicio, L., Gorelenkov, N., Finkenthal, M., Fredrickson, E., Kaye, S., Mazzucato, E., and Tritz, K. (2009). Correlation between electron transport and shear Alfvén activity in the National Spherical Torus Experiment, *Phys. Rev. Lett.* **102**, p. 115002, doi:http://dx.doi.org/10.1103/PhysRevLett.102.115002.

Sydora, R .D., Decyk, V. K., and Dawson, J. M. (1996). Fluctuation-induced heat transport results from a large global 3D toroidal particle simulation model, *Plasma Phys. Control. Fusion* **38**, A281, doi:10.1088/0741-3335/38/12A/021.

Tonetti, G., Heym, A., Hofmann, F., Hollenstein, C, Koechli, J., Lahlou, K., Lister, J. B., Marmillod, Ph., Mayor, J. M., Magnin, J. C., Marcus, F. B., and Rage, R. (1991). *Fusion Technology, 1990,* Proceedings of the 16th Symposium on Fusion Technology (North-Holland, London, U.K.) R. Hemsworth, ed., p. 587.

Wagner, F., Becker, G., Behringer, K., Campbell, D., Eberhagen, A., Engelhardt, W., Fussmann, G., Gehre, O., Gernhardt, J., Gierke, G. v., Haas, G., Huang, M., Karger, F., Keilhacker, M., Klüber, O., Kornherr, M., Lackner, K., Lisitano, G., Lister, G. G., Mayer, H. M., Meisel, D., Müller, E. R., Murmann, H., Niedermeyer, H., Poschenrieder, W., Rapp, H., Röhr, H., Schneider, F., Siller, G., Speth, E., Stäbler, A., Steuer, K. H., Venus, G., Vollmer, O., and Yü, Z. (1982). Regime of Improved Confinement and High Beta in Neutral-Beam-Heated Divertor Discharges of the ASDEX Tokamak, *Phys. Rev. Lett.* **49**, p. 1408, http://dx.doi.org/10.1103/PhysRevLett.49.1408.

Waelbroeck, F. L., Morrison, P. J., and Horton, W. (2004). Hamiltonian formulation and coherent structures in electrostatic turbulence, *Plasma Phys. Control. Fusion* **46**, 9, pp. 1331-1350, doi:10.1088/0741-3335/46/9/001.

Wei, X., Sokolov, V., and Sen, A. K. (2010). Experimental production and identification of electron temperature gradient mode, *Phys. Plasmas* **17**, 4, p. 042108, `http://dx.doi.org/10.1063/1.3381070`.

Weiland, J., and Holod, I. (2005). Drift wave transport scalings introduced by varying correlation length, *Phys. Plasmas* **12**, p. 012505, doi:10.1063/1.1828083.

Chapter 5

Operational Regimes
and their Properties

5.1 Ohmic Plasma Confinement Mode, H-mode, I-mode Plasmas

There is a rich variety of plasma confinement regimesound in tokamaks – each regime having its own characteristics and own name. The classic Ohmic heated plasma regime has two regimes, with the first having the energy confinement time increasing with the plasma density, which is called the linear Ohmic regime. The original Alcator machine with circular magnetic surfaces followed this regime with toroidal magnetic fields up to 7 T. Next the Saturated Ohmic Confinement (SOC) regime was discovered in which the energy confinement stopped increasing with the plasma density. Once auxiliary heating comparable to the ohmic heating power was applied to tokamaks the plasma confinement showed a new scaling called the L-mode confinement regime. This regime was discovered in the PLT machine by Goldston *et al.* (1987). The upgraded Alcator C and Tore Supra accumulated a large databases with linear ohmic and the L-mode discharges driven with RF heating. The superconducting coils in Tore Supra allowed long steady-state discharges in which the level of heating power was varied during a single discharge to measure the energy confinement time dependence on the heating power in a single discharge. The upgrade called Alcator C-Mod had a rectangular vacuum chamber with large toroidal coils used to create poloidal magnetic field configurations with a separatrix with either single null as in ITER or double (upper and lower) null points in the poloidal magnetic field structure. In the Alcator C-Mod the energy confinement time τ_E was found to increases linearly with the plasma density up to a certain maximum density that was then shown to be proportional to the total toroidal plasma current I_p.

A large number of parameter studies in a variety of different tokamaks have now found this density limit, which is called the Greenwald density limit. Since the thermonuclear power increases with the product of the deuterium and tritium density operating at the maximum density is required for producing the highest levels of fusion power. The density limit is complex so the plasma regime is regime is typically expressed ratio $f_{GW} = n_e/n_{GW}$ where the Greenwald density limit is $n_{GW} = C_{GW}I_p/a^2$ with the constant C_{GW} determined from the tokamak confinement databases [Greenwald, *et al.* (1988)]. The limit on the maximum current for a given minor radius a of the machine is the Kruskal-Shafranov in which the twist of magnetic field forms is such as to make one complete rotation in the poloidal direction during one turn about the major symmetry as of the torus. The rate of twisting of the magnetic field is given by the $q(r) = rB_\phi/RB_\theta$ which for high performance tokamak increases from a value close to unity at the core $r = 0$ and rises to large values at the chamber wall. The internal kink instability is strong and fast and starts when q drops below unity in the core $r = 0$ plasma. Thus, the name for the q parameter is the safety factor meaning that for MHD kink stable plasma the safety factor must be above unity at the core of the plasma. As the plasma is heated the core value will drop below unity and there are observed a quasi-periodic sequence of strong relaxation events called the sawtooth modes. Fusion tokamaks will operate with $q(0) \sim 1$ which translates into a maximum toroidal plasma current determined by the toroidal magnetic field B_ϕ and the geometry of the confinement system. The plasma geometry is controlled by limiters and divertors.

As described in Chapter 1 there was a breakthrough in confinement when the first divertor tokamak called ASDEX was operated in 1982 by the Wagner team at the Max-Planck Institute für Plasmaphysik. For the specified plasma current and machine parameters, the ASDEX machine found a transition to higher plasma confinement regimes now called the H-mode for high confinement. In the beginning, to obtain the transition from the L-mode found in PLT and numerous other tokamaks, to the H-mode sufficient auxiliary heating power P_{aux} needed to be added to the Ohmic heating power. Those experiments showed that there is critical amount of auxiliary heating power P_{L-H} to enter the H-mode. Now, we know that even without auxiliary heating power the plasma can make a transition to the H-mode when sufficient plasma shaping is used as in the Tokamak Configuration Variable described in Chapter 4. For typical tokamaks, the transition to the high-confinement regime requires that the auxiliary plasma heating exceed a threshold value known as the L-to-H power threshold $P_{L\to H}$ threshold [Greenwald, *et al.* (1988)]. To have a low threshold for the bifurcation or transition to the high-confinement mode, the plasma chamber needs to be shaped and the divertor chamber added below the X-point in the magnetic field surfaces for the lost plasma exhaust. For low power auxiliary heating to enter the H-mode confinement requires particular plasma shaping as explored in the Tokamak Configuration Variable illustrated by four H-mode regime discharge in the previous Chapter. Later essentially all tokamaks were found to be capable of finding the L- to-H-mode transition when sufficient auxiliary heating power is applied. Designing systems to minimize the required auxiliary power P_{aux} for the L- to H-transition occupied years of research in the 1990s and early 2000s. For the shaped divertor tokamak with the proper orientation of the grad-B and curvature drifts of the ions, discussed in Chapter 2, the L- to H-mode transition is achieved at a reasonably low level of auxiliary plasma heating power. Thus, the design of ITER was changed at some point in time, after these results were well documented, to operate in the H-mode rather than in the original L-mode plasma used in the earlier larger $R = 8\,\mathrm{m}$ design shown in Fig. 1.1. In the H-mode, the plasma energy confinement time is about a factor of 2 longer than in the L-mode. The profiles of density tend to be flat in the H-mode rather than peaked, as in the L-mode. Thus, it appears there is a partial transport barrier just inside the magnetic separatrix, where the density drops abruptly to low value and the temperature drops in crossing the magnetic separatrix into the outer region called the scrape-off layer (SOL) plasma. This is now the standard design for the $R = 6\,\mathrm{m}$ ITER under construction in France and shown in Fig. 1.2.

While the H-mode plasma has better global energy confinement times, there is an concomitant build-up of a steep density and pressure gradient across the last closed magnetic flux surfaces. These gradients drive the resistive ballooning g-modes unstable and at higher pressures compared to the poloidal magnetic field pressure – a limit given by the H-mode beta limit formula. In addition, the gradient of the toroidal plasma current density ∇j_ϕ adds the drive of a MHD ballooning-kink mode instability in this sharp gradient edge region. Thus, the optimal performance is obtained just below the onset of these two instabilities where the precise transitions point is a complicated function of the collisionality, the rotation of the plasma and shaping of the plasma from the shape of the magnetic separatrix created with large current carrying coils circling the symmetry axis. These are the poloidal coils shown in cross section in Fig. 1. The D-shaped separatrix seems to give the optimal plasma shape for fusion power. A "D" shaped plasma has more of the plasma volume with outward toroidal curvature giving g_{eff} pointing into the high pressure core plasma. Relatively simple nonlinear models for the L- to H-transition have been developed by combing the formulas for the transport of thermal energy with turbulence driven by the resistive g-instabilities and the collisionless pressure/temperature gradient driven modes against the stabilization effects from sheared plasma rotation driven by the concomitant turbulent Reynolds stress that drive cross-field momentum transport. The models have some qualitative successes in explaining the transitions from L- to H-mode confinement and the partially suppressed turbulent transport. Continuing to increase the auxiliary heating power leads to further bifurcations in the plasma state. The next state with increase core plasma temperature is one with bursts of turbulence and plasma loss across the magnetic separatrix. At the first power levels above the L-H transition the plasma transport has frequent small Type III Edge Localized Modes (ELMs). Adding further auxiliary heating the plasma makes a transition into the large Type I ELMs which give a higher level of high-temperature plasma flowing into the edge – SOL – of the confinement chamber that is of considerable concern and is a key reason for the need to build and test the ITER design before moving on to a demonstration tokamak fusion reactor that could have these non-stationary states with repeated nonlinear relaxation oscillations. This chapter describes the current research on the ELMs and ELM control mechanisms developed for ITER.

5.2   Control of Confinement Modes with External Sources of Momentum and Energy Injectors

The experimental control of the relaxation oscillations observed in the strongly-driven H-mode plasma are called ELMs. There are three varieties of ELMs observed and described by the H-mode Database Working Group [L-H-Mode Database (1994); Wolf, *et al.* (2000); Wolf, *et al.* (2001)]. There is an extensive review article [Wolf (2003)] on the types of transport barriers – both edge barriers and internal transport barriers in a wide range of plasma conditions.

With high auxiliary heating power the tokamak plasma develops both toroidal rotation and poloidal rotation. The rotational speeds or angular velocities have different values on neighboring magnetic surfaces which is called flow shear. Flow shear can in general be either stabilizing or destabilizing to the drift waves and thus changes the transport from that in a plasma with no sheared flows. There is experimental evidence that the transition from L-mode to H-mode is created by a rapid increase in the sheared flow in the edge plasma region, especially strong in divertor tokamaks. This H-mode regime was discovered in the first large shaped divertor tokamak called ASDEX [Wagner, *et al.* (1982)]. Subsequently, H-modes have been reported in all tokamaks that have sufficient heating power. Some machines such as the strongly-shaped TCV tokamak can operate exclusively in the H-mode. The ITER reference plasma design performance target was changed from the standard regime to the H-mode design in the late 1990s. The factor of two or more increase in the energy confinement time resulted in allowing the major radius of the design to be reduced from $R = 8\,\mathrm{m}$ to $R = 6\,\mathrm{m}$.

A model for the rotational shear flow necessary to produce a transport barrier is given by Diamond, *et al.* (1995) and Hahm and Burrell (1995). The model was developed to explain the transport barrier formed at the plasma edge in tokamaks after the L-(low) to H-(high) transition. Data on the rotational profiles is indirectly inferred from the spectroscopic measurements of the impurity ion transition lines. Then transport models are used to connect the impurity flow velocities to those of the hydrogen ions.

Simulations show that sheared flow significantly changes the cross-field transport rate. Often, the profile of the shear flow influences the effectiveness of the transport in atmospheres as well as plasmas. Sufficiently, strong sheared flows generate their own turbulent transport. However, the rules used for reducing the transport based on flow shear combined with the magnetic shear have been successful in explaining a variety of transport barriers in tokamaks. The name "transport barrier" denotes a region of reduced transport, not as in Hamiltonian chaos theory as a true barrier to transport. The rule that has wide success uses the ratio of the $\mathbf{E} \times \mathbf{B}$ shear to the magnetic shear for drift wave instabilities.

Various shear flow criteria are successfully used for the edge region and the core region of the tokamaks. These are the edge transport barrier (ETB) and internal transport barrier ITB in the plasma transport data. The rules have the universality needed to explain the edge transport barriers in both limiter and divertor tokamaks, stellarators, and mirror machines.

The shear flow transport barrier model is invoked to explain further the confinement improvement from the H-(high) mode to VH-(very high) mode seen in some tokamaks. The shear flow paradigm also applies to the core transport barriers formed in plasmas with negative or low-magnetic shear in the plasma core.

The reduction of cross-field transport associated with $\mathbf{E} \times \mathbf{B}$ velocity shear effects also has significant practical consequences for fusion research. The physics involved in transport reduction from sheared flow is (1) a reduction of the growth rate pressure gradient driven instabilities, (2) a reduction in the radial extent and phase correlation of turbulent eddies in the plasma. The same fundamental transport reduction process can be operational in various portions of the plasma because there are a number of ways to change the radial electric field with different heating and momentum injection schemes.

5.3 Bifurcations Models Describing Spontaneous Symmetry Breaking with Transitions to L, H and ELMy-H Modes

Transitions or bifurcations in the state of plasma are qualitatively similar to the changes that occur in heated neutral fluids that have been known and researched for decades with the Rayleigh-Benard convection experiments in heated fluids. Key parameters for the various fluids are their molecular viscosity and thermal diffusivity. Here the gravity is down and the heat flux is up driven by a high temperature lower plate whose temperature is steadily increased to produce the sequence of bifurcations. Some of the physics and mathematics for the sequence of bifurcations and changes in the thermal fluxes in the neutral fluid experiments has been successfully exported to the tokamak fusion program. Works following this approach include a set of nonlinear oscillations between stable and unstable configurations that are in the predator-prey models developed in works by Diamond, *et al.* (1995). Horton and Laval develop models following the fluid works of Krishnamurti and Howard (1981) with Rayleigh-Benard cells. Combined neoclassical-turbulence transport models are given in Sugama and Horton (1995), Horton, *et al.* (1996), Hu and Horton (1997) and Li and Kishimoto (2002).

Sugama and Horton (1995, 1997) and Sugama, *et al.* (2001) have developed a systematic combination of neoclassical and momentum transport moment equations from truncation of the fluid moments of the drift-kinetic equations including the finite gyroradius momentum stress tensor and heat fluxes. This produces complex coupled partial differential equations that in principle describe the similar change of events discussed in the simple fluid models described in the preceding paragraph. The complexity of the combined turbulence and neoclassical transport equations has essentially prevented the development of solutions to this full set of partial differential equations.

5.4   Hot Ion Mode Sets Record

ITER is designed to operate in the high-confinement mode called the H-mode. As discussed in Chapter 4, there is a factor of two or more longer energy-confinement time in the H-mode plasma than in the L-mode plasma. There are other confinement regimes described in Chapter 2 with improvements over the L-mode. Particularly interesting is the I-mode. The H-mode plasma confinement relies on edge confinement produced by a zone of reduced transport just inside the separatrix and called the "transport barrier". This term is to be understood as an acronym for a zone of reduced transport; not a true barrier to transport. There is no true barrier in the sense of a transport barrier as defined by the KAM surfaces in chaos theory.

The plasma density inside the magnetic separatrix is high and comparable to the core density level owing to the transport barriers achieved in the H-mode regimes. Discovery of this new confinement regime was a major achievement of the first toroidal machine, with a magnetic separatrix designed by K. Lackner and his team in the 1970s, and built as the ASDEX tokamak at the Max-Planck Institüt für Plasma Physics in 1980. The new confinement regime was discovered in ASDEX by an experimental team led by Wagner (1982). The team made numerous important discoveries through the 1980-1990s that are described in Wolf, *et al.* (2001). With this magnetic separatrix the experimental team made a fundamental discovery in the history of tokamaks. The high-pressure core plasma had a sharp edge just inside the magnetic separatrix. The plasma inside this sharp edge was shown to have an energy confinement about twice as long as given by the previously existing tokamak confinement laws derived from earlier circular flux surface machines. This regime is then defined as the high-confinement regime or H-mode. There is a steep pressure gradient at the edge which represents a radial width of a few percent of the plasma radius a_p on the equatorial plane. The plasma outside the magnetic separatrix is said to be in the scrape-off layer and rapidly flows in to the divertor chamber at the bottom of the machine in Figs. 1.1 and 1.2.

In these discharges the plasma starts as a typical ohmically-heated tokamak regime, now called the L-mode for low confinement, and then makes a transition to the steeper-edge profiles to the H-mode regime. There needs to be a sufficiently vigorous core heating for the transition to occur, giving rise to the empirical formula for the P_{LH} threshold heating power to achieve the H-mode confinement. Various sized and shaped tokamak experiments have documented the power $P_{threshold}$ required to make the transition to the H-mode providing a database that gives the empirical formula for the critical injected conditions required to produce the L to H bifurcation. Machines without the magnetic separatrix usually do not achieve the H-mode. There are both hot ion and hot electron plasmas with H-modes as reviewed by Wolf, *et al.* (2001).

Recent work on the L-H transition physics and the formation of edge transport barriers is presented in Dong, *et al.* (2013). The studies of the transition dynamics in HLAHL-2A and KSTAR are divided into four-stages using imaging diagnostics. Low collisionality is shown to give the transition from the trapped electron mode to ion temperature gradient transition by varying the plasma density. Fueling and wall coatings of lithium Li^3, beryllium Be^4 and boron B^5 are used to control the inward pinch of particles in tokamak. Carbon C^6 wall coatings are found to be unsuitable producing high impurity concentrations.

5.5 Discovery of Edge Localized Modes (ELMs)

Figure 5.3 shows a typical plasma pressure in the H-mode, which has a sharp drop in pressure over the narrow layer in the middle of the plasma edge. Once the H-mode is settled, this pressure gradient tends to increase with time until the peeling-ballooning instability leads to a turbulent state triggering the Edge Localized Modes (ELMs). There are three well-studied varieties of the ELMs defined as ELMs of Type I, II and III. The ELMs of Type III have the highest repetition rate and are least detrimental to steady-state operation. The Type I ELMS have a large, infrequent releases of the core plasma energy and are the most serious obstacle to steady-state operation. The energy released in the Type I ELM can exceed the capacity of the boron or beryllium coated metal walls to retain their integrity as discussed in Chapter 9. For wall loads of over $10\,MW/m^2$ the lifetime of the walls is too short – perhaps less than one year – which is now a key issue for the ITER machine to answer. For comparison the thermal energy flux at the solar surface is of order $60 MW/m^2$.

Articles on the understanding of the transition to the regimes with various types of transport barriers and the scaling of the required power for the transition are described in Connor (1998a) and Connor, *et al.* (1998b), Groebner, *et al.* (2009). The intermittent releases of the core plasma energy through the edge transport barriers are called Edge Localized Modes or ELMs. The ELMs are fast bursts of hot dense plasma on a fast time scale ($\sim 25\text{-}300\ \mu$ secs). Fast-frequency ELMs with low amplitude are called Type III ELMs and are less harmful to the plasma walls than the large Type I ELMs. Research on ways to eliminate or reduce the Type I ELMs to the smaller, more frequent ELMS is extensive. Currently, considerable research efforts is being performed by adding external magnetic field coils to create δB_r at the mode rational surfaces to trigger the release of the edge plasma energy before their stored energy is released by a large ELM. Thus, this approach has been to induce smaller, more frequent ELMs with these Resonant Magnetic Perturbations (RMPs) coils. The approach has been successful on JET and DIII-D and is, at the time of writing, being evaluated as a set of external coils to add to the ITER device on the outer mid-plane just outside the vacuum chamber wall. The second method to limit the growth of the ELMs to inject solid pellets with suitable composition

Summary of High-Performance TFTR Shots

	Supershot 80539A12	Li assisted 83546A15	High-I_i 95603A02	ERS 88170A51
I_p(MA)	2.7	2.3	2.0	1.6
B_t(T)	5.6	5.5	4.8	4.6
P_{NB}(MW)	39.6 (D-T)	17.4 (T only)	35.5 (D-T)	28.1 (D only)
$n_T/(n_D + n_T)(0)$	0.47	0.58	0.42	0
$n_e(0)(10^{19}/m^3)$	10.2	8.5	6.9	9.0
$n_{hyd}(0)(10^{19}/m^3)$	6.7	6.6	6.0	7.0
$Z_{eff}(0)$	2.4	2.0	1.6	2.1
$T_e(0)$(KeV)	13.0	12.0	8.0	8.0
$T_i(0)$(KeV)	36	43	45	25
W(MJ)	6.9	4.9	5.7	3.9
dW/dt(MW)	0.0	3.0	8.5	3.0
τ_E(s)	0.180	0.340	0.165	0.150
$\tau_E^* = W/P_{NB}$(s)	0.174	0.28	0.161	0.139
$\tau_{ITER-89P}$(s)	0.095	0.119	0.074	0.073
$\tau_E/\tau_{ITER-89P}$	1.89	2.86	2.23	2.05
$n_{hyd}(0)T_i(0)\tau_e$ $(10^{20}m^{-3} \cdot KeV \cdot s)$	4.3	9.6	4.5	2.6
$n_{hyd}(0)T_i(0)\tau_E^*$ $(10^{20}m^{-3} \cdot KeV \cdot s)$	4.2	8.0	4.4	2.4
P_{fus}(MW)	10.7	2.8	8.7	0
P_{fus}/P_{NB}	0.27	0.16	0.25	0
β_{norm}(mag) (% mT/MA)	1.83	1.35	2.50	1.95
β_{norm}(TRANSP)	1.83	1.5	2.40	1.95
β_{norm}^*(TRANSP)	2.99	3.0	3.9	3.7
κ	1.1	1.1	0.975	0.978
q_{cycl}	3.07	3.61	3.57	5.0
q^*	3.22	3.79	3.47	5.0

and velocity as discussed in Chapter 6.

5.6 Comparison of Four Confinement Modes in a Long Discharge

The TCV shot 29892 begins with a typical Ohmic H-mode with Type III ELMs. When the ECRH is turned on at $t = 0.7$ s, the plasma rapidly switches to a Type I ELMy H-mode. When the ELMs disappear, a quasi-stationary ELM-free H-mode plasma emerges. After a period of time extending over numerous energy confinement times, a single ELM event occurs, at $t = 0.9$ s, the quasi-stationary ELM-free H-mode is recovered but with a reduced energy confinement due to a magnetic island left as a residual the isolated ELM event. This magnetic island at the rational surface $q(r) = 3$ produces a local flattening of the temperature profile and keeps the energy confinement lower than in the previous phase before the isolated ELM event. The suppression of the ELMs in this case coincides with the onset of the ECRH power modulation. The transition to the ELM free state has also been observed to arise spontaneously [Porte, *et al.* (2006)]. The transition to an ELM-free H-mode is an interesting topic requiring further research.

This discharge exhibits a higher density ($\sim 6 \times 10^{19} \mathrm{m}^{-3}$) than what is commonly found with ECRH heated tokamaks. The use of the third harmonic (X3) heating system instead of the second harmonic system allows for strong electron heating at plasma density up to approximately $10^{20} \mathrm{m}^{-3}$. This in turn gives rise to considerable ion heating due to the faster thermal equilibration between ions and electrons with higher plasma density. Unfortunately there are no charge exchange recombination spectroscopy (CXRS) measurements during this shot that would provide profiles of carbon ion temperature, density and rotation. Such data is available for a sister shot which shows the higher ion temperature.

The core electron power deposition in this discharge varies through the four phases with an average of 1-2 MW/m^3 and has a maximum of 8 MW/m^3 on-axis. Thus, the power deposition is strongly peaked on-axis. With a core electron density, $n_e \sim 6 \times 10^{19} \mathrm{m}^{-3}$ this gives a heating rate of 800 keV/s per electron. The electron temperature is measured by Thomson scattering with 25 cords every 25 ms and the ECRH power deposition profile is calculated with the TORAY-GA ray-tracing code [Matsuda (1989)].

The electron heating is focused in the core plasma which gives rise to an accurately defined flow of thermal energy through the electron channel from the core plasma ($1 < T_e < 3$ keV) to the edge plasma. This large thermal flow yields a precise estimate of the radial electron heat flux $q_e(r)$ and the electron heat diffusivity χ_e as there is accurate electron temperature profile data (5% error). The profiles are stationary for more than five energy confinement times, τ_E, and therefore the power balance χ_e is accurately determined. The greatest uncertainty in q_e and χ_e arrives from the ion temperature, T_i, and the uncertainties of the measurements of the plasma gradients. The error analysis performed shows that allowing T_i to vary between $0.33 \, T_e$ to T_e only yields a 10% variation in the electron flux q_e in the ECRH heated phases. For the first Ohmic phase the error is considerably larger.

The TCV deuterium plasma analyzed has electron plasma beta $\beta_e > 2\%$ at the core such that $\beta_e m_i / m_e \simeq 73$, where m_i and m_e are the masses of ions and electrons. The TCV 29892 plasma has a toroidal magnetic field $B \simeq 1.27$ T, plasma current $I_p \simeq 400$ kA and Alfvén velocity $v_A \ll v_e = (T_e / m_e)^{1/2}$. This implies that we have an electromagnetic regime where the Alfvén wave falls in the low-parallel velocity $v_\parallel < v_e$ part of the electron energy distribution where electron Landau damping is significant. Hence, wave damping is important and the coupling to drift waves is strong. Due to the steep electron temperature gradient and the electron plasma pressure we expect the electromagnetic electron temperature gradient (ETG) drift-wave turbulence to be an important transport mechanism. The plasma also enters the trapped electron (TEM) and ion temperature gradient (ITG) modes in the electron energy spectrum.

The tools for the TEM-ITG analysis are an analytical model [Zhu, *et al.* (1999); Zhu, *et al.* (2000)] and the fluid-based Weiland model [Guo and Weiland (1997); Jarmen, *et al.* (1987); Nordman, *et al.* (1990); Weiland (2000)]. For the ETG analysis we used a code based on the three-coupled gyrofluid differential equations, the linear modes of which describes well the ETG instability [Horton, *et al.* (2005); Kim, *et al.* (2006)]. This ETG code has proven itself by accurately modeling a large database of electron thermal transport from ASDEX [Ryter, *et al.* (2001a); Ryter, *et al.* (2001b)], Tore Supra [Hoang, *et al.* (2003)] and the Frascati Tokamak Upgrade [Jacchia, *et al.* (2002)]..

To obtain a quantitative measure of how well the theoretical models explain the observed heat flux, the metric called the average relative variance (ARV) [Weigend, *et al.* (1990); Detman and Vassiliadis (1997)] is introduced. The ARV metric gives a dimensionless error measure for inter-model comparisons. The ARV increases quadratically with the magnitude of the model χ_e so it increases with a mismatch in the model's relative error with respect to the database χ_e. The model ETG thermal flux is also in good agreement with the power balance as shown in a figure for the thermal diffusivities in the ETG Transport section. The ARV parameters for the analytical TEM model and the local Weiland model show that these models fail to predict quantitatively the thermal transport. The ETG model, on the other hand, can, according to this metric, explain 70% of the variation of electron heat diffusivity.

5.6.1 *ECRH driven discharges*

A range of H-mode plasmas regimes from Ohmic to high power electron cyclotron resonance heating are produced in the tokamak with strong shaping of the magnetic flux surfaces called the Tokamak Configuration Variable, or TCV. We illustrate this and the different types of plasma confinement regimes in a single long pulse discharge in basic tokamak experiment with extensive diagnostics. The TCV discharge 29892 is described by Porte, *et al.* (2006) in the conference proceedings of the 21st IAEA Fusion Energy Conference [Porte, *et al.* (2006)]. This section gives the basic features of the shot (plasma quasi-steady state plasma discharge) needed to understand the underlying heat transport and energy confinement in a variety of plasma confinement states. TCV is a tokamak with a high-power density electron cyclotron resonance heating (ECRH) system in Table 5.1. In shot 29892, the third-harmonic X-mode (X3) heating system was used to enable significant electron heating at the plasma density $n_e \sim 6 \times 10^{19}\,\mathrm{m}^{-3}$ [Porte, *et al.* (2006)]. The electron heating gave rise to a significant ion heating $P_{(e \to i)}$ through thermal equilibration, $P_{e \to i} = 0.2 \pm 0.1\,\text{MW}$. Ion dynamics may directly or indirectly influence the electron transport. Ion heating from the electrons is also of fundamental importance for ITER and in nuclear fusion producing machines in which the ion heating is from the alpha particles through the electrons. Electron-to-ion heat transfer is vital to sustain the fusion reactions. In addition, this discharge features an ITER hybrid scenario q-profile with low magnetic shear.

X3 (third harmonic extraordinary mode polarized electron cyclotron wave) heating is applied at $t = 0.6\,\text{s}$ to a target plasma consisting of an Ohmic ELMy H-mode. The X3 ECH power is injected vertically from the top of the machine with the resonating surfaces close to the plasma axis which makes the heating peak in the inner core of the plasma as shown in Fig. 5.1. Moreover, the high temperature ($T_e \approx 14\,\text{KeV}$) caused by the high-energy confinement time, ($\tau_E \approx 37\,\text{ms}$), of the target plasma helps to increase the X3 absorption. The X3 heating does not produce a significant population of supra-thermal electrons that would have influenced the transport analysis. The ion temperature profile used for the calculation of the

Table 5.1   TCV machine and discharge 29892 parameters. Radial values are given at $r/a = 0.7$ in the Ohmic H-mode and the ELM-free H-mode.

Fixed parameters		
Plasma major radius, R (m)	0.89	
Plasma minor radius, a (m)	0.22	
Plasma elongation	1.75	
Plasma triangularity	0.51	
Plasma volume (m^3)	1.47	
Plasma current, I_p (kA)	400	
Toroidal field at the magnetic axis (T)	1.3	
Time dependent parameters	$t = 0.5\,\text{s}$	$t = 1.0\,\text{s}$
Total heating, P_{tot} (MW)	0.49	1.13
Global energy confinement time, τ_E (ms)	37	32
H-factor H$_{98(2,y)}$	0.72	1.1
Safety factor, q	1.29	1.30
Safety factor, q_{95}	2.18	2.16
Magnetic shear, s	0.80	0.65
Density gradient length, L_{ne} (m)	0.15	0.21
Temperature gradient length, L_{Te} m)	0.12	0.09
Electron temperature, T_e (KeV)	0.57	1.3
Electron diamagnetic drift velocity, v_{de} (km/s)	2.0	4.5
Electron magnetic drift velocity, v_{De} (km/s)	0.5	1.3
Ion Larmor radius at T_e, ρ_s (mm)	2.7	4.1
Power-balance diffusivity, χ^{PB} (m^2/s)	0.28	0.83
GyroBohm diffusivity, $D_{gB} = \rho_s v_{de}$ (m^2/s)	5.4	18.5
Drift wave frequency, $\omega = k_y v_{de}$ (kHz) at $k_y \rho_s = 0.3$	341	353
Electron collisionality, $\nu_*{}^1$	0.24	0.06
ω_{pe}/ω_{ce}	1.8	1.9

[1] refer to Eq. (18)

thermal exchange power density in Fig. 5.1 was taken from a sister shot where CXRS data was available. To validate the $T_i(r, t)$ profiles they were cross-checked with Neutral Particle Analyzer (NPA) [Karpushov, *et al.* (2006)] measurements of the energy spectra of the deuterium ions escaping the plasma.

The electron heat flux is proportional to the difference between the integrated electron power deposition P_e and the integrated electron-ion thermal equilibration $e \to i$, energy transfer power from electrons to ions through collisions. During the Ohmic phase at $t = 0.5$ s, the total thermal equilibration is comparable to the total power deposition $P_e = P_{oh} \sim 0.4$ MW. For the Ohmic regime, a 10% error in the ion temperature produces a 20% error in the electron heat flux. In the ECRH X3-heated phases of the discharge, a 10% error in the ion temperature leads to approximately a 10% variation in the electron heat flux.

The time history of the discharge TCV2989 is shown in Fig. 5.2 with five panels for plasma-neutral ionization $D_\alpha(t)$, mean density $\bar{n}_e(t)$, stored plasma energy $W_{dia}(t)$, plasma beta and electron temperature.

The D_α line is from electron transitions from the $n = 3$ to the $n = 2$ quantum (Bohr) levels of the deuterium atom. The signal is used to monitor the interaction of the hot plasma with neutral gas component that surrounds the plasma in edge and scrape-off layer. When there is a burst of brightness in the alpha line radiation, there is an excursion or a plume of hot plasma in the outer layer of the confinement chamber. Thus, the edge localized modes called ELMs give rise to bursts of the red $\lambda = 656$ nm radiation visible light from the plasma. For hydrogen working gas the line radiation is from the neutral hydrogen atom and for the deuterium working gas the line is from the neutral deuterium atoms. Owing to the higher nuclear mass of deuterium the D_α line has a slight shorter wavelength of $\lambda = 656$ nm than the hydrogen H_α line.

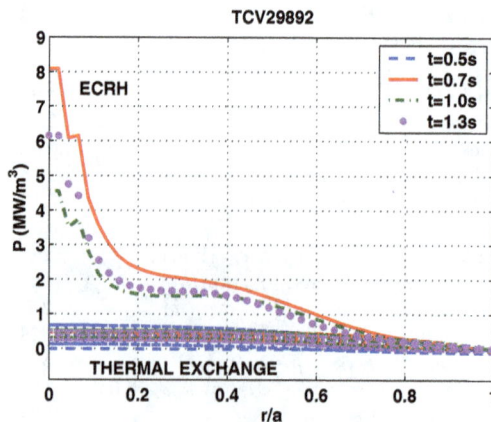

Fig. 5.1 Deposition of ECRH in TCV 29892 calculated by TORAY-GA [Matsuda (1989)] and the thermal equilibration power to the ions. The latter shows the range of values expected assuming a 5% error in n_e and T_e and assuming a 20% error in T_i.

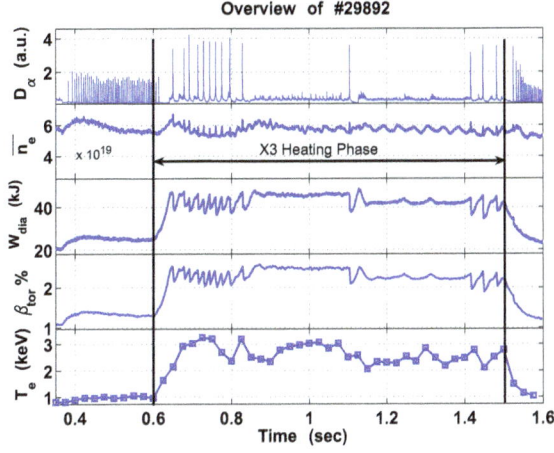

Fig. 5.2   Temporal evolution of TCV29892. From top to bottom are shown D_a emission, line-averaged density, total stored energy measured by the diamagnetic loop, toroidal β and on-axis electron temperature. At $t < 0.6\,\text{s}$, standard TCV H-mode with Type III ELMs. Full ECRH power for $0.6 \leq t < 0.8\,\text{s}$ and Type I ELMy H-mode. Modulated ECRH power for $t \geq 0.8\,\text{s}$ and quasi-stationary ELM-free H-mode. Single-ELM event at $t = 1.1\,\text{s}$ triggers MHD activity which reduces the confinement as shown by the decrease in W_{dia} at the D_α-spike at $t = 1.1\,\text{s}$.

The transport analysis is carried out in each of the four H-mode phases shown in Fig. 5.2 and Fig. 5.3. The first phase ($< 0.6\,\text{s}$) is a standard Ohmic H-mode with Type III ELMs. At $t = 0.6\,\text{s}$ the ECRH X3 heating is switched on at full power and the plasma enters into a Type I ELMy H-mode. The X3 heating is at full power ($\sim 1.35\,\text{MW}$ coupled to the plasma) until $0.8\,\text{s}$ when the power from one gyrotron is fully modulated at a frequency of $127\,\text{Hz}$ with a duty cycle of 50%. This modulation results in an average heating power reduction of approximately 17%. Although, in this case, there was a strong correlation between X3 power modulation and the onset of the ELM-free H-mode, power modulation is not necessary [Porte, *et al.* (2006)]. The ELM-free H-mode regime has been achieved without modulation. In contrast to other machines like DIII-D [Lohr, *et al.* (1988)] and Alcator C-Mod [Fielding, *et al.* (1996)] which have also produced ELM-free H-modes, the TCV tokamak accomplishes this with no direct ion heating, no active fueling and no cryo-pumping.

The ELM-free nature of the discharge is seen from the steady D_a signal in the top frame of Fig. 5.2 and Fig. 5.3 while the high-energy confinement is shown in the measurement of total energy content obtained from a diamagnetic loop, W_{dia} in the third frame of Fig. 5.2. The constant W_{dia} and line averaged density (second frame) show that this phase is quasi-stationary. The phase ends with a single ELM event at $1.1\,\text{s}$ at which time a $m/n = 3/2$ magnetic island is triggered by a core

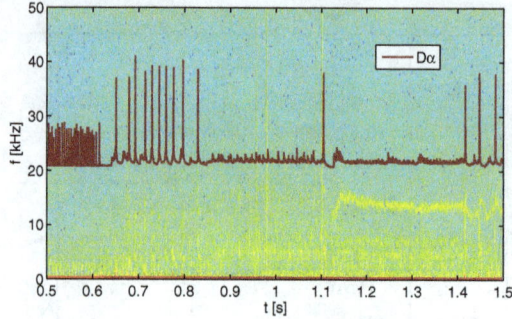

Fig. 5.3 Spectral temporal evolution of TCV29892 also including the D_a emission in red. The $m/n = 3/2$ mode with a frequency of 15 kHz emerges after the single ELM event at 1.1 s.

$m = 1$ sawtooth oscillation as seen in Fig. 5.3. This MHD mode has a frequency of 15 kHz and a magnetic island width of about 5 cm [Turri, *et al.* (2008)]. A magnetic island of this width located at a normalized minor radius around $r/a = 0.6$ gives rise to a decrease of total stored energy of about 25% according to the model of Chang and Callen (1990, 1992). The observed energy drop in Fig. 5.2 third frame is closer to 15%. There is therefore no doubt that it is the destabilization of the $m/n = 3/2$ MHD mode that causes the drop in confinement in this fourth and last H-mode phase of the discharge. Note that although affected by this MHD mode, this fourth phase is a quasi-stationary ELM-free H-mode. Moreover, $m/n = 1/1$ modes associated with sawteeth activity and intermittent $m/n = 2/1$ are present throughout the discharge [Turri, *et al.* (2008)].

Simulations of ELMs are performed with MHD codes like JOREK [Huysmans and Czarny (2007)]. Using models of the H-mode plasma profiles with sharp gradients in the pressure profile and the toroidal current profiles produce bursts of complex interchange modes called the peeling-ballooning mode that release up to 10% of stored plasma energy in a short burst. These events are monitored by the light emission from the ionization-recombination lines of the neutral deuterium atoms in the edge plasma. Examples of the ELMs are shown in Fig. 5.3.

For axisymmetric tokamak geometry without external toroidal momentum input the radial electric field E_r is arbitrary within the ordering of $|v_{E_r}| \sim \rho_i v_i/a$. The deviations from axisymmetry such as the toroidal field ripple lead to a particular E_r. The poloidal rotation velocity is of the order

$$v_\theta = C_{E_r} \frac{T_e}{eBL_{ne}} \tag{5.1}$$

with the numerical coefficient C_{E_r} bounded by 2. The standard observed tokamak core rotation is in the electron diamagnetic direction from $E_r < 0$. The Heavy Ion Beam Probe on TEXT (HIBP) [Hallock, *et al.* (1994)] measured this negative $E_r < 0$ directly. This magnitude of E_r is consistent with a collisional loss of ions in the edge or scrape-off layer plasma. In the outer edge region the sign of E_r flips to hold the electrons in resulting in a narrow shear poloidal flow shear layer at the edge of the plasma. Perez, *et al.* (2006) and Kim, *et al.* (2006) verified this picture with probes and the HIBP in the Ohmic heated L-mode TEXT tokamak and in LAPD.

In the single-null divertor configurations the inward E_r shear is enhanced by having the grad-B and curvature drifts directed toward the separatrix divertor chamber. This breaks up-down symmetry and results in a lower-threshold power for the transition from the L-mode to the H-mode presumably due to the stronger radial shear in the $E_r(r)$ scrape-off layer field.

This upper limit to E_r is used to analytically and numerically estimate the effect on the growth rates for ITG and ETG turbulence [Hamaguchi and Horton (1992); Sugama and Horton (1995)]. The effect is rather complicated since at first order there is a radial shift of the wave functions proportional to the E_r shear divided by the magnetic shear. The shear-flow effect is significant on weakly growing ITG modes and generally weak on the ETG modes. From the electron temperature T_e profile and the system parameters in Table 5.1, the range of the shearing frequency is $dv_{E_r}/dr = 5 \sim 8 \times 10^4 \text{ s}^{-1}$. This shearing rate may stabilize some of the slower growing TEM modes, but is too low to stabilize the turbulence considered here. For ETG this level of sheared rotation is negligible due to the fact that the electrons are interacting with much higher frequency fluctuations.

The electron temperature gradient drift wave frequency $\omega_{0,\text{ETG}}$ is given in Eq. (4.23) arising from the $\boldsymbol{E} \times \boldsymbol{B}$ convection of the electron density in the presence of adiabatic ions $\delta n_i = -eZ_i n_i \phi/T_i$. The frequency of the mode is shown in Fig. 4.7 for $k_y \rho_e = 1.0$. The smaller electron temperature gradient toward the edge gives the change of the wave direction from electron diamagnetic to ion diamagnetic drift direction.

A simple gyrokinetic simulation for the ETG was carried out with by Jenko and Dorland (2002) using the codes they developed called GENE and GS2. The results were that the ETG thermal flux for usual tokamak parameters is large and universal in nature. The comparison with the slab ETG simulation is given and the toroidal curvature-driven flux is about 30 to 40 times larger than the slab-model ETG thermal flux for their particular set of parameters. The slab ETG simulation with the GTC simulation code are given in Fu, *et al.* (2012) where a validation of the model and the simulations with measured plasma profiles and fluctuations in a steady-state hydrogen cylindrical plasma produced in the Columbia Linear Machine as described by Wei, *et al.* (2010). The simulations required large computing power owing to the small scale of the ETG simulation and the experiments require special microelectronic detectors for the megaHertz frequencies and 100 micron wavelengths. The GKV simulations show a strong inverse cascade from the high mode numbers to the lower modes with amplitudes peaking at $m = 11, 12, 13$ modes. The agreement with the laboratory data is impressive in Fu, *et al.* (2012).

The GENE simulations used teraflop computers with billions of effective particles. In this chapter we discuss the electromagnetic turbulence using gyrofluid equations for the electron dynamics introducing the electromagnetic field $A_\parallel$ to generate the associated magnetic turbulence. The gyrokinetic ETG simulations use several simplifications including an adiabatic ion response and electrostatic fields with a limited number of parameter variations.

Gyrofluid models of the ETG mode were developed to extend the convection fluid model for the study of drift wave turbulence [Horton, *et al.* (1988); Li and Kishimoto (2004); Li, *et al.* (2005); Holland and Diamond (2002)]. Here we describe the finite electron Larmor radius (gyrofluid) fluid model [Horton, *et al.* (1988); Horton, *et al.* (1990)]. The model was used for the analysis of the Tore Supra discharges [Horton, *et al.* (2004); Horton, *et al.* (2005)] and upgraded [Kim, *et al.* (2006)] with electron Landau damping effect [Hammett and Perkins (1990)] for NSTX. The model is constructed in a local toroidal geometry where the magnetic curvature and grad-B drifts are kept. The magnetic shear is low and positive in these discharges which allows the approximation of a constant $k_\parallel = |s|/qR$ in the analysis. This constant $k_\parallel$ approximation is acceptable when the growth rate is not near marginal stability. The higher-order finite Larmor radius (FLR) dynamics is included. The cross-field viscosities absorb the turbulence energy cascaded to high-k fluctuation components. The kinetic Landau damping physics is represented by the Hammett and Perkins closure model [Hammett and Perkins (1990)], $q_\parallel = -\sqrt{8/\pi}\, k_\parallel v_e \delta T_e(k)/|k_\parallel|$, for the fast parallel electron thermal dynamics.

The equations of the system can be described as

$$\widehat{T}\Phi = L\Phi + \widehat{N}(\Phi, \Phi), \tag{5.2}$$

where $\widehat{T}$ is the first-order time derivative operator, the $\widehat{L}$ and $\widehat{N}$ operators are the linear spatial operator and $\widehat{N}$ the nonlinear spatial operator, respectively.

To write out explicitly these nonlinear partial differential equations we use local (x, y) coordinates for radial and poloidal position and z for the toroidal position. The (x, y) variables are normalized by the electron gyroradius ρ_e, the toroidal z coordinate by L_n and time is normalized by L_n/v_{T_e} where L_n is the local electron density gradient length. The 3-component field vector field in the simulation is composed of $\Phi = (\phi, A_\parallel, \delta T_e)$ constructed from the electrostatic fluctuation ϕ, the parallel magnetic potential fluctuation $A_\parallel$ and the electron temperature fluctuation δT_e. All three field components are normalized by T_e/e, $\beta_e(T_e/e)v_e$ and T_e with the additional factor of ρ_e/L_n scaling the amplitude of the three fluctuating fields.

The time derivative operators are given by

$$\widehat{T}_{11} = \left(1 - \nabla_\perp^2\right)\frac{\partial}{\partial t},$$

$$\widehat{T}_{22} = \left(\frac{\beta_e}{2} - \nabla_\perp^2\right)\frac{\partial}{\partial t},$$

$$\widehat{T}_{33} = \frac{\partial}{\partial t} \quad \text{and} \quad \widehat{T}_{ij} = 0 \quad \text{where} \quad i \neq j. \tag{5.3}$$

The linear spatial derivative operators are given by a 3×3 matrix operator defined as L with the following components:

$$\widehat{L}_{11} = \left(1 - 2\epsilon_n + (1 + \eta_e)\nabla_\perp^2\right)\frac{\partial}{\partial y} - \mu\nabla^4, \quad \widehat{L}_{12} = \nabla_\perp^2\frac{\partial}{\partial z}, \quad \widehat{L}_{13} = 2\epsilon_n\frac{\partial}{\partial y} \quad (5.4)$$

$$\widehat{L}_{21} = -\frac{\partial}{\partial z}, \quad \widehat{L}_{22} = -\frac{\beta_e}{2}(1 + \eta_e)\frac{\partial}{\partial y} + \frac{\eta}{\mu_0}\nabla_\perp^2, \quad \widehat{L}_{23} = \frac{\partial}{\partial z} \quad (5.5)$$

$$\widehat{L}_{31} = -(\eta_e - 4\epsilon_n(\Gamma - 1))\frac{\partial}{\partial y}, \quad \widehat{L}_{32} = -(\Gamma - 1)\nabla_\perp^2\frac{\partial}{\partial z} \quad (5.6)$$

$$\widehat{L}_{33} = -2\epsilon_n(\Gamma - 1)\frac{\partial}{\partial y} + \chi_\perp\nabla_\perp^2 + \chi_\parallel\frac{\partial^2}{\partial z^2} + (\Gamma - 1)\sqrt{\frac{8}{\pi}}|k_\parallel| \quad (5.7)$$

where electron plasma beta $\beta_e = 2\mu_0 p_e/B^2$, the ratio of electron and ion temperature $\tau = T_e/T_i$ and the ratio of electron density gradient length L_n and magnetic field gradient R, $\epsilon_n = L_n/R$. Background viscosity μ, resistive diffusivity η/μ_0 and heat diffusivity $\chi_\perp, \chi_\parallel$ are included. Software and simulations for these simulation codes with this structure is maintained at `https://sites.google.com/site/turbulenttransport/`.

The linear operator matrix $\widehat{L}_{ij}$ is a function of five-dimensional parameter vector $\{\mu^5\} = \{\eta_e, \epsilon_n, T_e/T_i, \beta_e, \Gamma\}$ and on the three dissipation coefficients $\{\mu, \eta/\mu_0, \chi_\perp\}$. These high-$k$ damping parameters are adjusted according to the k-space spectral range being simulated so as to validate the truncation of the spectral representation of the fields for a given problem.

The nonlinear derivative terms in the dynamics are given by

$$\widehat{N}_1 = \widehat{N}_{111} + \widehat{N}_{122} = -\left[\phi, \nabla_\perp^2\phi\right] + \frac{\beta_e}{2}\left[A_\parallel, \nabla_\perp^2 A_\parallel\right] \quad (5.8)$$

$$\widehat{N}_2 = \widehat{N}_{212} + \widehat{N}_{221} + \widehat{N}_{223} = -\left[\phi, \nabla_\perp^2 A_\parallel\right] - \frac{\beta_e}{2}\left[A_\parallel, \phi - \delta T_e\right] \quad (5.9)$$

$$\widehat{N}_3 = \widehat{N}_{313} + \widehat{N}_{322} = \left[\phi, \delta T_e\right] - (\Gamma - 1)\frac{\beta_e}{2}\left[A_\parallel, \nabla_\perp^2 A_\parallel\right] \quad (5.10)$$

where the Poisson bracket $[f, g] = (\partial f/\partial x)(\partial g/\partial y) - (\partial f/\partial y)(\partial g/\partial x)$.

The energy transfer equations without viscosity, resistivity and thermal diffusion derived from the dynamics are

$$\frac{\partial W_E}{\partial t} = \left\langle \phi \hat{\boldsymbol{L}}_{11} \phi \right\rangle + \left\langle \phi \hat{\boldsymbol{L}}_{12} A_\| \right\rangle + \left\langle \phi \hat{\boldsymbol{L}}_{13} \delta T_e \right\rangle - \left\langle \phi \widehat{\boldsymbol{N}}_{122} \right\rangle$$

$$= \left\langle \phi \nabla_\perp^2 \frac{\partial A_\|}{\partial z} \right\rangle + 2\epsilon_n \left\langle \phi \frac{\partial \delta T_e}{\partial y} \right\rangle - \frac{\beta_e}{2} \left\langle \phi \left[A_\|, \nabla_\perp^2 A_\| \right] \right\rangle \tag{5.11}$$

$$\frac{\partial W_B}{\partial t} = -\left\langle \nabla_\perp^2 A_\| \hat{\boldsymbol{L}}_{21} \phi \right\rangle - \left\langle \nabla_\perp^2 A_\| \hat{\boldsymbol{L}}_{22} A_\| \right\rangle - \left\langle \nabla_\perp^2 A_\| \hat{\boldsymbol{L}}_{23} \delta T_e \right\rangle$$

$$+ \left\langle \nabla_\perp^2 A_\| \left(\widehat{\boldsymbol{N}}_{221} + \widehat{\boldsymbol{N}}_{223} \right) \right\rangle$$

$$= -\left\langle \phi \nabla_\perp^2 \frac{\partial A_\|}{\partial z} \right\rangle + \left\langle \delta T_e \nabla_\perp^2 \frac{\partial A_\|}{\partial z} \right\rangle + \frac{\beta_e}{2} \left\langle \phi \left[A_\|, \nabla^2 A_\| \right] \right\rangle$$

$$- \frac{\beta_e}{2} \left\langle \delta T_e \left[A_\|, \nabla^2 A_\| \right] \right\rangle \tag{5.12}$$

$$\frac{\partial W_{\delta T_e}}{\partial t} = \frac{1}{\Gamma - 1} \left(\left\langle \delta T_e \hat{\boldsymbol{L}}_{31} \phi \right\rangle + \left\langle \delta T_e \hat{\boldsymbol{L}}_{32} A_\| \right\rangle + \left\langle \delta T_e \hat{\boldsymbol{L}}_{33} \delta T_e \right\rangle - \left\langle \delta T_e \widehat{\boldsymbol{N}}_{322} \right\rangle \right)$$

$$= \left(\frac{\eta_e}{\Gamma - 1} - 4\epsilon_n \right) \left\langle \phi \frac{\partial \delta T_e}{\partial y} \right\rangle - \left\langle \delta T_e \nabla_\perp^2 \frac{\partial A_\|}{\partial z} \right\rangle + \frac{\beta_e}{2} \left\langle \delta T_e \left[A_\|, \nabla_\perp^2 A_\} \right] \right\rangle \tag{5.13}$$

where the three components of the total turbulent energy are

$$W_E = \frac{1}{2} \left\langle \tau |\phi|^2 + |\boldsymbol{\nabla}_\perp \phi|^2 \right\rangle \tag{5.14}$$

$$W_B = \frac{1}{2} \left\langle \frac{\beta_e}{2} |\boldsymbol{\nabla}_\perp A_\||^2 + |\nabla_\perp^2 A_\||^2 \right\rangle \tag{5.15}$$

$$W_{\delta T_e} = \frac{1}{2(\Gamma - 1)} \left\langle |\delta T_e|^2 \right\rangle \tag{5.16}$$

and the integration $\langle \cdot \rangle$ is done over $dxdydz$. The total energy evolves as

$$\frac{\partial}{\partial t} \left(W_E + W_B + W_{\delta T_e} \right) = -\left(\frac{\eta_e}{\Gamma - 1} - 4\epsilon_n \right) \left\langle \delta T_e \frac{\partial \phi}{\partial y} \right\rangle - \left\langle \sqrt{\frac{8}{\pi}} |k_\|| |\delta T_e|^2 \right\rangle \tag{5.17}$$

which in terms of the turbulent thermal flux

$$q_e = \langle v_{Ex} \delta T_e \rangle = -\left\langle \frac{\partial \phi}{\partial y} \delta T_e \right\rangle$$

becomes

Energy Balance Theorem

$$\frac{d}{dt} \left(W_E + W_B + W_{\delta T_e} \right) = q_e \left(\frac{\eta_e}{\Gamma - 1} - 4\epsilon_n \right) - \left(\frac{8}{\pi} \right)^{1/2} \left\langle |k_\|| |\delta T_e(k)|^2 \right\rangle. \tag{5.18}$$

Thus, the heat flux times electron temperature gradient beyond the compressional temperature gradient threshold drives up the turbulent fluctuations. The $\boldsymbol{E} \times \boldsymbol{B}$-induced turbulent energy flux $q_e = -\langle \delta T_e \partial \phi / \partial y \rangle = \langle v_x \, \delta T_e \rangle$, determines the total energy evolution.

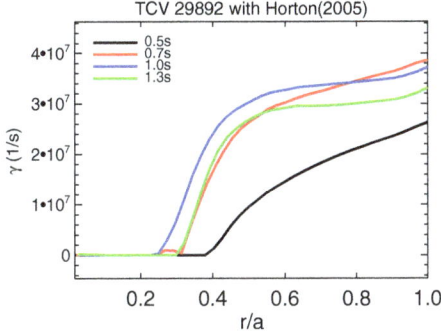

Fig. 5.4   The ETG mode growth rates used to calculate the χ_{ETG} in the four regimes of the TCV discharge 29892.

The linear stability analysis of the gyrofluid model shows a significant increase of the growth rates from $t = 0.5$ s to 0.7 s Fig. 5.4 for the TCV discharge. In the inner region $r/a < 0.3$ the mode is almost linearly stable. The strong ETG instability and associated turbulence spectra given by the model is consistent with the plasma data. The large turbulent heat diffusivities derived from the turbulent fluctuations is consistent with that transport derived from the power balance analysis with the heating power deposited by ECH RF waves and the accurately measured electron temperature profiles.

The nonlinear simulations in Fig. 5.5(a) for $r/a = 0.7$ for the TCV transport data show that the electron heat fluxes are overestimated by a factor of two. By increasing the damping in the pdes, the turbulence level can be brought down to the experimental level. The simulations predict the sharp increase of the electron heat fluxes observed between the Ohmic phase and the later ECRH phases and comparable heat fluxes during the three $t = 0.7$ s, 1.0 s, 1.3 s ECRH phases. The error bars in Fig. 5.5 are the standard deviation at the nonlinear stage of the simulation. In the nonlinear stage of the simulation, the vortices and radially extended streamer structures with size of 10-$20\rho_e$ are produced as shown in Fig. 5.5(b). Thus, the turbulent energy cascades from the source region of scale ρ_e to the ion scale $(m_i/m_e)^{1/2} \sim 40\rho_e$ through the formation of vortex merging and streamer formation.

Gyrokinetic simulations of the ETG turbulence show similar results as the gyrofluid simulations described here, but require extensive computer simulation resources [Nakata, *et al.* (2010)]. An example giving similar comparisons of the power balance χ_e with the simulations from a gyrokinetic code are given in Joiner, *et al.* (2006) and Brizard and Hahm (2007).

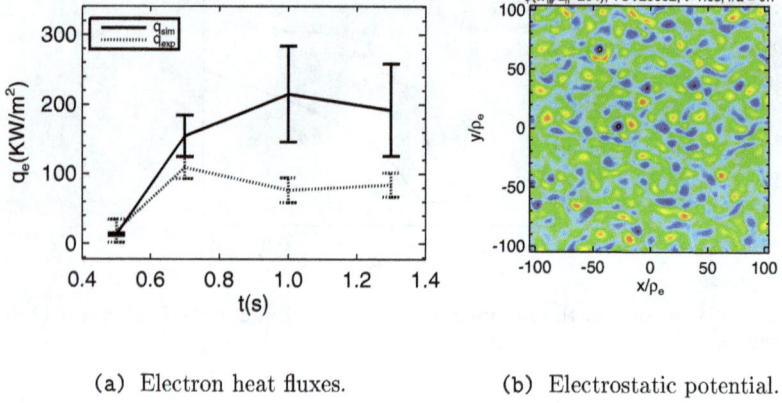

(a) Electron heat fluxes.   (b) Electrostatic potential.

Fig. 5.5   Electron heat fluxes (a) from nonlinear ETG fluid simulations. The solid and dashed line represent the simulations and the experiment. The contours of electrostatic potential $\phi(x,y)$ (b) from the nonlinear simulation at TCV29892 $r/a = 0.7$ and $t = 1.0$ s where $\rho_e = 75\mu$ m and the box corresponds to 1.5 cm × 1.5 cm.

Validation Analysis of the Electron Transport Modeling

Since there has historically been a multitude of models for the anomalous electron thermal transport problem a validation method is needed to rank the success of the various models. Here we briefly describe the use of one metric commonly used in weather systems to compare the TEM-ITG model with the ETG model for the discharge analyzed in detail in this chapter. There will be a need for a community wide method of comparing models in the future as the number of models and the associated number of computer simulations codes is continually expanding.

Validation Analysis of the ETG Model

When the nonlinear simulations are too heavy numerically to give the heat transport over the whole minor radius, one may use the theoretical scaling law to estimate the ETG-driven transport in the discharge. The gyroBohm heat diffusivity with length scale $q\rho_e$ and time scale L_{T_e}/v_{T_e} [Horton, et al. (2004)] is

$$\chi_e = \alpha_{\text{ETG}} \, q^2 \rho_e^2 v_{T_e} \left(\frac{1}{L_{T_e}} - \frac{1}{L_{T_e}^{\text{crit}}} \right) \tag{5.19}$$

where q is the safety factor and α_{ETG} is a scaling coefficient obtained from comparison with experimental data or with more rigorous theoretical calculations.

In the region $r/a > 0.4$, we calculated χ_e in Eq. (5.19) using the inverse of the critical temperature gradient length,

$$\frac{1}{L_{T_e}^{\text{crit}}} = \left(1 + Z_{\text{eff}} \frac{T_e}{T_i} \right) \left(\frac{1.33 + 1.9|s|}{q} \right) /R \tag{5.20}$$

[Horton, *et al.* (2004)]. The critical temperature length turns out to be five to ten times larger than the electron temperature gradient length, L_{T_e} at $r/a > 0.4$ derived from the measure electron temperature profiles.

The average relative variance (ARV) [Weigend, *et al.* (1990); Detman and Vassiliadis (1997)] is the degree to which the theoretical models for χ_e explain the power balance data χ^{pb} and is given in terms of the data and the model by

$$\text{ARV}_j = \frac{\sum_i \left(\chi_{e,i} - \chi_i^{pb}\right)^2}{\sum_i \left(\chi_i^{pb} - \left\langle\chi_i^{pb}\right\rangle\right)^2} \tag{5.21}$$

with the spatial and temporal indices i and j, respectively. For TCV 29892, the power balance diffusivities χ_i^{pb} are accurately known from the plasma data and the theoretical heat diffusivities $\chi_{e,i}$. Here $\left\langle\chi_i^{pb}\right\rangle$ is the average power balance diffusivity over all radial points for each temporal index j. For each temporal slice with index j, there is a total of 41 points whereof there are 30 radial points for $r/a > 0.4$.

The ARV is a widely-used metric measure for evaluating the how well a model's prediction follows the behavior of experimental data. When the averages of $\chi_{e,i}$ and χ_i^{pb} are comparable, ARV < 1 indicates that a model behaves in the same way as the experiment as shown in Fig. 5.6 where χ_e^{pb} increases as χ_e^{ETG}. A smaller ARV means better agreement or higher predictive power of the model.

For the turbulent electron fluxes shown by TCV, the model with $\alpha_{ETG} \sim 0.8$ shows the best agreement beyond $r/a = 0.4$ in Fig. 5.6. The gyroBohm ETG heat diffusivity formula with a free parameter α_{ETG}, Eq. (5.19) explains not only the electron heat diffusivity increase with the H-mode transition but also gives ARV ~ 0.3 for $t = 0.7\,\text{s}$ and $1.0\,\text{s}$. For $t = 0.5\,\text{s}$, the flat experimental heat flux profile gives a small denominator in Eq. (5.21) producing the large ARV. With $\alpha_{ETG} \sim 1$ and the assumption that the dominant time scale is $\omega \sim v_{T_e}/L_{T_e}$, we can estimate the correlation length $l_c \sim 2\pi q \rho_e \sim 10\rho_e$, consistent with the nonlinear simulation result.

Comparison of T_e and $q_{\|e}$ radial gradients

The ratios of the contributions of the radial gradients of T_e and $q_\|$ to the growth rate are

$$\frac{dT_e}{dx} : \frac{k_\|}{\omega + i\varepsilon} \frac{1}{n_{0e}} \frac{dq_\|}{dx}. \tag{5.22}$$

Thus, the comparison involves the two gradient scale lengths

$$\frac{1}{L_{q_\|}} = \frac{1}{q_\|} \frac{dq_\|}{dx} \gg \frac{1}{L_{T_e}} \tag{5.23}$$

and the magnitude of the heat flux parameter

$$\alpha_{q_\|} = \frac{q_\|}{n_e T_e v_e} \equiv \frac{u_{\|q}}{v_e} \tag{5.24}$$

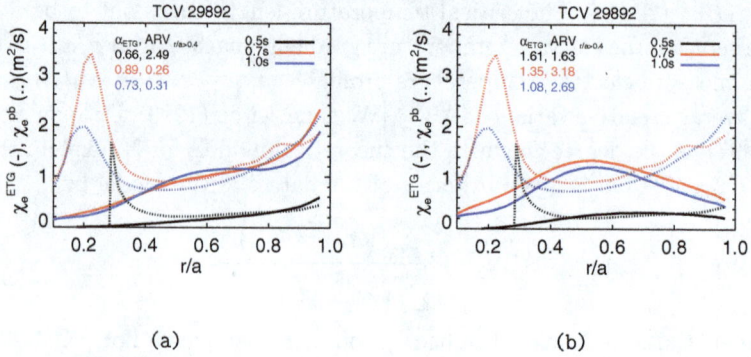

(a)   (b)

Fig. 5.6 The comparison between the electron gyroBohm heat diffusivity, χ_e in Eq. (5.19), with ETG time and spatial scales (the lines with *) and the power balance heat diffusivity χ_e^{pb} (no mark) for each time slice $t = 0.5$ (solid), 0.7 (dotted) and 1.0 (dashed) are shown.

of the parallel thermal flux. For RF current drive with lower hybrid waves (LHCD) the electron distributions have $u_{\|q} \gtrsim u_{\|d} \cong j_\|/n_e e$ where $j_\| = I_p/\pi\Delta^2$ is the driven plasma current density. The tokamak plasma current I_p is over the region $\Delta < a$ and produces the plasma confinement.

The dimensionless α_q parameter measures the skewness of the $v_\|$ of the electron parallel velocity distribution function. For example, the LHCD phase space distribution function f_e may have a radial profile such that

$$\frac{L_{T_e}}{L_{q_\|}} \gtrsim 5. \tag{5.25}$$

The combined ray tracing/Fokker-Planck code DELPHINE of Imbeaux and Peysson (2005) is used to compute the driven electron distributions $f(v_\|, v_\perp, x)$. A new interactive combination of the ray tracing code and Fokker-Planck code for the electron distribution function is LUKE-C3PO [Peysson and Decker (2007)] gives the radial profiles of the parallel electron current and parallel heat flux profiles.

Kinetic dispersion relation for LHCD steady state plasmas

The electrostatic modes are given by

$$D_k = k^2 + \sum_i \frac{n_i e_i^2}{\varepsilon_0 T_i} + \frac{n_e e^2}{\varepsilon_0 T_{\|e}} \left[1 + \left(\frac{T_{\|e}}{T_{\perp e}} - 1 \right) (1 - \Gamma_0(b_e)) - P_e^{LHCD} \right] \tag{5.26}$$

where $b_e = k_\perp^2 T_{e\perp}/m_e \omega_{ce}^2$ and

$$P_e^{LHCD}\left(\omega, k_\|, k_\perp \right)$$

$$= -T_{e\|} \int \frac{\frac{\partial f_e}{\partial E_\|} J_0^2 d^3 p \left[\omega - \frac{k_y T_{e\|}}{eBLn_e} \left(1 + \eta_{e\|} \left(\frac{\epsilon_\|}{T_\|} - \frac{1}{2} \right) \right) + \eta_{e\perp} \left(\frac{\epsilon_\perp}{T_\perp} - 1 \right) \right]}{\omega - k_\| v_\| - k_y \hat{v}_D(\epsilon_\| + \epsilon_\perp/2)}. \tag{5.27}$$

Hydrodynamic-FLR limit of P^{LHCD}-function

To gain insight into the form of the unstable modes driven by the LHCD distribution function we take the hydrodynamic or fluid limit of the integrals. Thus, we obtain the dispersion relation for the LHCD plasma

$$
D_{f\ell}^{\text{LHCD}} = k^2 + \sum_i \frac{n_i e_i^2}{\varepsilon_0 T_i} + \frac{ne^2}{\varepsilon_0 T_{\|e}} \left[1 + \left(\frac{T_\|}{T_\perp} - 1 \right) k_\perp^2 \rho_{\perp e}^2 - 1 + \frac{\omega_{*ne}}{\omega} \right]
$$

$$
- \left(1 - \frac{\omega_{*pe}}{\omega} \right) \frac{k_\|^2 T_\|}{m\omega^2} - \frac{k_y v_{De}}{\omega}
\tag{5.28}
$$

which has the form

$$
= A + \frac{B}{\omega} + \frac{C}{\omega^2} + \frac{D}{\omega^3}.
\tag{5.29}
$$

Slab Modes $k_y q \rho_{e\parallel} < 1$

The slab regime ETG modes are given by $k_\parallel v_{e\parallel} \sim (v_{e\parallel}/qR)(1 + |s|) > k_y v_{De}$ with the growth rate given by $\gamma_{\text{slab}} = |k_\parallel| v_{e\parallel} (L_{ne}/L_{T_e} - \eta_{e,\text{crit}})^{1/2}$.

Toroidal ETG Modes $k_y q \rho_{e\parallel} > 1$

The toroidal regime modes are given by

$$\gamma_{\text{toroidal}} = \frac{v_{e\parallel}}{R} \left(\frac{R}{L_{T_e}} - \frac{R}{L_{T_e,\text{crit}}} \right)^{1/2} \frac{|k_y| \rho_{e\parallel}}{\sqrt{1 + k_y^2 \rho_{e\parallel}^2 (1 + \omega_{ce}^2/\omega_{pe}^2)}}.$$

$$- \frac{k_\parallel^2 v_{e\parallel}^2}{\omega^2} \left(1 - \frac{\omega_{*e}(1 + \eta_{e\parallel})}{\omega^2} \right) + \frac{2k_y^2 \epsilon_n}{\omega^2} \left[1 + \eta_{\parallel e} + \frac{T_e}{T_\parallel}(1 + \eta_{\perp e}) \right]. \tag{5.30}$$

The maximum χ_e arises from longest wavelength with $\lambda_\perp \lesssim \rho_i$

$$\chi_e \leq \rho_i^2 \frac{v_{\parallel e}}{R} \left(\frac{R}{L_{T_e}} - \frac{R}{L_{T_r,\text{crit}}} \right)^{1/2}.$$

There are two useful limiting forms of the LHCD-ETG waves from the fluid expansion of the kinetic P function. For small $k_\parallel L_{T_e}$ and $\bar{\omega} = \omega - k_\parallel u_\parallel$ we find from the fluid equations

$$\bar{\omega}(\bar{\omega} - \omega_{*i}) = \frac{k_\parallel k_y T_i}{eB Z_{\text{eff}}} \frac{du_\parallel}{dr} + \frac{k_\parallel^2 T_i}{m_e Z_{\text{eff}}} \left[1 - \frac{k_y T_e}{\bar{\omega} e B n_e} \frac{dn_e}{dr} - \frac{k_y T_e}{eB} \frac{1}{\bar{\omega} - k_\parallel^2 v_e^2/(\bar{\omega} + i\nu)} \right.$$

$$\left. \times \left[\left(\frac{1}{T_e} \frac{dT_e}{dr} - \frac{k_\parallel dq_\parallel/dr}{(\nu - i\omega)p_e} \right) \right] \right]. \tag{5.31}$$

The drift wave frequency ω_{*i} in Eq. (5.31) arises from the $\boldsymbol{E} \times \boldsymbol{B}$ convection of the electron density in the presence of the adiabatic ions $\tilde{n}_i = -e Z_i n_i \phi/T_i$. Due to the electron temperature gradient η_e, for frequency formula is approximately

$$\omega(k) = \frac{-\omega_{*e} S(b, \eta_{e\perp})}{\frac{Z_{\text{eff}} T_e}{T_i} + k_\perp^2 \rho_e^2 + k^2 \lambda_{De}^2}$$

with $S(b, \eta_e) = e^{-b} I_0 - \eta_{e\perp} b e^{-b}(I_0 - I_1) \simeq 1 - k_\perp^2 \rho_e^2(1 + \eta_{e\perp})$.

For larger $k_\parallel L_{T_e} \gtrsim 1$ the waves are given by

$$\omega^2 = \frac{k_\parallel^2 T_i}{Z_{\text{eff}} m_e} \left[1 - \frac{k_y T_e}{eB\omega} \left[\frac{1}{T_e} \frac{dT_e}{dr} + \frac{k_\parallel}{\omega p_e} \frac{dq_\parallel}{dr} \right] \right]. \tag{5.32}$$

Often the temperature gradient is subdominant to the gradient of the parallel heat flux and the instability becomes

$$\omega_k = \omega_{*i} + i \left(\frac{k_\parallel^2 T_i}{2m_e} \right)^{1/3} \left| \frac{k_y T_{\parallel e}}{eB L_{T_e\parallel}} \frac{dq_\parallel}{dx} \right|^{1/3}.$$

Table 5.2   Parameters for LHCD Driven Plasmas.

Drift Wave Parameters in LHCD Driven Plasmas

$$\omega_{*e} = \frac{k_y T_e}{q_e B n_e}\frac{dn_e}{dr} = \frac{k_y T_e}{eBL_{n_e}} \quad \text{and} \quad \omega_{*eT_e} = \eta_{e\|}\omega_{*e} = -\frac{k_y}{eB}\frac{dT_{e\|}}{dr} = \frac{k_y T_{e\|}}{eBL_{T_e}},$$

where $\quad \eta_{e\|} = \dfrac{\partial_r \ell n T_{e\|}}{\partial_r \ell n n_e} = \dfrac{L_{n_e}}{L_{T_e}} \quad$ and $\quad \eta_{e\perp} = \dfrac{\partial_r \ell n T_{e\perp}}{\partial \ell n n_e}.$

The dangerous instabilities have: $\omega_k \leq |\omega_{*e}| \leq \dfrac{v_e}{L_{T_{e\|}}} \simeq (3-5)k_\| v_e$

and thus $\quad k_\| L_{T_{e\|}} \leq 1/3 \quad$ and $\quad k_\perp \rho_e \lesssim 1.$

The nonuniform parallel thermal flux introduces the electron drift-gradient term

$$\frac{k_y \phi}{B}\frac{dT_e}{dx} + \frac{k_\| k_y \phi}{n_e(\omega + i\varepsilon)B}\frac{dq_\|}{dx} \quad \text{where} \quad \boldsymbol{E} = -\nabla\phi.$$

Comparison of T_e and $q_{\|e}$ gradient terms $\quad \dfrac{dT_e}{dx} : \dfrac{k_\|}{\omega + i\varepsilon}\dfrac{1}{n_{0e}}\dfrac{dq_\|}{dx}.$

Thus, the comparison involves the scale length $\quad \dfrac{1}{L_{q\|}} = \dfrac{1}{q_\|}\dfrac{dq_\|}{dx} \gg \dfrac{1}{L_{T_e}}$

and the magnitude of the heat flux parameter $\quad \alpha_q = \dfrac{q_\|}{n_e T_e v_e} \equiv \dfrac{u_{\|q}}{v_e}.$

LHCD electron distributions have:

$\quad u_{\|q} \gtrsim j_\|/n_e e \quad$ where $\quad j_\| = I_p/\pi\Delta^2$

is the driven plasma current density required for confinement.
The dimensionless α_q parameter measures the skewness of the $v_\|$-distribution.

These heat-flux gradient driven modes have

$$\omega_k = \left(\frac{k_\|^2 T_i}{Z_{\text{eff}} m_e}\right)^{1/4} |k_\| u_\| q\omega_{*i}|^{1/4} \propto k_\|^{3/4} k_y^{1/4} u_\|^{1/4} v_e^{1/2}. \tag{5.33}$$

The LHCD driven modes have a mean parallel wavenumber $\bar{k}_\|$ so the eigenmodes are shifted off the rational magnetic surfaces. Thus, the modes form magnetic islands in their nonlinear state when the plasma electron beta is appreciable and contribute to the radial transport of $j_\|$ and $q_\|$.

Analytic quasilinear RF velocity diffusivity

The intense LH-waves produce the parallel velocity diffusion coefficient $D_\|^{\text{LH}}$ that is modeled by a box function with

$$D_\|^{\text{LH}}(v_\|) = \hat{D}\nu_e v_e^2 \quad \text{for} \quad v_1 \lesssim v_\| \lesssim v_2 \tag{5.34}$$

with $D_\|$ vanishing outside the interval $[v_1, v_2]$. For CIEL and other LHCD systems the upper parallel velocity $v_\|$ may be extended to $v_2 \lesssim c/2 = 1.5 \times 10^8\,\text{m/s} \sim 13 v_e$. The dimension measure of the LH power is given by Stix parameter for RF power density

$$\hat{D} = \frac{P^{\text{RF}}}{n_e T_e v_e} \tag{5.35}$$

where ν_e is the Coulomb collision frequency for electron collisions that restore the distribution to the local Maxwellian.

In the region $[v_1, v_2]$ a useful analytic model is obtained by taking $\widehat{D} \gg 1$ so that the collisional effects are negligible and the absorbed RF power forces $\partial f / \partial v_\parallel = 0$ in this $v_\parallel$-interval. This model is referred to as the plateau LHCD model. In reality the electron Coulomb collisions will always be non-negligible due to the formation of boundary layers at the transition across the $v_\parallel = v_1$ and $v_\parallel = v_2$ layers.

The electron distribution function, in the absence of the mirror force and toroidal velocity drifts, is

$$\frac{\partial f}{\partial t} = \frac{\partial}{\partial v_\parallel}\left[D_\parallel(v_\parallel)\frac{\partial f}{\partial v_\parallel}\right] + \frac{\nu_e}{v^2}\frac{\partial}{\partial v}\left[v^2 g(v)\left(fv + \frac{T_e}{m_e}\frac{\partial f}{\partial v}\right)\right]$$
$$+ \nu_{ei}Z_{\text{eff}}\left(\frac{v_e}{v}\right)^3\frac{\partial}{\partial \mu}\left[(1-\mu^2)\frac{\partial f}{\partial \mu}\right] \tag{5.36}$$

where ν_e is the electron collision frequency, $g(v)$ is the coefficient of drag (which we take as unity in the following discussion), ν_{ei} is the electron-ion (working gas) collision frequency and $Z_{\text{eff}} = \Sigma_I Z_I^2 n_I / n_e$ gives the enhancement of the electron pitch-angle scattering rate from impurities I over that in the pure hydrogenic plasma for which $Z_{\text{eff}} = 1$.

High-power LHCD plateau model F_e

In the limit that $D_\parallel / \nu_e v_e^2 \gg 1$ in the interval (v_1, v_2), we may obtain the solution of thorough geometrical construction. The key point is that one may map the values of $f(v_\parallel = v_1, v_\perp) \to f(v_\parallel = v_2, v_\perp)$ since $\partial f / \partial v_\parallel = 0$ in this interval. The coordinates of the locus of points $(v_2, v_\perp)$ parameterized by the pitch-angle α is $v_2 = v\cos\alpha$ and $v_\perp = v\sin\alpha = v_2\tan\alpha$. Thus, the value of f along the high-velocity resonant plane $v_\parallel = v_2$ is given by $f(v_2, v_\perp) = f(v_2, v_2\tan\alpha)$ with $0 \le \alpha \le \pi/2$ parameterizing the position along the $v_\parallel = v_2$ surface.

Now the value of $f(v_2, v_\perp)$ is taken as the same as the value of $f(v_1, v_\perp)$ in this high-power limit due to $\partial f / \partial v_\parallel = 0$. So, on the surface $v_\parallel = v_1$ one has $v_\perp = v\sin\alpha$ and $v_\parallel = v\cos\alpha = v_1$. For parallel velocities below $v_\parallel = v_1$ the distribution is Maxwellian since this is the solution of $Cf = 0$ in the low-velocity region up to v_1 and for all $\pi/2 < \alpha \le \pi$ (co-current moving electrons). Thus, f is known through the region $[v_1, v_2]$ and on the surface $v_\parallel = v_2$. Pitch-angle collisions dominate for $v_\parallel > v_1$.

Typical values of the temperatures in the core LHCD plasma are $T_\parallel \sim 750\,\text{keV}$ to $1\,\text{MeV}$, and $T_\perp \sim 150 - 300\,\text{keV}$ with lower-energy Maxwellian with temperature of 50-100 keV. The analytic model [Stevens, *et al.* (1985)] used in the literature is:

$$
f_{3T} =
\begin{bmatrix}
C_n \exp\left(-\dfrac{p_\perp^2}{2T_\perp} - \dfrac{p_\parallel^2}{2T_{\parallel F}} \right) & \text{for} & p_\parallel > 0 \\[2em]
C_n \exp\left(-\dfrac{p_\perp^2}{2T_\perp} - \dfrac{p_\parallel^2}{2T_{\parallel B}} \right) & \text{for} & p_\parallel < 0
\end{bmatrix}
$$

and $f_{3T} = 0$ for $E(p) = mc^2\gamma > E^* \sim 2\,\text{MeV}$. The temperatures are determined by the photon spectrum from the Bremsstrahlung spectroscopy

The Nyquist stability analysis gives the marginal stability condition

$$
-\frac{1}{T_{e\parallel}} \frac{dT_{e\parallel}}{dr} > \left[\frac{1}{L_{ne}} + \frac{1 + 2|s|/q}{R} \right]
$$
$$
\times \left[1 + \frac{Z_{\text{eff}} T_{e\parallel}}{T_i} + \left(\frac{T_{\parallel e}}{T_{\perp e}} - 1 \right)(1 - \Gamma_0(b_e)) + k_\perp^2 \lambda_{De\parallel}^2 \right]. \tag{5.37}
$$

In conclusion the LHCD can stimulate the (a) fast-growing toroidal modes $T_\parallel > T_{\perp i}$ and (b) fast-growing slab-like modes $k_\perp \rho_{e\parallel} \sim 1$. Secondly, one finds that (c) short wavelengths $b_e = k_\perp^2 \rho_{e\perp}^2 > 1$ are stabilized by $T_{\parallel e} > T$ and finally (d) that the effect of $Z_{\text{eff}} T_{e\parallel}/T_i$ is important in controlling the threshold.

Formulas for the anomalous thermal flux associated with the $\boldsymbol{E} \times \boldsymbol{B}$ motion and the magnetic $\delta B_\perp$ fluctuations for LHCD driven tokamaks remain an unsolved problem. The turbulence for the c/ω_{pe} scale electromagnetic turbulence is typical of the ∇T_e-driven short wavelength drift modes.

The design for LHCD for ITER is given in Hoang, *et al.* (2009). Lower-hybrid current drive produce record-breaking steady-state tokamak discharges in Tore Supra [Saoutic, *et al.* (1994)] and Alcator-C [Bonoli, *et al.* (1988)]. In divertor tokamaks, there is a turbulent plasma outside the last closed flux surface that produces refraction and scattering of the LHCD waves [Peysson, *et al.* (2011)].

More laboratory studies of ETG include extensive power balance studies on RF heating in NSTX in Kaye, *et al.* (1997) and third harmonic electron cyclotron heating in TCV by Asp, *et al.* (2008). Finally, a basic plasma physics experiment for the ETG instability and nonlinear saturation with validation with the gyrokinetic simulations is given in Fu, *et al.* (2012) for the Columbia Linear Machine where steady-state ETG turbulence is measured and controlled.

5.7   Edge Localized Modes and Plasma Pedestals

This high-temperature plasma is insulated from the vessel walls which must remain below 1000 K to retain their integrity. Thus, steep temperature gradients of order 10 keV/m must be stably maintained in the thermonuclear fusion device. The associated average ion thermal diffusivity χ_i is then of order or less than $1\,\text{m}^2/\text{s}$ in devices with minor radius a few meters.

Such magnetic thermal insulation is consistent with the results from the latest generation of tokamak experiments under favorable conditions. The thermal flux across the magnetic surfaces is large $q_i = -n_i \chi_i dT_i/dr$ reaching $0.1\,\mathrm{MW/m^2}$. Large heating rates of many MW are required to maintain the plasma against turbulent losses. The alpha particle decay products from the fusion reactions provide a fraction of the power to maintain the ion temperature and the electron temperature $T_i > T_e$ in the $10\,\mathrm{keV}$ range. This section is devoted to instabilities driven by the temperature gradient and the associated turbulent transport of thermal energy across the magnetic flux tubes. The electron thermal transport problem is the most difficult to control in the magnetic confinement experiments.

In this chapter we consider plasmas that are stable to ideal MHD motions and study the remaining smaller-scale drift wave instabilities. The drift waves have dispersive frequencies and finite parallel electric fields which means the associated waves and vortices have resonant wave-particle interactions. Figure 5.5 shows the growth of the ion temperature gradient driven drift wave vortices starting on the outside of the torus in frame (a) and spreading throughout the torus in frames (b), (c), and (d) from gyrokinetic simulation.

The drift wave dispersion relation in a collisionless plasma need not assume a local thermal Maxwell-Boltzmann distribution function. In the high temperatures of fusion plasmas the particle distributions in energy typically have high energy tails and different effective temperatures across and parallel to the magnetic field. In low-beta $\beta = p/(B^2/2\mu_0)$ plasmas (thermal plasma pressure to magnetic pressure) the ion temperature gradient instability, known as the ITG mode, has nearly an electrostatic polarization for the coupled drift waves and the ion acoustic waves. In higher-pressure plasmas with $\beta > 0.01$ the ITG mode couples with the shear Alfvén wave becoming a dispersive oscillation with an electromagnetic polarization. As the plasma pressure increases the inductive electric field from the time changing δB_x cancels part of the electrostatic component of the parallel electric field. This cancellation reduces the energy transfer rate $\langle j_\parallel E_\parallel \rangle$ and reduces the growth rate of the instability.

There are numerous theoretical studies of the magnetic ITG modes including Hong, *et al.* (1989ab), Kim, *et al.* (1993), and Rewoldt, *et al.* (1998). At certain intermediate values of plasma beta both the ITG and the kinetic ballooning mode of the MHD polarization are present with different mode frequencies and growth rates.

A thermodynamic description of the energy source for the turbulence from the temperature gradient. The kinetic theory describes the essence of turbulent thermal flux of the ion temperature gradient driven stability gives the precise properties of the turbulent fluxes.

5.8 Thermodynamics of the ITG Instability

For the Maxwellian phase space distributions the electrostatic ITG (ion temperature gradient) instability starts when the ratio of the ion temperature gradient $d \ln T_i/dr$ exceeds the density gradient $d \ln n_i/dr$ by a number that depends on the compressibility of the ion gas. This critical gradient is best understood by considering the maximum work that can be extracted from the plasma temperature profile with a Carnot cycle operating between the high- and low-temperature regions. The "engine" for the cycle is the drift wave convection between the high-temperature and low-temperature regions and the working gas is the ion component of the plasma with an adiabatic gas constant $\Gamma = (d+2)/d$ where $d = $ number of degrees of freedom active in the dynamics of the ion gas. The situation is shown in Figs. 5.7 and 5.8, where the radially-extended vortex cells, tilted to follow the twist of the magnetic field lines, cycle the hot interior plasma to the lower-temperature plasma.

The Carnot cycle is used to calculate the maximum energy W_{ITG} that can be extracted per convection cycle from the drift wave convection from T_1 to T_2. The plasma is adiabatically compressed and expanded between the high- and low-ion temperature regions so that the adiabatic gas constant Γ controls the stability condition. For $\Gamma = 5/3$ one finds that the condition for a net energy release from the temperature gradient is given by

$$\eta_i > \eta_{\text{crit}} = \frac{2}{3}. \tag{5.38}$$

For f degrees of freedom gas equation of state the result is $\eta_{\text{crit}} = \Gamma - 1$ where $\Gamma = (f+2)/f$.

This η parameter is widely used in the plasma literature and defined for a charge species s by

$$\eta_s = \frac{d \ln T_s}{d \ln n_s}. \tag{5.39}$$

In the early plasma literature the instabilities driven by ion temperature gradient were often called the η_i modes.

The temperature gradient instability condition follows from a calculation of the work done, $W = Q_1 - Q_2$, from the thermal energies Q taken from the hot-ion region at $T_i(x)$ and adiabatically expanding the gas to lower the temperature to the low-temperature regions at $T_i(x + \Delta x)$. Then the convection recompresses the gas to return to the high-temperature region at $T_i(x)$. Subtracting the work done in the compression minus the work gained in the expansion and adding the work done by the gas in the two isothermal strokes gives the net energy W released for one cycle. The energy W is proportional to the differences $\Delta T/T(\Delta T/T - 2/3\Delta n/n)$. For the general value of the adiabatic gas constant the factor $2/3$ is replaced by $\Gamma - 1$. Thus for the slab model of the ITG threshold, where there is only the parallel compressible motion, there is a one-degree-of-freedom compression and the threshold is raised to the value of $\Gamma - 1 = 2$. In toroidal systems there are three-degrees-of-freedom and the threshold starts at $2/3$.

The corresponding changes in the wave-particle resonances are shown in Fig. 5.7.

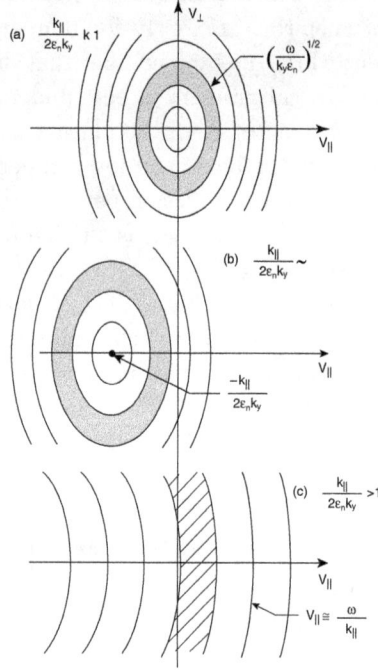

Fig. 5.7   The band of resonant ion velocities for a given drift-wave fluctuation $(\omega, k_y, k_\parallel)$. (a) the guiding center drift delta function resonance curve $\omega_k = \omega_D$, (b) the shifted contours of resonant $\omega_k - k_\parallel v_\parallel = \omega_D$, and (c) the slab Landau resonance $\omega_k = k_\parallel v_\parallel$.

These threshold formulas agree with the results from Nyquist diagrams from the dispersion function $D_\mathbf{k}(\omega/\eta_i)$, which give the stability condition of the electrostatic drift waves when other complications such radiative damping are absent.

Here some intermediate steps in the Carnot cycle construction are given for clarity. The different temperatures $T_i(x + \Delta x)$ and $T_i(x)$ are the driving force of the convection. In the ideal isentropic limit the work W done from the release of energy from the hot $T_2(x)$ plasma to reservoir $T_1(x + \Delta x)$ to the lower temperature $T_1(x + \Delta x)$. Using conservation of energy (first law of thermodynamics) the $W = Q_2 - Q_1$ and $Q_2 = T_2 \Delta S_2$ and $Q_1 = -T_1 \Delta S_1$, with $\Delta S_1 + \Delta S_2 = 0$ gives the maximum work W that can be extracted as per

$$W = N\left(T_2 - T_1\right) \underbrace{\Delta\, \Delta\ell n\left(T^{3/2}/N\right)}_{\Delta S_2} = N\Delta T\left(\frac{3}{2}\frac{\Delta T}{T} - \frac{\Delta N}{N}\right) \tag{5.40}$$

from the gradients of $T(x)$ and $N(x)$. The complete dispersion relation from solving the eigenvalue problem in the toroidal geometry has been developed into several computer codes. The fastest growing eigenmode from this numerical solution with

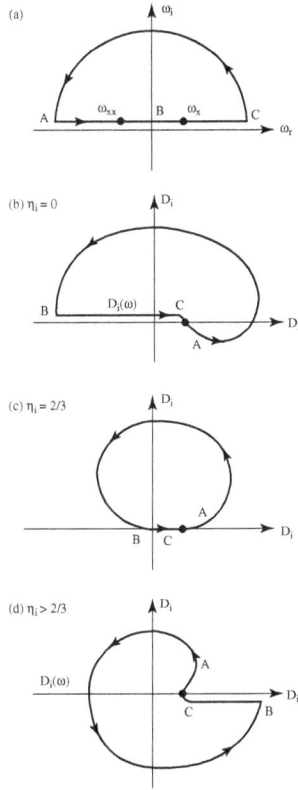

Fig. 5.8 The Nyquist diagram for the ion temperature gradient driven instability. Frame (a) is the contour enclosing the unstable upper-half plane that is mapped by the dispersion function in frames (b), (c) and (d) for three values of the stability parameter η_i. For the first case in (b) there is no temperature gradient and the mapped contour does not encircle the origin, so there is no unstable root. In frame (c) with $\eta_i > 2/3$ there is a marginally stable root. For frame (c) with $\eta_i > 2/3$ the D-function contour encircles the origin proving that there is an unstable root within the A-B-C contour in frame (a) of the complex omega plane.

the ITG and TEM physics for typical discharge parameters in TFTR is shown in Fig. 5.9 from Rewoldt, *et al.* (1998). One sees in Fig. 5.9(a) that for the collisionality of $\nu_{*e} = 1$ the mode onsets with a well-defined ion temperature gradient of $d \ln T_i / d \ln n_i = 1$ and rotates in the plasma rest frame in the ion diamagnetic direction. For higher temperatures where there is a substantial fraction of trapped electrons Fig. 5.9(b) shows that the mode is unstable with vanishing ion temperature gradient $d \ln T_i / d \ln n_i = 0$ from the eigenvalue code. This instability is the trapped electron mode (TEM) and has growth rate with $\gamma_k / \omega_{*e} = 0.25$ with even for no ion temperature gradient.

In the next section we describe how the electrons respond and destabilize the drift wave in this weakly collisional regime where the ions and electrons with large

pitch angles are trapped by the mirror force $F_\parallel = -\mu\nabla_\parallel B$ in the low-B field region
on the outside of the torus. These types of modes exist in all toroidal confinement
devices and produce drift wave instabilities that are universal to toroidal confinement geometries. The growth rate in Fig. 5.7a for $d\ln T_i/d\ln n_i = 0$ is from the
mirror trapped electrons resonantly interacting with the drift wave.

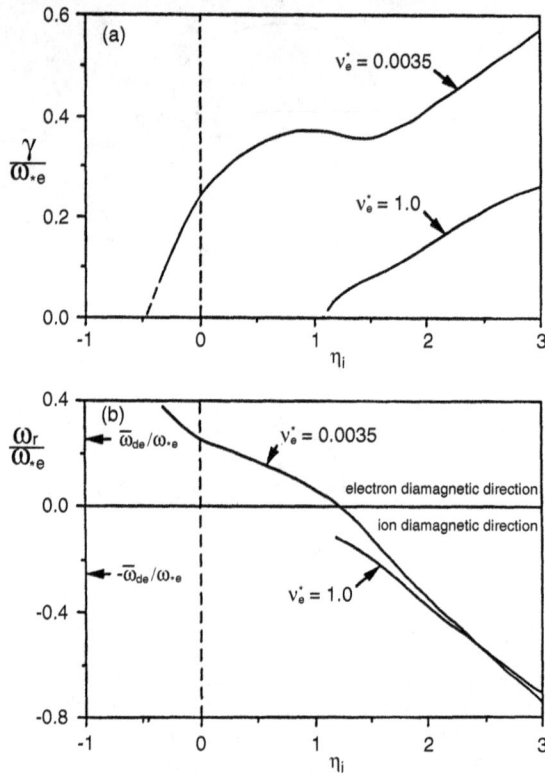

Fig. 5.9   The normalized (a) growth rate and (b) frequency of the complex kinetic toroidal eigenvalue problem. At high collisionality only the ITG mode appears for $\eta_i > 1.2$ while at low
collisionality the trapped electron mode appears for $\eta_i < 1.5$ [Rewoldt (1998)].

The change in the unstable region for $k_\parallel L_n$ versus $k_y\rho_i$ in a stability diagram
with varying values of $\eta_\perp$ and $\eta_\parallel$ is worked out in Hojo (1995). The result shows
that at low values $\eta_\perp$, the critical condition is a value of $\eta_\parallel$ equal to 3 as given by
the thermodynamics of the one-degree of freedom gas. The change in the unstable
region for $k_\perp L_n$ versus $k_y\rho_i$ in a stability diagram with varying values of $\eta_\perp$ and $\eta_\parallel$
is worked out in Hojo (1995). The result shows that at low $\eta_\parallel$ the critical condition
is value of $\eta_\perp > 1$ to 2 also consistent with the thermodynamics.

For the critical ion temperature gradient is also a function of T_i/T_e. Plasmas with high T_i/T_e can remain stable to the ITG modes for $\eta_\perp$ moderately large. The analysis [Kim, *et al.* (1992)] shows why these hot ion plasmas are stable to ITG modes.

In hot ion mode plasmas, one finds the ITG instability is stable. A stability to ITG at high T_i/T_e ratios was reported in the TFTR supershots [Horton, *et al.* (1992)].

The kinetic stability analysis of the temperature gradient driven modes is determined by three kinetic plasma response functions $P(\omega, k)$, $Q(\omega, k)$, $R(\omega, k)$ which are integrals of the gradients of the phase space density function f weighted by the Bessel function $J_0^2, v_\perp J_1 J_0$ and $v_\perp^2 J_1^2$ where $J_n = J_n(k_\perp v \sin \alpha/\Omega)$ with α the pitch angle and $v = (2H_0/m)^{1/2}$ the particle speed. Here we use the symbol H_0 for the kinetic energy of the ion of mass m and charge e. To start we take $\sin \alpha$ as a parameter by using $\lambda = k_\perp v \sin \alpha/\Omega$ as the dimensionless variable for the mean or characteristic speed v_0 in the $f(H_0, x)$ ion distributions.

The simplest, and most studied response function $P(\omega/k_y, \{p\})$ describes the electrostatic response of the plasma. Here $\{p\}$ denotes the set of system parameters. For electrostatic modes, the modes are given by dispersion relation

$$D_{ES}k, \omega = 1 + \frac{T_e}{T_i}[1 - P(k, \omega)] = 0. \tag{5.41}$$

The function $P(\omega, k_y, \{p\})$ may be thought of as the generalization of the plasma dispersion function $Z(\omega/k_\parallel v_{th})$ when the grad-B and curvature drifts dominate over the parallel phase velocity resonance.

The $P(\omega, k_y, \{p\})$ function is defined by

$$P(\omega, k_y, \{p\}) = -\int \frac{dH_0 H_0^{1/2} \left(\omega \dfrac{\partial f}{\partial H_0} + \dfrac{k_y}{eB_z} \dfrac{\partial f}{\partial x} \right) J_0^2}{\omega - k_y \bar{v}_D H_0 + i0^+}. \tag{5.42}$$

We may define an effective temperature T (or $k_B T$) from the mean ion energy in the distribution function $f(H_0, x)$ such that T becomes the thermodynamic temperature when $f(H_0, x) \Rightarrow f_{\text{Max-Boltzmann}}$.

For a Maxwellian $f_{\text{Max-Boltzmann}} = N \exp(-H_0/T)/(2\pi T/m)^{3/2}$ the P-function is thoroughly studied in the literature [Similon, *et al.* (1984)].

The $\bar{v}_D$ in equation Eq. (5.42) is the sum of the grad-B and curvature drift frequencies bounce-averaged and taken at some representative pitch angle (typically the variation with the pitch angle α is weak). The transition from the toroidal response function P to the slab model response function Z is given in Kim and Horton (1991).

The gradient $\partial/B\partial x$ is with respect to the flux function ψ with $Bdx = d\psi$ in the magnetic flux coordinate (per unit of dy where ∇y is the bi-orthogonal direction to the magnetic field vector and the radial gradient direction).

The spatial gradient of the plasma density is given

$$\frac{\partial f}{\partial x} \equiv -\frac{1}{L_n} f(H_0) \tag{5.43}$$

defining the length L_n. The local approximation is adequate for the first stage of analysis. With temperature gradients the driving term becomes

$$\frac{\partial f_M}{\partial x} = \left[\frac{1}{M}\frac{\partial N}{\partial x} + \frac{1}{T}\frac{\partial T}{\partial x}\left(\frac{H_0}{T} - \frac{3}{2}\right)\right]f_M \equiv \frac{f_M}{L_n}\left[1 + \eta\left(\frac{H_0}{T} - \frac{3}{2}\right)\right]. \qquad (5.44)$$

The energy dependence of $\eta(H_0/T - 3/2)$ arises describing that higher-energy ions have a stronger gradient scale length than that of the density itself.

The instabilities from this effect are sometimes called η_i-modes due to the historical definition of η as

$$\eta_i = \frac{d_x \ell n\, T_i}{d_x \ell n\, N}. \qquad (5.45)$$

For the Maxwellian distribution the electrostatic instability starts at

$$\eta_i > \eta_{\text{crit}} = \frac{2}{3}. \qquad (5.46)$$

This critical ratio of the temperature gradient instability threshold condition is understood by constructing a Carnot cycle between the different temperatures $T_i(x + \Delta x)$ and $T_i(x)$ and asking for the ideal isentropic limit of the work W done from the release of energy from the hot reservoir.

The analysis also shows that the case of inverted temperature and density gradients is strongly unstable. This regime of negative η_{crit} occurs when the pressure gradient is negligible across a region with a temperature and density gradient. The same situation occurs in the electron temperature gradient driven modes where the negative η_e instability has been demonstrated in basic laboratory experiments. The situation is common when two different plasmas across a transition layer as the plasma pause and the chromosphere-corona boundary.

Real-time stability monitoring

The thermodynamic formulas may be used with diagnostics for the density and temperature profiles to monitor in real-time the stability of the profiles. Using conservation of energy (first law of thermodynamics) the $W = Q_2 - Q_1$ and $Q_2 = T_2 \Delta S_2$ and $Q_1 = T_1 \Delta S_1$, with $\Delta S_1 + \Delta S_2 = 0$ gives the maximum work W that can be extracted by the turbulence per convection cycle as given in Eq. (5.47). The formula leads to the maximum turbulence energy density formula

$$\frac{W^{\text{turb}}(t)}{nT} = \frac{3}{2}\frac{\Delta T}{T}\left(\frac{\Delta T}{T} - \frac{2}{3}\frac{\Delta n}{n}\right) \qquad (5.47)$$

that can be a useful real-time diagnostic for feedback control of the auxiliary power $P_{\text{aux}}(t)$.

Thus, without Landau damping and other loss processes we have a bound, valid nonlinearly, on the maximum turbulent energy W that can be extracted from the distributions. We will return to the generalization of Eq. (5.47) for non-Maxwellian $f_0(H, x)$ distributions.

Full 3 × 3 electromagnetic ion temperature gradient modes

Low-frequency drift wave fluctuations satisfy the matrix equation $\mathbf{A} \cdot \mathbf{X} = 0$ defined by

$$\begin{bmatrix} a & b & c \\ b & d & e \\ c & e & f \end{bmatrix} \begin{bmatrix} \phi \\ \psi \\ \dfrac{\delta B_\parallel}{B} \end{bmatrix} = 0 \tag{5.48}$$

where the six complex kinetic response functions are

$$a = -1 + \frac{T_e}{T_i}(P-1)$$

$$b = 1 - \frac{\omega_{*e}}{\omega}$$

$$c = Q \tag{5.49}$$

$$d = \frac{k_\perp^2 \rho_s^2 \omega_A^2}{\omega^2} - \left(1 - \frac{\omega_{*e}}{\omega}\right) + \frac{\omega_{De}}{\omega}\left(1 - \frac{\omega_{*p_e}}{\omega}\right)$$

$$e = -\left(1 - \frac{\omega_{*p_e}}{\omega}\right)$$

$$f = \frac{2}{\beta_e} + \frac{T_i}{T_e}R.$$

The polarization of the mode is given by the ratio of the components of the vector $X^T = (\phi, \psi, \delta B_\parallel/B)$.

The physics contained in Eq. (5.48) is the following: The first row of the matrix equation is the condition of quasi-neutrality $\delta \tilde{n}_e = \delta \tilde{n}_i$. The second row is from Ampére's law, $\nabla_\perp^2 A_\parallel = -\mu_0 \delta j_\parallel$, where $k_\perp^2 \hat{\omega}_A^2 = -\partial_s v_A^2 k_\perp^2(s)\partial_s$ is the line-bending operator. The third row is from the "radial" ($\nabla \Psi$) component of Ampére's law.

Here the three ion kinetic response functions for ($|\omega| \gg k_\parallel v_i$) are given by P, Q, and R with analytic continuations from the upper-half of the complex ω plane

$$P = \left\langle \frac{\omega - \omega_{*i}(\epsilon)}{\omega - \omega_{Di}} J_0^2 \right\rangle, \tag{5.50}$$

$$Q = \left\langle \frac{\omega - \omega_{*i}(\epsilon)}{\omega - \omega_{Di}} \left(\frac{m_i}{b_i T_i}\right)^{1/2} v_\perp J_0 J_1 \right\rangle, \tag{5.51}$$

$$R = \left\langle \frac{\omega - \omega_{*i}(\epsilon)}{\omega - \omega_{Di}} \cdot \frac{m_i v_\perp^2}{b_i T_i} J_1^2 \right\rangle. \tag{5.52}$$

Here $\omega_{*i}(\epsilon) = \omega_{*i}[1 + \eta_i(\epsilon - 3/2)]$ with $\eta_i = d\ell n\, T_i/d\ell n\, n_i$, and likewise of ω_{*e} with $\eta_e = d\ell n\, T_e/d\ell n\, n_e$. The quantities $\omega_{*pj} = \omega_{*j}(1 + \eta_j)$, where p stands for the pressure gradient and j denotes the species. Similon, *et al.* (1984) describe the calculation and properties of the analytic P, Q, R functions.

Limiting cases of the electromagnetic ITG dispersion relation

Let us consider various limits of $D(\omega) = \det|A| = 0$ from Eq. (5.48). For a low-β plasma, $f \gg 1$ and the determinant D of Eq. (5.48) is

$$D = \left(ad - b^2\right) f \simeq 0. \tag{5.53}$$

For this system the MHD modes have $a \simeq -b$ and $d \cong -b$. Equation (5.53) for $f \neq 0$ gives the kinetically modified MHD modes $\omega^2 - \omega\omega_{*pi} - k_\parallel^2 v_A^2 + \gamma_{mhd}^2 = 0$ with small $E_\parallel$. There are electron drift modes with finite $E_\parallel$ given by

$$\omega^2 \left(1 + k_\perp^2 \rho_s^2\right) - \omega\omega_{*e} - k_\parallel^2 v_A^2 k_\perp^2 \rho_s^2 = 0. \tag{5.54}$$

The $\delta B_\parallel \neq 0$ compressional mode has $f(\omega) \simeq 0$ and is stable until the mirror mode instability condition $\beta(p_\perp/p_\parallel - 1) > C_m \approx 1$ is satisfied.

The turbulent electron transport is accurately defined in ECH plasmas where the location of the deposited power is well defined. High-power plasmas at ITER-like plasma conditions were carried out in the TCV Tokamak. For the TCV 29892 discharge the drift-waves and TEM modes at the radial position $r/a = 0.7$ at $t = 1.0\,\text{s}$, $(T_e, T_i) = (1.30, 0.76)\,\text{KeV}$, $n_e = 5.74 \cdot 10^{19}\,\text{m}^{-3}$, $B_T = 1.09\,\text{T}$, $\eta_e = 2.3$ and $\hat{s} = 0.65$ were driven by the high-T_e and dT_e/dr values. The relative plasma pressure is $\beta_e = 2\mu_0 p_e/B_T^2 = 0.025$ so that the $\omega \ll k_\parallel v_A$ and the modes are approximately electrostatic. Figure 5.10 shows the comparison of the growth rates

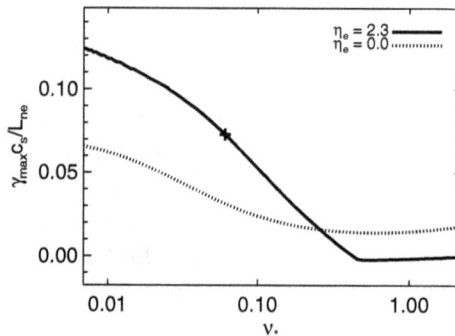

Fig. 5.10   The comparison of the maximum TEM growth rate versus the collisionality ν_* between the $\eta_e = 2.3$ (solid data) and $\eta_e = 0$ (dotted). The variation of collisionality ν_* is obtained by varying $T_e = 200 \sim 3{,}000\,\text{eV}$ and with the fixed temperature ratio T_e/T_i. Example for TCV 29892 ECH data in Porte, et al. (2006).

between the case $\eta_e = 2.3$ (solid) and the reference case $\eta_e = 0$ (dotted) where in the absence of an electron temperature gradient where the density gradient drives turbulence. The comparison shows that the electron temperature gradient η_e leads to the distinct transition around $\nu_* = 0.45$ between a low growth rate collisional drift wave mode and the high growth rate trapped electron mode. At ν_* lower than 0.45, where the data point (+), $\nu_* = 0.06$, corresponds, the electron temperature gradient destabilizes the trapped electron mode by a factor 2 over the density driven TEM mode.

The magnetic shear determines the parallel wavelength $k_\| L_n = k_y \rho_s \hat{s}^{1/2}$. Both the collisional and collisionless drift waves are affected by the magnetic shear through collisional dissipation and Landau damping respectively. The magnetic shear dynamics has a weak influence over the trapped electron mode. The electron response functions for $\eta_e = 2.3$ and 0 with $\nu_* = 0.06$ of the experiment are compared in Figs. 5.11 and 5.12. The wave/guiding-center drift resonances occur at $E/T_e \sim \omega_0/\omega_{De}$. With the temperature-driven case $\eta_e = 2.3$, the resonant electrons accord in the range $E/T_e = 1$ to 2.

For all time slices and for all the radial profiles, the mode frequency, growth rate and electron response functions are calculated. The mode frequencies and the maximum growth rates for each radial point are shown at each time slice in Fig. 5.11.

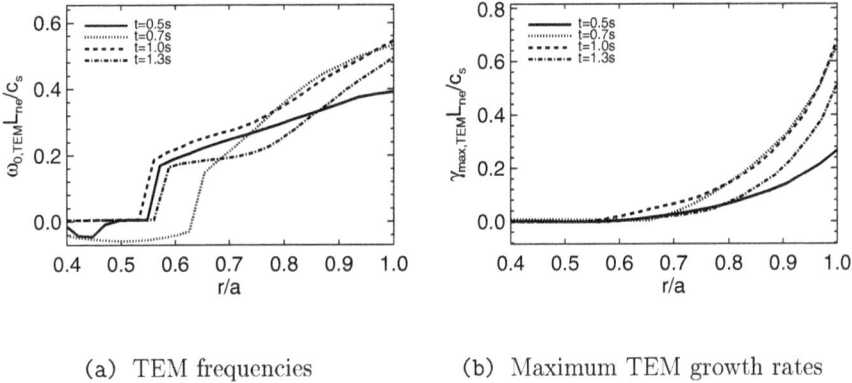

(a) TEM frequencies (b) Maximum TEM growth rates

Fig. 5.11 Time evolution of the TEM mode frequencies and the maximum growth rates of TCV 29892 at $t = 0.5$ (solid), 0.7 (dotted), 1.0 (dashed) and 1.3 s (dashed-dot). The maximum growth in panel (b) rate is obtained among the different k_y values.

At $r/a = 0.45 - 0.55$, there is a transition between ion drift waves and trapped electron modes. The ion drift waves propagate in the ion diamagnetic direction with low frequencies $\omega_0 \ll 0.1 c_s/L_{ne}$. The TEM modes with higher frequencies $\omega_0 \sim 0.1 c_s/L_{ne}$ at $r/a \gtrsim 0.5$ propagate in the electron drift wave direction.

For each time slice, the ion drift wave modes are weakly destabilized. Beyond the transition radial position $r/a = 0.5$, the trapped electron mode is destabilized by η_e and the same temperature gradient is stabilizing the ion drift wave mode. Towards the edge, both the density and temperature gradients contribute to the instability with comparable contributions.

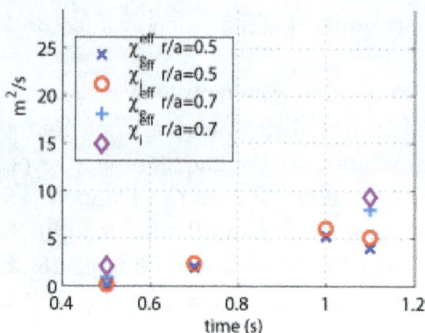

Fig. 5.12 Space and time evolution of the electron and ion diffusivities calculated by the Weiland model at $r/a = 0.5$ and 0.7 [Asp, *et al.* (2008)].

5.9   Isotope Scaling of Energy Confinement Time

The global energy confinement time and the local turbulent diffusivity depend on the mass of the work gas ions. The heavier the work gas ions the slower the turbulent energy loss rate. This behavior is called the isotope scaling of the confinement times and is well validated comparing toroidal confinements in plasma experiments with different isotopes of hydrogen as the working gas.

The first detailed validation study of the isotope effect is reported in a team work on the TFTR plasma in Scott, *et al.* (1995). In these discharges designed specially to test the isotope scaling after numerous earlier toroidal experiments appeared to have longer time times for the deuterium than for hydrogen discharges. In the Scott, *et al.* (1995) experiments the confinement for otherwise similar discharges were compared changing the working gas from hydrogen, to deuterium-tritium mixtures and to helium. Using the atomic number $A = 1, 2, 3$ and 4 for these gases the dependence both the local ion thermal diffusivity from local (r, t) power balance analysis and the global energy confinement time was determined to be strongly varying with A. The global energy confinement time and the local ion thermal diffusivity is reported to vary as $\tau_E = \tau_1 \langle A \rangle^{-0.9}$ and $\chi_i = \chi_1 \langle A \rangle^{-2.6+-0.5}$ where the range of uncertainty $+, -$ is from 2Σ from both diagnostic errors and the shot-to-shot variations.

Scott, *et al.* (1995) report that the local ion thermal transport is a factor of 2 lower in the D-T discharges compared with the pure deuterium discharges.

This rather strong dependence requires both the linear and nonlinear variations of the drift wave turbulence with ion mass. The ion sound wave speed is lower by the $A^{-0.5}$ and the ion sound gyroradius is larger by $A^{+0.5}$. Clearly, the simple gyroBohm form for χ_i has the wrong variation with ion mass. To understand this variation one first examines the maximum ITG growth rates scaling with A and observes that the growth rate decreases with increasing A. This behavior is also verified in the CLM basic cylindrical plasma experiments. Secondly, in the nonlinear simulations and theory the small value of the turbulence driving measure by $\Gamma_{max}/\omega(k_{max})$ produces a regime of weaker turbulence. Finally, using either turbulence theory or the simulations [Wang, *et al.* (2006); Lin, *et al.* (1999); Lee and Santoro (1997] one derives that there is a strong decrease of the turbulence and the turbulent thermal diffusivity comparable to that measured in the TFTR experiments [Dong, *et al.* (2013)].

Confinement enhancement over that of ELM free H-mode scaling law was obtained by the DIII-D team by coating the vacuum vessel walls with a film of 100 nm average thickness of boron. The coatings were analyzed with Auger electron spectroscopy and shown to be about 90% B and 10% C put on the walls by repeated glow discharges at pressure of 5×10^{-3} mbar of helium-diboranre gas. The Ohmic discharges after the boron coating of the walls showed the impurity concentrations of B:C:O with ratios of 2:1:1. The new series of shots with plasma parameters $B_T = 2.1$ to $2.5\,\mathrm{T}$, $I_P = 0.75$ to $4.5\,\mathrm{MA}$ and NBI injected power $P = 2$ to $10\,\mathrm{MW}$ with double null configuration of the separatrix.

The energy confinement time in the Jackson, *et al.* (1991) work is compared with that from the earlier DIII-D and JET data that scaled as $\tau_E = 0.106\,\mathrm{P_{nib}^{-0.46}I_p^{1.03}R^{1.48}}$ without boronization. After boronization the confinement time improved by a factor of 1.8 following the same scaling in the range of injected power below 10 MW. For injected power higher that 10 MW[12.5 MW] the confinement dropped owing to the onset of ELMs. The ELM-free states with enhanced confinement was defined as the VH-mode for "very high confinement" mode of operation.

The boronization is a type of chemical vapor deposition (CVD) process that is widely used in pre-process to obtain enhanced plasma confinement, owing mostly to the reduction of the impurity levels of carbon and oxygen in the working gas. The change in the stability and ion thermal transport with impurities for the tokamaks with various hydrogen isotopes for the working gas is analyzed in Dong, *et al.* (1994). The reductions in the growth rate with impurity fractions and with hydrogen isotope mass explains the variation of confinement times for a number of tokamak experiments.

The Weiland ITG/TEM Transport Model

The Weiland model [Jarmen, *et al.* (1987); Nordman, *et al.* (1990); Nordman, *et al.* (2005); Guo and Weiland (1997); Weiland (2000); Weiland and Holod (2005)] is often used to predict the radial global transport. Here we show how the model is used as a local stability tool to analyze the instabilities, their thresholds, and to calculate the local turbulent transport. The version of the Weiland model used here includes electromagnetic effects, collisions, impurity and magnetic shear effects but neglects varying correlation length [Weiland and Holod (2005)] and toroidal and poloidal rotation effects [Eriksson, *et al.* (2007)]. The two latter effects are included in a more refined version of the model.

In principle, the TEM (ITG) mode mainly drives electron (ion) thermal transport. The Weiland model (W) gives the following gyroBohm mixing-length formula

[Weiland and Nordman (1988)] for the thermal diffusivities

$$\chi_{e,i}^{W} = \frac{R/L_{T_e,i} - R/L_T^{\text{crit}}(\widehat{w}_{\text{re},i}, L_n)}{R/L_{T_e,i}}$$

$$\times \left\{ f_{e,i} \frac{m_i^{1/2}}{k_\theta \rho_s e^2 B^2 L_{ne}} T_e^{3/2}(eV) \sum_{g>0} \frac{\widehat{g}^3}{(\widehat{w}_{\text{r}} \mp F_{e,i})^2 + \widehat{g}^2} \right\} \quad (5.55)$$

where

$$F_{e,i} = \frac{10}{3} \frac{L_{ne}}{R} \frac{T_{e,i}}{T_e}. \quad (5.56)$$

Above e and i denote electron and ion, respectively. Here $f_e = f_t$, is the trapped electron fraction and $f_i = 1$. The mode frequency, $\widehat{w}_{\text{r}}$, and growth rate, $\widehat{g}$, are normalized to the diamagnetic drift frequency of the electrons ($w_* = K_\theta \rho_s c_s / L_{ne}$). The model assumes that the radial correlation length is of the same order as the poloidal one, i.e. $k_r^2 \approx k_\theta^2$. A heat pinch arises in Eq. (5.55) from the threshold function $p(\widehat{w}_{\text{re},i}, L_n) = R/L_T^{\text{crit}}$. The part of Eq. (5.55) within the curly brackets gives the mixing length diffusivity.

In Eq. (5.55), the summation over g is over all unstable modes. Hence, TEM modes can drive ion transport and ITG modes the electron transport depending on the relative sizes of the resonance factor, $(\widehat{w}_{\text{r}} \mp F_{e,i})$, and the growth rate. In the case of ion transport, the resonance factor will decrease with F_i if both R/L_{ne} and T_e/T_i are much larger than unity. In such a case, a significant TEM drive of χ_i can be expected [Asp, *et al.* (2007)].

The results of applying the Weiland model to the TCV discharge 29892 are shown in Fig. 5.12. The trend of the diffusivities at $r/a = 0.7$ agrees with the trend of the experimental heat fluxes shown in Fig. 5.12 up to $t = 1.0$ s. At $t = 1.3$ s the MHD activity reduces the gradients in the outer part of the plasma, which results in a marked drop in the diffusivities at $r/a = 0.7$. This drop was also seen in the nonlinear ETG simulations for the same reason.

The diffusivities in Fig. 5.12 for $r/a = 0.5$ follows the same trend as for $r/a = 0.7$ with the exceptions that between $t = 1.0$ s and 1.3 s, the diffusion only decreases by 20% and that the diffusivities at 0.7 s are not of the same magnitude as at 1.0 s.

During the Ohmic phase at $t = 0.5$ s, the transport is much lower, as the Ohmic heating of 0.49 MW is much smaller than the total heating with ECRH, $P_{\text{tot}} \sim$ 1.2 MW. This Ohmic phase is also the only phase in which the ITG instability is about twice as strong as the TEM. As a consequence, the ITG modes drive most of the ion transport and the TEM most of the electron. In the following phases the TEM dominates both the ion and electron transport.

The TEM electron heat diffusivities are twice as high as those from the power-balance calculations at $r/a = 0.7$ two time values $t = 0.7$ s and 1.0 s. The ETG turbulence gives closer agreement with the power-balance diffusivity. In the next section we introduce a metric called the Average Relative Variance to measure the degree to which a model and data agree.

An error analysis shows that the diffusivities can be reduced by, on average, $\sim 75\%$ for the ions and $\sim 65\%$ for the electrons by decreasing all gradients (R/L_{ne}, R/L_{T_e} and R/L_{T_i}) by 30%, which is within the error bars of the ion gradients. This would give diffusivities of similar magnitude to the experiment in inner part of the plasma, whereas further out they would still be overestimated. The local theoretical diffusivities version can therefore give a good idea of which modes are present and how the transport behaves qualitatively but not quantitatively.

Dorland and Hammett (1993) show a way to implement periodic radial boundary conditions in a sheared magnetic field to overcome (or reduce) the difficulty introduced by boundary effects where there is an absence of rational surfaces with Dirichlet boundary conditions. Periodic boundary conditions allow the heat flux leaving the box on the cold side to re-enter the simulation domain on the warm side. This recycling may not represent adequately the steady injection from heat sources on the high-temperature side of the box and a sink on the low-temperature side. The benefit of this method over simply zeroing out the ($k_y = 0, k_z = 0$) modes is that these modes are important in the nonlinear dynamics with respect to the generation and decay of the sheared flows and this mechanics is lost if these modes are completely removed from the simulation.

Electron Temperature Gradient-Driven Instabilities Producing Anomalously Low-Electron Temperatures and Regions of Ergodic/Stochastic Magnetic Field Lines

Without the ion-wave resonances, the dispersion relation $D(\boldsymbol{k}, \omega, \boldsymbol{P})$ with temperature and density gradient driven drift-waves ($\omega \sim \omega_* = k_y T_e/eBL_n$) is given [Zhu, *et al.* (1999); Zhu, *et al.* (2000)] by

$$D(\boldsymbol{k}, \omega, \boldsymbol{P}) = 1 + \frac{T_e}{T_i} - \left(\frac{T_e}{T_i} + \frac{\omega_*}{\omega}\right)\Gamma_0 - \eta_i \frac{\omega_*}{\omega} b_i(\Gamma_1 - \Gamma_0)$$

$$- \int_0^\infty 2\pi v^2 dv F_e^M(v) \int_{-1}^{+1} d\mu h_e(v, \mu, \omega) = 0 \qquad (5.57)$$

where the perturbed electron phase-space distribution fraction is $\tilde{f} = (e\phi/T_e)$ $F_e^M(1 - h_e)$ with $F_e^M(v)$ the Maxwell distribution for the electron velocity. Moreover, $\eta_{i,e} = d\log T_{i,e}/d\log n_{i,e}$ and the usual finite ion gyroradius Bessel function $\Gamma_n(b_i) = e^{-b_i} I_n(b_i)$ where $b_i = k_y^2 \rho_i^2$. The non-adiabatic part of the electron distribution $h_e(v, \mu)$ satisfies the kinetic equation

$$\nu(v)\frac{\partial}{\partial \mu}\left(1 - \mu^2\right)\frac{\partial h_e}{\partial \mu} + i\left[\omega - \omega_{De}\frac{v^2}{v_e^2} - k_\parallel v\mu\right] h_e$$

$$= i\left[\omega - \omega_*\left[1 + \eta_e\left(\frac{v^2}{v_e^2} - \frac{3}{2}\right)\right]\right] \qquad (5.58)$$

where $\nu(v) = \nu_e(v_e/v)^3$. $\omega_{De} \simeq \omega_*(2L_n/R) = 2w_*\epsilon_n$, $\eta_e = d\log T_e/d\log n_e \neq 0$, $v_e = (2T_e/m_e)^{1/2}$, and $\mu = v_\parallel/v$.

With the assumption of $\gamma/\omega_0 \ll 1$ and $h_e \ll 1$, the drift frequency ω_0 is given by

$$\omega_0(k, \boldsymbol{P}) = \omega_* \frac{[\Gamma_0 + \eta_i b_i(\Gamma_1 - \Gamma_0)]}{1 + (T_e/T_i)(1 - \Gamma_0)} \qquad (5.59)$$

and with $\Delta = 1 - \omega_0/\omega_*$ the growth rate given by

$$\frac{\gamma \omega_*}{\omega_0^2} [\Gamma_0 + \eta_i b_i(\Gamma_1 - \Gamma_0)] = \Delta \operatorname{Im} h^0 + \eta_e \operatorname{Im} h^1. \qquad (5.60)$$

For even n, the $\operatorname{Im} h^n$ are positive definite and determine the electron particle diffusivity D and the heat diffusivity $\mathcal{X}_e$. For odd n, the $\operatorname{Im} h^n$ are indefinite in sign. From Eq. (5.60) the contributions of Δh^0 and $\eta_e h^1$ to the growth rates determines which is the more critical parameter between the density gradient and the temperature gradient.

The electron temperature gradient in Eq. (5.60) drives drift wave instability at smaller space-time scales than the ion temperature gradient. This higher k, ω turbulence is called the ETG mode and as with the ITG mode occurs both in cylindrical and toroidal plasmas under conditions with $\eta_e > \eta_{\text{crit}}$. The frequency and growth rate now scale with v_e/L_{T_e} where $v_e = \sqrt{(T_e/m_e)}$ and the electron gyroradius ρ_e or $\rho_{e,i}$ where the later gyroradius is the analog of the ρ_s of the ITG modes that has the mixed measure of electron temperature and the ion mass. Here the $\rho_{e,i} = (m_e T_i)^{1/2}/eB$ so that there is an interchange of the roles of electrons and ions in-going from the ITG to the ETG instability.

An important difference is that owing to the small scales of the ETG the quasi-neutral approximation that is well satisfied for the ITG drift waves is marginal and often not sufficiently well satisfied for the ETG modes. The condition for the ETG to be quasi-neutral is that $k_\perp \lambda_{De} \ll 1$ which requires that $\omega_{pe} \gg \omega_{ce}$ which is not satisfied in high-performance regimes of ITER.

Comparison of the two types of temperature-gradient driven instabilities are shown in Fig. 5.13. The ETG modes derived in the next section, Sec. 5.10, plotted along with the TEM results for a TCV discharge driven strongly into the low-ν_{*e} regime with third harmonic electron cyclotron heating in the core plasma.

Carefully note the change in the frequency scale on the vertical axis for the frequencies and the wavenumber scale on the horizontal scale for the high-frequency ETG and the lower-frequency TEM waves in Fig. 5.13. Also note the reversal of the direction of rotation of the waves. The longer wavelength modes of the TEM may cause a larger transport when present with sufficient amplitude, however the smaller scale ETG modes grow much faster. The TEM modes live in a bath of the ETG modes. The ETG modes have an inverse cascade of their energy to wave lengths comparable to the ion gyroradius where they can feed the lower-frequency drift waves. Both forms of turbulence exist in plasmas with steep electron temperature gradients however the smaller scale modes are not sensitive to the details of the geometry such as the fraction of trapped electrons and are well above the usual electron collision frequencies. Thus, the universal nature of the anomalous

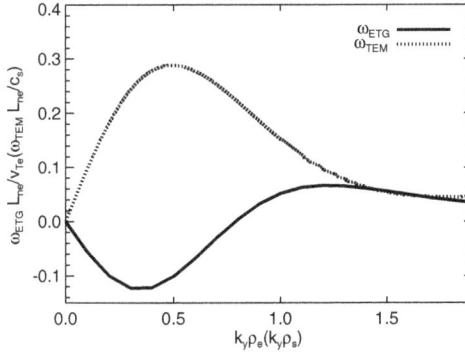

Fig. 5.13 The ETG (solid) and TEM (dotted) mode frequencies with increasing $k_y \rho_s$ ($k_y \rho_e$) at $r/a = 0.7$ and $t = 1.0\,$s for TCV discharge 29892 in Table 5.3. The frequencies are respectively normalized by v_{T_e}/L_n and c_s/L_n. The high frequency-small scale ETG mode, analyzed in Chapter 14, has many features similar to the TEM-ITG as shown here.

electron transport across all magnetic confinement geometries is likely due to the small scale ETG modes. When the trapped electron mode appears at low collisionality with significant trapped electrons is present, then the associated transport, TEM, will mask that due to the ETG turbulence. The role of zonal flows [Itoh, *et al.* (2008)] in controlling the level of transport from the trapped electron mode is explored in Ernst, *et al.* (2009). The work shows that the zonal flows can reduce the level of the electron thermal transport and the fluctuation level in general for the both of the trapped electron modes and the modes driven by the ion temperature gradients. The method of investigation uses and compares the results of the particle-in-cell gyrokinetic codes and the continuum codes. The works [Nevins, *et al.* (2005); Nevins, *et al.* (2006)] characterize the role of the interaction of the microturbulence and the sheared flows.

Table 5.3 JT-60U High-β_p Experiment 17110 with Internal Transport Barrier: parameters in Phase I of shot.

R/a	$3.1\,$m$/0.7\,$m
B_ϕ	$4.4\,$T
I_p	$2\,$MA
P_{NBI}	$27\,$M
$n_D(0)$	$4.1 \times 10^{19}\,$m^{-3}
$T_i(0)/T_e(0)$	$38\,$KeV$/12\,$KeV
$n_D \tau_E T_i$	$1.1 \times 10^{21}\,$m$^{-3} \cdot s \cdot$KeV
$v_\phi(0)$	$-100\,$km/s

5.10   Visualization of the Coherent Structures in ELMy Discharges

The simulations show that plasma vortex structures, called coherent structures, form in the steep density gradient regions produced by the magnetic separatrix. On the outside mid-plane of a torus these density and pressure gradients are in a region of unfavorable magnetic curvature (Ch. 1.4.1) such that there is an unstable effective accelerating force described by $g_{\text{eff}} = 2c_s^2/R$ producing vortex or "blobs" of plasma accelerating outward from the g_{eff} which is of order 10^{10}m/s^2 and higher, for high temperatures. This force in the steep gradient region produces high density blobs that move outward in the SOL and bubbles that move into the core plasma across the separatrix. This transport process limits both the density gradient and the jump in the density going from the scrape-off layer in the open separatrix region to the closed magnetic surfaces inside the separatrix. Experiments have been constructed to observe these structures by a complex method of splitting an RF or microwave beam and using interferometer methods analogous to the method used in the Michelson-Morely experiment to measure the speed of light waves. Here the waves are lower-frequency millimeter wavelength waves launched by infrared or RF beams and the method is called Phase Contrast Imagining [Terry, *et al.* (2005)]. Figure 5.10 shows

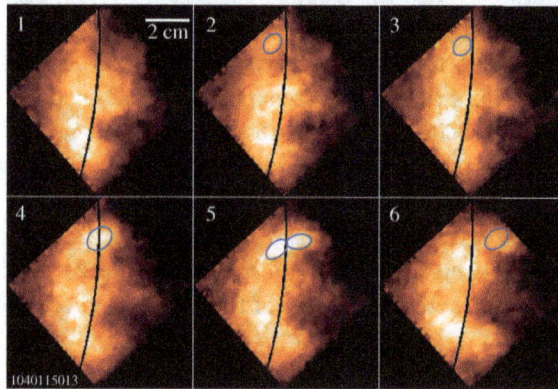

Fig. 5.14   Movie frames of edge turbulence in Alcator C-Mod from Terry, *et al.* (2005). The ovals locate the 'birth' and the subsequent motion of a blob, where the heavy vertical lines show the location of the magnetic separatrix. The emission is from He I and the time between frames is $4\,\mu s$.

the propagation of coherent vortex structures that transport plasma density down the steep density gradient that occurs across a magnetic separatrix. A qualitatively similar sharp density gradient occurs across the produces coherent structures called blobs and bubbles that give large transport $\Gamma = \langle \delta n v_r \rangle$ of plasma into the scrape off-layer. A simple model of the process is simulated using a the separatrix magnetic field in Meyerson, *et al.* (2014). A review of the experiments on C-Mod, DIII-D and

NSTX for the stability of the the steep pressure gradient and the current density gradient just inside the last closed magnetic surface is given in Groebner, *et al.* (2013). The work is a review work of the data and simulation with 65 co-authors reporting from the three tokamak teams. The work gives data for the boundaries between stable and unstable plasmas in the machines for various regimes, principally defined by the onset of the Type-I ELM events, dimensionless parameters from the MHD ballooning mode α and a normalized edge current profile. The work is comprehensive but technical, with numerous complex diagnostic techniques and computer simulations used in the data analysis.

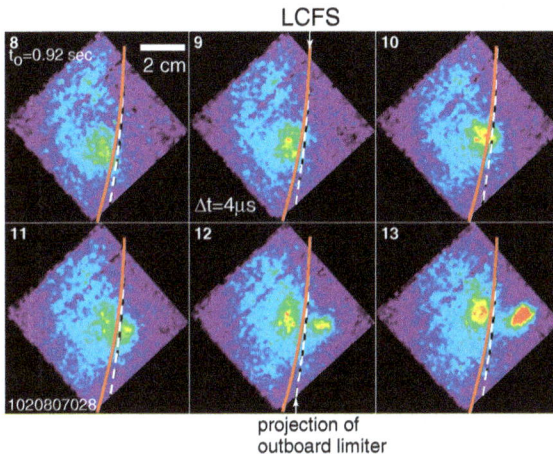

Fig. 5.15 Sequence of six images taken at 250 kHz from emission of the 667 nm He I line after He is puffed into the plasma on the out-board side of the C-Mod machine. The images show the outward transport of density blobs (trapped in vortex structures) with excess density from the plasma inside the separatrix, which is shown by the heavy vertical black curve to the plasma outside the magnetic separatrix. This outer-side plasma is in scrape-off layer (SOL). As the core density rises to the density limit, the plasma transport increases both to the outer wall and to the divertor plates at the bottom of the confinement chamber. The edge plasma temperature is measured to drop rapidly at the density limit. One of these quasi-coherent vortex structures is highlighted and traced with light from the He I emission. One of the plasma blobs is highlighted within the circles for the six frames separated by 4μ sec. The effective gravity g_{eff}, defined in Chapter 1.6, produces the $E \times B$ plasma drift velocity driving the bulge outward and the density holes inward across the separatrix. A simple mathematical model for this transport across a realistic magnetic field structure around the magnetic separatrix is given in Meyerson, *et al.* (2014). The plasma density and pressure transport across the magnetic separatrix is driven by the local electric field arising from the ions and electrons drifting in opposite directions in the curved toroidal magnetic field lines.

Experiments investigating the mechanisms responsible for the widely-observed density limit (the Greenwald density limit) were carried out in Alcator C-Mod with measurements of the change in edge temperature and the change in fluctuations that occur as the plasma approaches the density limit [LaBombard (2001);

Greenwald (2002); Greenwald (2014)]. The experiments showed that before the limit was reached, changes in the time-averaged SOL density profiles were observed, with progressive increases in the density outside the magnetic separatrix rising, giving, an overall flattening of the profiles. The flattening occurs even with modest increases in the separatrix density. At the same time, the amplitude, frequency, and birth rate of the plasma blobs increased [LaBombard (2002); Garcia, *et al.* (2013)]. This picture is supported by fluid models which predict strong transport under these conditions [Rogers, *et al.* (1998)]. At still higher densities, the boundary between the near-SOL and far-SOL moved inward, with the region of the colder plasma with intermittent fluctuations eventually crosses the separatrix entering the regions of closed field lines as shown in Fig. 5.10. When that boundary reaches roughly to the position of $\rho = 0.85$ of the normalized flux (a movement of about 3 cm on C-Mod), a density limit disruption is triggered. As the density limit is approached, the cross-field transport of plasma energy increases rapidly. The result is to create the density limit owing to upstream core plasma rushing to supply more density into the scrape off layer. The fast transport is partially across the field and partially parallel to the divertor chamber. This contrasts with the situation at lower density where all power is lost via the parallel channel to the divertor. In that case, the upstream temperature is pinned to a narrow range, typically to 60-100 eV, at the boundary between open and closed field lines. At densities close to the limit, perpendicular transport dominates on the open field lines and the plasma temperatures can drop to low values.

The phase contrast imaging diagnostic is described further in Chapter 8 and is used in both the tokamak and in the Large Helical Device [Tanaka, *et al.* (2012)] for understanding the cross-field plasma transport.

These expulsions of density blobs become more intense as the core plasma density is raised toward the Greenwald density limit, and are an important part of the physics of density limit. Further aspects of the density limit are described in Greenwald (2002). Further evidence of this process is described in LaBombard, *et al.* (2002).

References

Bonoli, P. T., Porkolab, M., Takase, Y., and Knowlton, S. F. (1988). Numerical modeling of lower-hybrid RF heating and current drive experiments in the Alcator-C tokamak, *Nucl. Fusion* **28**, p. 991, doi:10.1088/0029-5515/28/6/004.

Brizard, A. J., and Hahm, T. S. (2007). Foundations of Nonlinear Gyrokinetic Theory, *Rev. Mod. Phys.* **79**, p. 421, doi:http://dx.doi.org/10.1103/RevModPhys.79.421.

Chang, Z., and Callen, J. D. (1992). Unified fluid/kinetic description of plasma microin-stabilities. Part II: Applications, *Phys. Fluids B: Plasma Phys.* **4**, pp. 1182-1192, `http://dx.doi.org/10.1063/1.860126`.

Chang, Z., and Callen, J. D. (1990). Global energy confinement degradation due to macro-scopic phenomena in tokamaks, *Nucl. Fusion* **30**, pp. 219-233, doi:10.1088/0029-5515/30/2/003.

Connor, J. W. (1998a). A review of models for ELMs, *Plasma Phys. Control. Fusion* **40**, 2, p. 191, doi:10.1088/0741-3335/40/2/003.

Connor, J. W., Hastie, R. J., Wilson, H. R., and Miller, R. L. (1998b). Magne-tohydrodynamic stability of tokamak edge plasmas, *Phys. Plasmas* **5**, p. 2687, doi:10.1063/1.872956, `http://dx.doi.org/10.1063/1.872956`.

Detman, T., and Vassiliadis, D. (1997). Review of techniques for magnetic storm forecast-ing, *Geophysical Monograph, Am. Geophys. Union* **98**, pp. 253-266, ISSN:0065-8448.

Diamond, P. H., Lebedev, V. B., Newman, D. E., and Carreras, B. A. (1995). Dynamics of spatio-temporally propagating transport barriers, *Phys. Plasmas* **2**, p. 3685, `http://dx.doi.org/10.1063/1.871068`.

Dong, J. Q., Si, Y. J., Tamura, N., Jhang, H., Watanabe, T.-H., and Ding, X. T. (2013). (IOP Publishing and International Atomic Energy Agency, Nuclear Fusion) *Nucl. Fusion* **53**, p. 027001 (11pp), doi:10.1088/0029-5515/53/2/027001.

Dong, J. Q., Horton, W., and Dorland, B. (1994). Isotope scaling and η_i mode with impurities in tokamak plasmas *Phys. Plasmas* **1**, p. 3635, `http://dx.doi.org/10.1063/1.870942`.

Eriksson, A., Nordman, H., Strand, P., Weiland, J., Tala, T., Asp, E., Corrigan, G., Giroud, C., de Greef, M., and Jenkins, I. (2007). Predictive simulations of toroidal momentum transport at JET, *Plasma Phys. Control. Fusion* **49**, p. 1931, doi:10.1088/0741-3335/49/11/012.

Fielding, S. J., Ashall, J. D., Carolan, P. G., Colton, A., Gates, D., Hugill, J., Morriss, A. W., Valovic, M., and the COMPASS-D and ECRH Teams. (1996). *Plasma Phys. Control. Fusion* **38**, p. 1091, doi:10.1088/0741-3335/38/8/002.

Garcia, O. E., Czieglera, I., Kubea, R., LaBombard, B., Terry J. L. (2013). Burst statistics in Alcator C-Mod SOL turbulence, *J. Nucl. Mat.* **438**, Supplement, pp. S180-S183, doi:10.1016/j.jnucmat.2013.01.054.

Goldston, R. J., Kaita, R., Beiersdorfer, P., Gammel, G., Herndon, D. L., McCune, D. C., and Meyerhofer, D. D. (1987). Charge exchange measurements of MHD activity during neutral beam injection in the Princeton Large Torus and the Poloidal Divertor Experiment, *Nucl. Fusion* **27**, p. 921, doi:10.1088/0029-5515/27/6/004.

Greenwald, M. (2014). 20 Years of Research on the Alcator C-Mod Tokamak, *Phys. Plasmas.*

Greenwald, M. (2002). Density limits in toroidal plasmas, *Plasma Phys. Control. Fusion* **44**, pp. R27-R80, doi:10.1088/0741-3335/44/8/201.

Greenwald, M., Terry, J. L., Wolfe, S. M., Ejima, S., Bell, M. G., Kaye, S. M., and Neilson, G. H. (1988). A new look at density limits in tokamaks, *Nucl. Fusion* **28**, 2199, doi:10.1088/0029-5515/28/12/009.

Groebner, R.J., Chang, C. S., Hughes, J. W., Maingi, R., Snyder, P. B., Xu, X. Q., Boedo, J. A., Boyle, D. P., Callen, J. D., Canik, J. M., Cziegler, I., Davis, E. M., Diallo, A., Diamond, P. H., Elder, J. D., Eldon, D. P., Ernst, D. R., Fulton, D. P., Landreman, M., Leonard, A. W., Lore, J. D., Osborne, T. H., Pankin, A. Y., Parker, S. E., Rhodes, T. L., Smith, S. P., Sontag, A. C., Stacey, W. M., Walk, J., Wan, W., Wang, E. H.-J., Watkins, J. G.,White, A. E., Whyte, D. G., Yan, Z., Belli, E. A., Bray, B. D., Candy, J., Churchill, R. M., Deterly, T. M., Doyle, E. J., Fenstermacher, M. E., Ferraro, A. E. Hubbard, I. Joseph, J. E. Kinsey, B. LaBombard, C. J. Lasnier, Z. Lin, B. L. Lipschultz, C. Liu, Y. Ma, G. R. McKee, D. M. Ponce, J. C. Rost, N. M., Schmitz, L., Staebler, G. M., Sugiyama, L. E., Terry, J. L., Umansky, M. V., Waltz, R. E., Wolfe, S. M., Zeng, L., and Zweben, S. J. (2013). Improved understanding of physics processes in pedestal structure, leading to improved predictive capability for ITER, *Nucl. Fusion* **53** p. 09302, doi:10.1088/0029-5515/53/9/093024.

Groebner, R. J., Leonard, A. W., Snyder, P. B., Osborne, T. H., Maggi, C. F., Fenstermacher, M. E., Petty, C. C. and Owen, L. W. (2009). Progress towards a predictive model for pedestal height in DIII-D, *Nucl. Fusion* **49** p. 085037, doi:10.1088/0029-5515/49/8/085037.

Guo, S., and Weiland, J. (1997). Analysis of eta i mode by reactive and dissipative descriptions and the effects of magnetic q and negative shear on the transport, *Nucl. Fusion* **37**, pp. 1095-1107, doi:10.1088/0029-5515/37/8/105.

Hahm, T. S., and Burrell, K. H. (1995). Flow shear induced fluctuation suppression in finite aspect ratio shaped tokamak plasma, *Phys. Plasmas* **2**, p. 1648, http://link.aip.org/link/doi/10.1063/1.871313.

Hallock, G. A., Hickok, R. L., and Hornady, R. S. (1994). *The TMX heavy ion beam probe, Plasma Science, IEEE Transactions on Plasma Science* **22**, 4, pp. 341-349, ISSN:0093-3813, doi:10.1109/27.310639.

Hamaguchi, S., and Horton, W. (1992). Effects of sheared flows on ion-temperature-gradient-driven turbulent transport, *Phys. Fluids B* **4**, p. 319, http://dx.doi.org/10.1063/1.860280.

Hammett, G. W., and Perkins, F. W. (1990). Fluid moment models for Landau damping with application to the ion-temperature-gradient instability, *Phys. Rev. Lett.* **64**, pp. 3019-3022, doi:10.1103/PhysRevLett.64.3019.

Hoang, G. T., Horton, W. Bourdelle, C., Hu, B., Garbet, X., and Ottaviani, M. (2003). Analysis of the critical electron temperature gradient in Tore Supra, *Phys. Plasmas* **10**, pp. 405-412, doi:10.1063/1.1534113.

Holland, C., and Diamond, P. H. (2002). Electromagnetic secondary instabilities in electron temperature gradient turbulence, *Phys. Plasmas* **9**, p. 3857, http://dx.doi.org/10.1063/1.1496761.

Horton, W., Bekki, N., Berk, H. L., Hong, B.-G., LeBrun, M. J., Mahajan, S. M., Tajima, T., and Zhang, Y. Z. (1988). *12th Intl. Conf. Plasma Phys. Control. Nucl. Fusion Res.* (IAEA, London) CN-50/D-4-3.

Horton, W., Hong, B.-G., and Tang, W. M. (1988). Toroidal electron temperature gradient driven drift modes, *Phys. Fluids* **31**, p. 2971-2983.

Horton, W., Hong, B.-G., Tajima, T., and Bekki, N. (1990). *Comm. Plasma Phys. Control. Fusion* **13**, pp. 207-217.

Horton, W., Lindberg, D., Kim, J. Y., Dong, J. Q., Hammett, G. W., Scott, S. D., Zarnstorff, M. C., and Hamaguchi, S. (1992). Ion-temperature-gradient-driven transport in a density modification experiment on the Tokamak Fusion Test Reactor, *Phys. Fluids B* **4**, p. 953, http://dx.doi.org/10.1063/1.860112.

Horton, W., Hu, G., and Laval, G. (1996). Turbulent transport in mixed states of convective cells and sheared flow, *Phys. Plasmas* **3**, pp. 2912-2923, http://dx.doi.org/10.1063/1.871651.

Horton, W., Zhu, P., Hoang, G. T., Aniel, T., Ottaviani, M., and Garbet, X. (2000). Electron transport in Tore Supra with fast wave electron heating, *Phys. Plasmas* **7**, pp. 1489-1510, doi:10.1063/1.873969.

Horton, W., Hoang, G. T., Bourdelle, C., Garbet, X., Ottaviani, M., and Colas, L. (2004). Electron transport and the critical temperature gradient, *Phys. Plasmas* **11**, 5, p. 2600, http://dx.doi.org/10.1063/1.1690761.

Horton, W., Wong, H. V., Morrison, P. J. Wurm, A., Kim, J.-H., Perez, J. C., Pratt, J., Hoang, G. T., LeBlanc, B. P., and Ball, R. (2005). Temperature gradient driven electron transport in NSTX and Tore Supra, *Nucl. Fusion* **45**, pp. 976-985, doi:10.1088/0029-5515/45/8/025.

Hu, G., and Horton, W. (1997). Minimal model for transport barrier dynamics based on ion-temperature gradient turbulence, *Phys. Plasmas* **4**, pp. 3262-3272, http://dx.doi.org/10.1063/1.872467.

Huysmans, G. T. A., and Czarny, O. (2007). MHD stability in X-point geometry: Simulation of ELMs, *Nucl. Fusion* **47**, pp. 659-666, doi:10.1088/0029-5515/47/7/016, stacks.iop.org/NF/47/659.

Itoh, K., Itoh, S.-I., Diamond, P. H., *et al.* (2008). *Physics of Zonal Flows*, pp. 106-126.

Jacchia, A., Luca, F. D., Cirant, S., Sozzi, C., Bracco, G., Bruschi, A., Buratti, P., Podda, S., and Tudisco, O. (2002). Gradient length driven electron heat transport study in modulated electron cyclotron heating FTU tokamak, *Nucl. Fusion* **42**, p. 1116, doi:10.1088/0029-5515/42/9/310.

Jarmen, A., Andersson, P., and Weiland, J. (1987). Fully toroidal ion temperature gradient driven drift modes, *Nucl. Fusion* **27**, p. 941, doi:10.1088/0029-5515/27/6/006.

Karpushov, A. N., Duval, B. P., Schlatter, C., Afanasyev, V. I., and Chernyshev, F. V. (2006). Neutral particle analyzer diagnostics on the TCV tokamak, *Rev. Sci. Instrum.* **77**, p. 033503, doi:10.1063/1.2185151, http://dx.doi.org/10.1063/1.2185151.

Kaye, S. M., Greenwald, M., Stroth, U., Kardaun, O., Kus, A., Schissel, D., DeBoo, J., Bracco, G., Thomsen, K., Cordey, J., Miura, Y., Matsuda, T., Tamai, H., Takizuda, T., Hirayama, T., Kikuchi, H., Naito, O., Chudnovskij, A., Ongena, J., and Hoang, G. T. (1997). ITER L-mode confinement database, *Nucl. Fusion* **37**, pp. 1303-1328, doi:10.1088/0029-5515/37/9/I10.

Kim, J.-H., Perez, J. C., Horton, W., Chagelishvili, G. D., Changishvili, R. G., Lominadze, J. G., and Bowman, J. C. (2006). Self-sustaining vortex perturbations in smooth shear flows, *Phys. Plasmas* **13**, 6, p. 062304, http://dx.doi.org/10.1063/1.2209229.

Kim, J-Y., Horton, W., Choi, D-I., Migliuolo, S., and Coppi, B. (1992). Temperature anisotropy effect on the toroidal ion temperature gradient mode, *Phys. Fluids B* **4**, 152, http://dx.doi.org/10.1063/1.860428.

Kim, Jin-Yong, and Horton, W. (1991). Transition from toroidal to slab temperature gradient driven modes, *Phys. Fluids B* **3**, p. 1167, http://dx.doi.org/10.1063/1.859808

Krishnamurti, R., and Howard, L. N. (1981). Large-scale flow generation in turbulent convection, *Proc. Nat. Acad. Sci.* **78**, p. 4.

LaBombard, B., Greenwald, M., Boivin, R. L., Carreras, B. A., Hughes, J. W., Lipschultz, B., Mossessian, D., Pitcher, C. S., Terry, J. L., Zweben, S. J.,and Alcator C-Mod Research Team. (2002). Density Limit and Cross-Field Edge Transport Scaling in Alcator C-Mod, *Fusion Energy* (Vienna: IAEA, Lyon, 2002).

LaBombard, B., Boivin, R. L., Greenwald, M., Hughes, J. W., Lipschultz, B., Mossessian, D., Pitcher, C. S., Terry, J. L., Zweben, S. J., and Alcator Group. (2001). Particle transport in the scrape-off layer and its relationship to discharge density limit in Alcator C-Mod, *Phys. Plasmas*, **8**, p. 2107, doi:10.1063/1.1352596, http://dx.doi.org/10.1063/1.1352596.

Lee, W. W., and Santoro, R. A. (1997). Gyrokinetic simulation of isotope effects in tokamak plasmas, *Phys. Plasmas* **4**, p. 169, http://dx.doi.org/10.1063/1.872128.

L-H-Mode Database Working Group. (1994). *Nucl. Fusion* **34**, pp. 131-167.

Li, J. Q., Kishimoto, Y., Miyato, N., Matsumoto, T., and Dong, J. Q. (2005). Dynamics of large-scale structure and electron transport in tokamak microturbulence simulations, *Nucl. Fusion* **45**, 11, pp. 1293-1302, doi:10.1088/0029-5515/45/11/010.

Li, J., and Kishimoto, Y. (2004). Numerical study of zonal flow dynamics and electron transport in electron temperature gradient driven turbulence, *Phys. Plasmas* **11**, pp. 1493-1511, doi:10.1063/1.1669397, http://dx.doi.org/10.1063/1.1669397.

Lin, Z., Hahm, T. S., Lee, W. W., Tang, W. M., and Diamond, P. H. (1999). Effects of Collisional Zonal Flow Damping on Turbulent Transport, *Phys. Rev. Lett.* **83**, 18, pp. 3645-3648, http://link.aps.org/doi/10.1103/PhysRevLett.83.3645.

Lohr, J., Stallard, B. W., Prater, R., Snider, R. T., Burrell, K. H., Groebner, R. J., Hill, D. N., Matsuda, K., Moeller, C. P., Petrie, T. W., St. John, H., and Taylor, T. S. (1988). Observation of H-mode confinement in the DIII-D Tokamak with electron cyclotron heating, *Phys. Rev. Lett.* **60**, pp. 2630-2633, doi:http://dx.doi.org/10.1103/PhysRevLett.60.2630.

Matsuda, K. (1989). Ray tracing study of the electron cyclotron current drive in DIII-D using 60 GHz, *IEEE Transactions on Plasma Science* **17**, p. 6, doi:10.1109/27.21664.

Meyerson, D., Michoski, C., Waelbroeck, F., and Horton, W. (2014). Effect of chaos on plasma filament dynamics and turbulence in the scrape-off layer, *Phys. Plasmas* **21**, p. 072310, http://dx.doi.org/10.1063/1.4890349.

Nakata, M., Watanabe, T.-H., Sugama, H., and Horton, W. (2010). Formation of coherent vortex streets and transport reduction in electron temperature gradient driven turbulence, *Phys. Plasmas* **17**, p. 042306, http://dx.doi.org/10.1063/1.3356048.

Nevins, W. M., Candy, J., Cowley, S., Dannert, T., Dimits, A., Dorland, W., Estrada-Milla, C., Hammett, G. W., Jenko, F., Pueschel, M. J., and Shumaker, D. E. (2006). Characterizing electron temperature gradient turbulence via numerical simulation, *Phys. Plasmas* **13**, p. 122306, http://dx.doi.org/10.1063/1.2402510.

Nevins, W.M., Hammett, G.W., Dimits, A. M., Dorland, W., and Shumaker, D. E. (2005). Discrete particle noise in particle-in-cell simulations of plasma microturbulence, *Phys. Plasmas* **12**, p. 122305, http://dx.doi.org/10.1063/1.2118729.

Nordman, H., Strand, P., Eriksson, A., and Weiland, J. (2005). *Plasma Phys. Control.*

Fusion **47**, p. L11, doi:10.1088/0741-3335/47/6/L01.

Nordman, H., Weiland, J., and Jarmén, A. (1990). Simulation of toroidal drift mode turbulence driven by temperature gradients and electron trapping, *Nucl. Fusion* **30**, p. 983, doi:10.1088/0029-5515/30/6/001.

Perez, J. C., Horton, W., Bengtson, R. D., and Carter, T. (2006). Study of strong cross-field sheared flow with vorticity probe in the large plasma device, *Phys. Plasmas* **13**, p. 057701.

Peysson, Y., Decker, Joan, Morini, L., and Coda, S. (2011). RF current drive and plasma fluctuations, *Plasma Phys. Control. Fusion* **53** p. 124028, doi:10.1088/0741-3335/53/12/124028.

Peysson, Y., and Decker, J. (2007). Advanced Lower-Hybrid Current Drive Modeling, Radio Frequency Power in Plasmas: 17th Topical Conference on Radio Frequency Power in Plasmas, *AIP Conference Proceedings* **933**, pp. 293-296.

Porte, L., Coda, S., Alberti, S., Arnoux, G., Blanchard, P., Bortolon, A., Fasoli, A., Goodman, T. P., Klimanov, Y., Martin, Y., Maslov, M., Scarabosio, A., and Weisen, H. (2006). Plasma Dynamics with Second and Third Harmonic ECRH on TCV, *21st International Atomic Energy Agency Fusion Energy Conf.* (Vienna) EX/P6-20.

Rewoldt, G., Beer, M. A., Chance, M. S., Hahm, T. S., Lin, Z. and Tang, W. M. (1998). Sheared rotation effects on kinetic stability in enhanced confinement tokamak plasmas, and nonlinear dynamics of fluctuations and flows in axisymmetric plasmas. *Phys. Plasmas* **5**, 1815, `http://dx.doi.org/10.1063/1.872851`.

Rogers, B. N., Drake, J. F., and Zeiler, A. (1998). Phase Space of Tokamak Edge Turbulence, the L-H Transition, and the Formation of the Edge Pedestal, *Phys. Rev. Lett.* **81**, p. 4396, doi:http://dx.doi.org/10.1103/PhysRevLett.81.4396.

Ryter, F., Imbeaux, F., Leuterer, F., Fahrbach, H.-U., Suttrop, W., and ASDEX Upgrade Team. (2001a). Experimental characterization of the electron heat transport in low-density ASDEX Upgrade Plasmas, *Phys. Rev. Lett.* **86**, pp. 5498-5501, doi:http://dx.doi.org/10.1103/PhysRevLett.86.5498.

Ryter, F., Leuterer, F., Pereverzev, G., Fahrbach, H.-U., Stober, J., Suttrop, W., and ASDEX Upgrade Team. (2001b). Experimental evidence for gradient length-driven electron transport in tokamaks, *Phys. Rev. Lett.* **86**, pp. 2325-2328, doi:http://dx.doi.org/10.1103/PhysRevLett.86.2325.

Saoutic, B., Beaumont, B., Becoulet, A., Bizarro, J. P., Fraboulet, D., Garbet, X., Goniche, M., Guiziou, L., Hoang, G. T., Hutter, T., Joffrin, E., Kuus, H., Litaudon, X., Mollard, P., Moreau, D., Nguyen, F., Pecquet, A. L., Peysson, Y., Rey, G., van Houtte, D., and Zabiego, M. (1994). High-power ICRF and LHCD experiments on Tore Supra, *AIP* **289**, pp. 24-31, `http://link.aip.org/link/?APC/289/24/1`.

Similon, P., Sedlak, J. E., Stotler, D., Berk, H. L., Horton, W., and Choi, D-I. (1984). Guiding-center dispersion function, *J. Comp. Phys.* **54**, 2, pp. 260-277, doi:10.1016/0021-9991(84)90118-9.

Stevens, J., Von Goeler, S., Bernabei, S., Bitter, M., Chu, T. K., Efthimion, P., Fisch, N., Hooke, W., Hosea, J., Jobes, F., Karney, C., Meservey, E., Motley, R., and Taylor, G. (1985). Modeling of the electron distribution based on Bremsstrahlung emission during lower-hybrid current drive on PLT, *Nucl. Fusion* **25**, p. 1529, doi:10.1088/0029-5515/25/11/002.

Sugama, H., Watanabe, T. H., and Horton, W. (2001). Collisionless kinetic-fluid closure and its application to the three-mode ion temperature gradient driven system, *Phys. Plasmas* **8**, 6, pp. 2617-2618, `http://dx.doi.org/10.1063/1.1367319`.

Sugama, H., and Horton, W. (1997). Transport Processes and Entropy Production in Toroidally Rotation Plasmas with Electrostatic Turbulence, *Phys. Plasmas* **4**, pp. 405-418, http://dx.doi.org/10.1063/1.872099.

Sugama, H., and Horton, W. (1995). L-H confinement mode dynamics in three-dimensional state space, *Plasma Phys. Control. Fusion* **37**, p. 345, doi:10.1088/0741-3335/37/3/012.

Tanaka, H., Ohno, N., Tsuji, Y., Kajita, S., Masuzaki, S., Kobayashi, M., Morisaki, T., Komori, A., and the LHD Experimental Group. (2012). Blob/Hole Generation in the Divertor Leg of the Large Helical Device, *Plasma Fusion Res.* **7**, p. 1402152, http://dx.doi.org/10.1585/pfr.7.1402152.

Terry, J. L., Basse, N. P., Cziegler, I., Greenwald, M., Grulke, O., LaBombard, B., Zweben, S. J., Edlund, E. M., Hughes, J. W., Lin, L., Lin, Y., Porkolab, M., Sampsell, M., Veto, B., and Wukitch, S. J. (2005). Transport phenomena in the edge of Alcator C-Mod plasmas *Nuc. Fusion* **45** p. 1321, doi:10.1088/0029-5515/45/11/013.

Turri, G., Sauter, O., Porte, L., Alberti, S., Asp, E., Goodman, T. P., Martin, Y. R., Udintsev, V., and Zucca, C. (2008). The role of MHD in the sustainment of electron internal transport barriers and H-mode in TCV, *J. Phys.* **123**, p. 1, doi:10.1088/1742-6596/123/1/012038.

Wagner, F., Becker, G., Behringer, K., Campbell, D., Eberhagen, E., Englehardt, W., Fussmann, G., and Gehre, O. (1982). Regime of improved confinement and high beta in neutral beam heated divertor discharges of the ASDEX Tokamak, *Phys. Rev. Lett.* **49**, p. 1408, doi:http://dx.doi.org/10.1103/PhysRevLett.49.1408.

Wang, W. X., Lin, Z., Tang, W. M., Lee, W. W., Ethier, S., Lewandowski, J. L. V., Rewoldt, G., Hahm, T. S., and Manickam, J. (2006). Gyrokinetic simulation of global turbulent transport properties in tokamak experiments, *Phys. Plasmas* **13**, p. 092505, http://dx.doi.org/10.1063/1.2338775.

Weigend, A. S., Huberman, B. A., and Rumelhart, D. E. (1990). Predicting the Future: A connectionist Approach, *Int. J. Neur. Syst.* **1**, pp. 193-209, doi:10.1142/SO129065790000102.

Weiland, J., and Holod, I. (2005). Drift wave transport scalings introduced by varying correlation length, *Phys. Plasmas* **12**, p. 012505, doi:10.1063/1.1828083.

Weiland, J. (2000). *Collective modes in inhomogeneous plasma* (IOP Publishing Ltd.).

Weiland, J., and Nordman, H. (1988). Theory of fusion plasmas: Proceedings of the Joint Varenna-Lausanne International Workshop, eds. A. Guthrie and R. K. Wakerling (Editrice Compositori for Societa Italiana di Fisica, Bologna, 1988).

Wolf. R. C., Günter, S., Leuterer, F., Peeters, A., Pereverzev, G., Gruber, O., Kaufmann, M., Lackner, K., Maraschek, M., McCarthy, P. J., Meister, H., Salzmann, H., Schade, S., Schweinzer, J., Suttrop, W., and ASDEX Upgrade Team (2000). Response of internal transport barriers to central electron heating and current drive on ASDEX Upgrade, *Phys. Plasmas* **7**, p. 1839, http://dx.doi.org/10.1063/1.874006.

Wolf, R. C., Hobirk, J., Conway, G. D., Gruber, O., Gude, A., Günter, S., Kirov, K., Kurzan, B., Leuterer, F., Maraschek, M., McCarthy, P. J., Meister, H., Pereverzev, G. V., Poli, E., Ryter, F., Treutterer, W., Yu, Q., and ASDEX Upgrade Team (2001). Performance, heating and current drive scenarios of ASDEX Upgrade advanced tokamak discharges, *Nucl. Fusion* **41**, 1259, doi:10.1088/0029-5515/41/9/315.

Wolf, R. C. (2003). Internal transport barriers in tokamaks plasmas, *Plasma Phys. Control. Fusion* **45**, pp. R1-R91, doi:10.1088/0741-3335/45/1/201.

Zhu, P., Bateman, G., Kritz, A. H., and Horton, W. (2000). Predictive transport simulations of internal transport barriers using the Multi-Mode model, *AIP* **7**, pp. 2898-2908, http://ojps.aip.org/pop/popcr.jsp.

Zhu, P., Horton, W., and. Sugama, H. (1999). The radial electric field in a tokamak with reversed magnetic shear, *Phys. Plasmas* **6**, p. 2503, doi:10.1063/1.873522.

Chapter 6

Transport Barriers and ELM Control

6.1 Record DT Fusion Power Discharges in the Joint European Torus (JET)

In 1999 the JET team announced the production of DT discharge approaching $Q_{\text{fus}} \lesssim 1$ within the error bars of the measurement of the fusion power of 16.1 MW driven by a combination of NBI and RF heaters programmed so as to create both an H-mode edge barrier and an internal transport barrier with weakly reversed magnetic shear. The plasma was in an ELM-free H-mode state and produced 1.3 MW of alpha particle heating power.

The high power was correlated with the discharge having a high-electron temperature. The DT fuel mixture was programmed to optimize the fusion power. The tokamak parameters where $I_p = 4.2$ MA of toroidal current in a $B_t = 3.6$ T toroidal magnetic field.

The fusion power triple product confinement measure reached the record value of $n_{\text{DT}}^{(0)} \tau_{\text{Edia}} T_i^{(0)} = 8.7 \times 10^{20} \pm 20\% \, \text{m}^{-3}\text{s keV}$ and was maintained for one-half an energy confinement time.

Magnetic fluctuation spectra showed no evidence of Alfvénic instabilities driven by alpha particles. Internal transport barriers were established in the DT plasma with optimized reversed magnetic shear. The ion thermal conductivity in the plasma core was a few times the neoclassical levels presumably from drift wave turbulence driven by the steep radial gradients.

Real-time power control maintained the plasma core close to limits set by pressure gradient driven MHD instabilities, allowing 16 MW of DT fusion power with $n_{\text{DT}}^{(0)} \tau_{\text{Edia}} T_i^{(0)} \sim 10^{21} \, \text{m}^{-3} \, \text{s keV}$.

Other discharges at that time were produced in quasi-steady-state discharges with simultaneous internal and edge transport barriers with a longer lower-level of external heating that producing average fusion power of up to 7 MW.

6.2   Radial Electric Field E_r in H-mode Transport Barriers

Measuring the electric field in the tokamak is a difficult problem that requires complex atomic physics modeling described in Chapter 8. Using a charge exchange recombination spectroscopy (CXRS) data from the edge transport barrier in ASDEX with He^{+2} ions the profiles of the poloidal rotational velocities were measured [Viezzer, *et al.* (2014)]. The rather complex analysis using the electron density, ion temperature and the corresponding ion pressure profiles derived from atomic physics data, gives the data necessary profiles to compare the rotational velocities (toroidal and poloidal components) of the fully ionized hydrogenic working gas ions with the formulas in neoclassical collisional transport theory.

The ASDEX team gives the profiles obtained in an H-mode ASDEX plasma with $B_\phi = -2.5\,\text{T}$, $I_p = 1\,\text{MA}$ driven by $P_{\text{ECH}} = 0.5\,\text{MW}$ and deuterium neutral beam with $P_{\text{NBI}} = 9.2\,\text{MW}$ heating. The discharge 27879, in $t = 2.55\text{-}2.85\,\text{s}$ with central line averaged density is analyzed with line density of $10^{20}/\text{m}^3$ for its transport properties in the barrier region of $\rho = 0.94$ to 1.00. The poloidal impurity rotational velocity of N^{7+} from the $n = 9 \rightarrow 8$ spectral line (567 nm) is used to compare with the neoclassical predictions of Kim, *et al.* (1991). Good agreement is shown for the region from $\rho = 0.9$ to 1.00. Several neoclassical codes NEOART, NEO, and HAGIS are used in the comparison. All three models agree and are consistent with the measured profiles. In ASDEX the T_e edge profiles are measured with electron cyclotron emission (ECE) with a sampling rate of 31 kHz and a spatial resolution of 1 cm. Integrated data analysis (IDA) combined with laser interferometry data yield n_e profiles with spatial resolution of 5 mm in the edge transport barrier (ETB). Edge ion temperature profiles and the toroidal velocity profiles were measured by charge exchange recombination spectroscopy (CXRS) diagnostics with a temporal resolution of 1.9 ms and a spatial resolution of between 3 and 8 mm. Radial electric field profiles were inferred from line integrated He II emission profiles. For the analysis, all frames which contain an ELM are omitted, so that only frames in between ELMs are used for the reconstruction of E_r profiles. The profiles are mapped to the magnetic midplane and n_e and T_e data from the Thomson scattering diagnostic are used in the profile reconstruction. From this data Wolfrum, *et al.* (2009) conclude that the electron pressure gradient is the dominant controlling factor in the onset of the ELMs. Their analysis suggests that the MHD ballooning instabilities limit the pressure gradient, but do not trigger the ELM. Providing the radial electric field behaves given by neoclassical transport formulas, namely that the E_r minimum is positioned where p_i/n_i is maximal, the data analysis suggests that it is the electron temperature gradient that correlates with the triggering of the ELMs.

In the steepest gradient region the gradient scale length approaches the poloidal ion gyroradius which is approximately 1-2 cm for the deuterium ions. This steep gradient breaks the small expansion parameter used in the analytic models. In addition the collisionality at this steep gradient region is at the banana to plateau transition point for deuterium. The impurity ions of nitrogen are in the Pfirsch-Schlüter regime. The toroidal rotation velocity of the deuterium ions is estimated to be between 0.05 to 0.2 of the ion thermal velocity. The conclusion drawn by Viezzer, *et al.* (2014) is that in the edge transport region the radial electric field is well described by the neoclassical formulas. The deuterium poloidal rotation velocity is inferred to be 1-3 km/s which is less than 0.03 the thermal velocity which means that the main ion poloidal velocity is driven by the ion temperature gradient in the transport barrier region. The data indicates the ion orbit-squeezing effect is negligible.

In DIII-D experiments where the ion collisionality parameter varied from 0.1 to 0.3 the main ion poloidal velocity was positive in contrast to that in the ASDEX experiment where the ion poloidal velocity is negative. Viezzer, *et al.* (2014) propose that this difference is due to the higher-ion collisionality parameter in the ASDEX data, which is an order of magnitude higher than in the DIII-D data.

In the DIII-D data the ion poloidal velocity is of order +1 m/s to +20 km/s in the ion diamagnetic direction [Grierson, *et al.* (2013)]. In contrast, in the ASDEX data the poloidal velocity is −1 to −2 km/s in the electron diamagnetic direction.

The ASDEX data is found to agree with a variety of models from Kim, *et al.* (1991), and HAGIS, NEOART and NEO, which is a drift kinetic code of Belli and Candy (2008).

The ASDEX data show that the fully developed, highly-collisional edge transport barrier of an H-mode has a radial electric field given by the neoclassical model with impurities playing an important role [Rogister (1994)]. This is connected to the finding that the main ion poloidal rotation velocity is at the neoclassical value. The impurity rotation velocity is the dominant contribution to the radial electric field E_r determining the depth of the E_r well in the H-mode transport barrier. A further implication is that the impurity ion transport is adequately described by neoclassical transport as reported in Pütterich, *et al.* (2011).

The mechanism for damping the toroidal rotational velocity to small speeds requires more understanding of the toroidal momentum transport in the edge pedestal region as discussed in Peeters (2011).

6.3 Internal Transport Barriers from ITG/TEM Turbulence

The turbulent transport from the ion temperature gradient driven drift waves and the trapped electron modes transport produces an ion momentum flow in the plasma fluid in addition to the transport of thermal energy and particles. The turbulence produces a flux of momentum from the turbulent $\boldsymbol{E} \times \boldsymbol{B}$ drift wave fluctuations [Connor and Wilson (2000)]. In the neutral fluid literature the calculations of the momentum flux $\langle v_i v_j \rangle$ from turbulence is an long standing research area and the turbulent flux is called the Reynolds stress tensor given by $\rho_m \langle v_i v_j \rangle$ where ρ_m is the mass density and the components of the fluctuating velocity are v_i. Calculations of Hamaguchi and Horton (1992) for the ITG turbulence included the calculation of the momentum stress tensor and showed that this flux can generate a mean sheared background flow velocity commonly called a zonal flow.

A clear indication of the formation of a diffusion barrier is shown by integrating ensembles of test particles in a specified spectrum of drift waves existing on top of the background plasma with flow shear or reversed magnetic field. This method gives a direct picture of how the radial transport is reduced by the sheared flows and magnetic fields. The reversed magnetic field breaks the Chirikov overlap condition so that the stochastic diffusion across the minimum q_{min} surface is not possible [Beklemishev and Horton (1992); Horton, *et al.* (1998)]. There are trajectories that cross the q_{min} surface but these are better described by long-flight orbits than by diffusion. Some works giving the details of the orbits in realistic fluctuation fields are Horton, *et al.* (1998), Evstatiev, *et al.* (2003), and Batista, *et al.* (2006).

Perpendicular shear flows are responsible for both symmetry breaking and suppression of turbulence, resulting in a shearing rate at which there is a maximum contribution to the momentum transport. The $\boldsymbol{E} \times \boldsymbol{B}$ momentum transport is shown to be quenched by increasing flow shear more strongly than the standard linear quench rule for turbulent heat diffusivity.

The analogous fluid turbulence calculations and experiment are given by Howard and Krishnamurti (1986). In their experiment in a Bernard convection chamber with water, the low thermal gradient driving the Bernard convection cells – analogous to the drift waves in the plasma – interact with one another creating a mean circulation in the chamber. A simple mathematical model with a few key k_x, k_y modes was derived by Howard and used to explain the bifurcation with increasing temperature gradient to a state of turbulence with the global mean sheared flows.

The next two subsections describe first a simple model for the transport barrier dynamics and then full computer simulation models.

6.3.1   *Predator-prey models*

Let us consider the plasma physics analog of this simple fluid experiment in which an increasing ion temperature gradient drives plasma turbulence which in turn drives a mean sheared $\boldsymbol{E} \times \boldsymbol{B}$ flow that reacts back on the turbulence. Since the reaction of the turbulence-driven sheared flow on the ITG turbulence is stabilizing, this situation has been characterized as a "predator-prey" situation. The predator is the sheared flow living off the prey which is the turbulence. Here the sheared flow feeds on the turbulence and at some point when sufficiently strong sheared flows are present the ITG turbulence level – as the prey – is reduced. This process can go through cycles as often seen in the simulations and in the dithering of plasmas between the two states of low-mode L and high-mode H of confinement [Diamond and Kim (1991)].

To understand the mathematical model of this problem we consider the dynamics of the ITG driven drift wave turbulent energy $W(t)$ in the toroidal magnetic geometry. The standard representation for the toroidal axisymmetric magnetic field in the torus is

$$\boldsymbol{B} = I(\psi)\nabla\phi + \nabla\phi \times \nabla\psi \qquad (6.1)$$

where the toroidal angle is ϕ and the poloidal magnetic flux function is ψ and the toroidal field is $I(\psi)/R$ where $\nabla\phi$ is the unit vector in the toroidal direction divided by R the distance from the axis of symmetry to the spatial point x, y, z under consideration for the $\boldsymbol{B}(\boldsymbol{x})$ field.

From the analysis of the ITG turbulence earlier in this chapter we can know that the thermal ion flux q_i and the turbulent wave energy density $W(r,t)$ are coupled by

$$q_i = -n_i \chi_i(W)\left(\frac{dT_i}{dr}\right) \qquad (6.2)$$

$$\gamma(W) = \gamma^\ell - \gamma^{n\ell}W \qquad (6.3)$$

$$W = \sum_k \omega_k \left(\frac{\partial \epsilon_k}{\partial \omega_k}\right)\frac{|E_k|^2}{8\pi n_e T_e} \simeq \frac{1}{2}\sum_k \frac{e^2|\phi_k|^2}{T_e^2} \qquad (6.4)$$

with the dynamics driven by

$$\frac{dW}{dt} = \left[2\gamma_0\left(\frac{R}{L_{T_i}} - \left(\frac{R}{L_{T_i}}\right)_{\text{crit}}\right) - 2\gamma^{n\ell}W\right] \qquad (6.5)$$

$$\frac{3}{2}\frac{\partial}{\partial t}(n_i T_i) = \frac{1}{r}\frac{\partial}{\partial r}r\left(n_i \chi_0 W\frac{\partial T_i}{\partial r}\right) + P_E(r,t). \qquad (6.6)$$

The gradient of the heating power per ion is

$$P_E' = \frac{\partial}{\partial r}\left(\frac{2P_E(r)}{3n_i(r)}\right). \qquad (6.7)$$

The radial gradient of the power deposited per ion is of order keV/m per ion per second from Eq. (6.7).

The ion temperature profile is determined by the combination of the injected heating power $P_E(r,t)$, the turbulence $W(r,t)$ profile, and the turbulent thermal flux $q_i(W, dT_i/dr)$ defined in Eqs. (6.2)-(6.6). Clearly the radial profile of the absorbed heating power $P_E(r,t)$ is important in determining the temperature gradient. In the following analysis, the spatial dependence of the heating power is reduced to the local gradient defined in Eq. (6.7).

The transport problem is now reduced to a system of spatially local ODEs for the turbulence level dW/dt and the temperature gradient dT_i/dr. To this we add the equation for the momentum transport. Already at this simple stage there is interesting dynamics of the evolution of the turbulence and temperature gradient. There is a stable fixed point of the system given by the formulas in Eqs. (6.3)-(6.8).

The energy confinement time is then determined from the fixed points of Eqs. (6.5) and (6.6) given by

$$W^*(P_E') = \left(\frac{\gamma^0}{2\gamma^{n\ell}}\right)\left[\sqrt{\left(\frac{T_i'}{T_i}\right)_c^2 + \frac{6\Delta^2\gamma^{n\ell}P_E'}{\gamma^0\chi_0 T_i}} - \left(\frac{T_i'}{T_i}\right)_c\right] \qquad (6.8)$$

$$\frac{T_i'(P_E')}{T_i} = \frac{1}{2}\left[\sqrt{\left(\frac{T_i'}{T_i}\right)_c^2 + \frac{6\Delta^2\gamma^{n\ell}P_E'}{\gamma^0\chi_0 T_i}} - \left(\frac{T_i'}{T_i}\right)_c\right]. \qquad (6.9)$$

From the fixed point one calculates the energy confinement time as

$$\tau_E = a\left(\frac{\gamma^{n\ell}}{\chi_0\gamma^0}\right)^{1/2} \bigg/ [W^*(P_E')]^{1/2}. \qquad (6.10)$$

The plasma flow velocity is

$$\boldsymbol{u} = \frac{\boldsymbol{B}}{B_0}\left(u_{\parallel} + \frac{B_T}{B_p}u_{\perp}\right) - \frac{u_{\perp}B}{R_0 B_p}R^2\nabla\phi.$$

Typically, one breaks the flow velocity into poloidal (p) and toroidal (t) components using

$$u_p = \left(\frac{B_t}{B}\right)u_{\perp} - \left(\frac{B_p}{B}\right)u_{\parallel} \qquad (6.11)$$

$$u_t = \left(\frac{B_t}{B}\right)u_{\parallel} + \left(\frac{B_p}{B}\right)u_{\perp}. \qquad (6.12)$$

The stability of the ITG turbulence is strongly controlled by the ratio $L_s u_{\perp}'/c_s$. The critical value of the flow shear divided by magnetic shear is

$$\left(\frac{L_s u_{\perp}'}{c_s}\right)_c = \gamma_{\text{crit}} = 2\sqrt{\frac{T_i}{T_e}(\eta_i - \eta_{ic})}. \qquad (6.13)$$

For the JT-60U data the balance in Eq. (6.13) occurs at

$$\left(\frac{L_s u_{\perp}'}{c_s}\right)_c = 2\sqrt{\frac{4\,\text{KeV}}{2\,\text{KeV}}(2-1)} = 2.8 \approx 3. \qquad (6.14)$$

With local gradients of the momentum injection profiles given by

$$P_t' = \frac{\partial}{\partial r}\left(\frac{P_t}{m_i n_i}\right) \quad \text{and} \quad P_p' = \frac{\partial}{\partial r}\left(\frac{P_p}{m_i n_i}\right), \qquad (6.15)$$

one may model the amount of flow shear required to produce an ion transport barrier (ITB). The nonlinear dynamics of the coupled heating and flow shear are given by the four coupled ODEs:

$$\frac{dW}{dt} = 2\gamma_0\left[(\mu - \mu_c) + c_1 F_{\parallel} - c_2 F_{\perp} - \gamma^{n\ell}W\right]W \qquad (6.16)$$

$$\frac{d\mu}{dt} = -\chi_x W_{\mu} + P_E' \qquad (6.17)$$

$$\frac{dF_{\perp}}{dt} = -\nu^{\text{nc}}F_{\perp} + \bar{c}_2 W F_{\perp} + \sqrt{F_{\perp}}\,P_{\perp}' \qquad (6.18)$$

$$\frac{dF_{\parallel}}{dt} = -\bar{c}_1 W F_{\parallel} + \sqrt{F_{\parallel}}\,P_{\parallel}'. \qquad (6.19)$$

A typical model for the NBI power deposition profile is given by

$$P_{bi}(r) = \frac{P_T n \sigma}{(2\pi r)(2\pi r)\sin\theta} \left[e^{-(a-r)n\sigma/\sin\theta} + e^{-(r+a)n\sigma/\sin\theta} \right] \tag{6.20}$$

where θ is the angle of injection with tangential injection having $\theta_{\min} \simeq (a/R)^{1/2}$ and σ is the charge exchange cross-section at the neutral beam injection energy. The condition of depositing the power in the core plasma puts a narrow range on $a/\lambda_{\text{eff}} = an\sigma(E_0)/\sin\theta$ so that we can estimate $a/\lambda_{\text{eff}} \approx 2$ in calculations with Eq. (6.20). Using this estimate for $\lambda_{\text{eff}} \simeq a/2$ yields the deposit heating power from Eq. (6.20) is

$$\frac{P_{\text{tot}}}{(2\pi R)(\pi ra)} \simeq \frac{20 \times 10^6 \,\text{Watts}}{2 \times 10^7 \,\text{cm}^3} \simeq 1\,\text{MW/m}^3.$$

The ion heating rate at $n = 3 \times 10^{13}\,\text{cm}^{-3}$ is calculated as

$$\frac{2P_{bi}}{3n_i} = 150\,\text{keV/s}. \tag{6.21}$$

For JT-60U this is sufficient to produce the internal transport barrier (ITB) found in the experiments.

6.3.2 *Computer simulations for interaction of the zonal flows and the drift wave turbulence*

Plasma turbulence is controlled by the local zonal flows as shown by the predator-prey model in the previous section, 6.3.1. Extensive toroidal computer simulations have been reported for the interaction of the turbulence and the zonal flows [Dimits, et al.(2000); Garbet, et al. (2010); Watanabe, et al. (2007)]. Experimental studies confirm some of the predictions of the theories [Rhodes, et al. (2011); Tanaka, et al. (2010)].

A detailed comparison between the data obtained with phase contrast imaging (PCI) and 3D gyrokinetic Vlasov simulations is made by the team of Nunami and Tanaka on the Large Helical Device (LHD) in a variety of quasi-steady-state discharges. The simulations that shaped magnetic flux surfaces of the LHD machines for high ion temperature gradient discharges in the ITG turbulence is strong. The helical machines have more control and detailed data on the radial electric field $E_r, (r,t)$ profiles than in tokamaks. Thus, these comparisons of simulations and data in the LHD system are considered as validation studies of the zonal flow-drift wave turbulence modes of internal transport barriers.

The strong correlations between the turbulence and zonal flows is demonstrated by introducing the energies $\tau(t)$ of the turbulence and the average zonal flow potential $Z(t)$ [Nunami, et al. (2012)] and constructing the Lissajous figures for $Z(t)$ versus $\tau(t)$. Clear correlations are founded giving a validation of the models of the control of turbulence with sheared zonal flows through the correlations developed in theory and simulations.

6.4   ELM Control with Resonant Magnetic Perturbation

Experiments reporting the results of transport barriers controlled through the externally-induced resonant-magnetic perturbations (RMPs) are reported in toka-mak DIII-D [Evans, *et al.* (2004); Burrell, *et al.* (2005)] and confirmed by JET [Liang, *et al.* (2007)] and TEXTOR [Finken, *et al.* (2007)].

The current I_{RMP} in the external coils is increased to find the values required to trigger small bursts of stored core energy without degrading the net energy confine-ment time excessively. In the next two subsections this research is described in some detail since the problem of controlling the large ELMs is one of high priority now for envisioning steady-state operation of a tokamak. The RMP coils are shown to induce a layer of stochastic magnetic fields near the last resonant magnetic surface, typically $q(r_{m/l}) = 3$ in the tokamak. To induce the stochastic magnetic field at $q = 3$, one applies external coils with currents flowing in the pattern of the topology of mode numbers $m = 12$ and $n = 4$. To give a width to the stochastic magnetic field layer, a typical design is to use external coils with toroidal mode number $n = 4$ and a set of poloidal mode numbers $m = 10, 11, 12, 13$ and 14. In the next section we follow the analysis of Beyer, *et al.* (2011) and Marcus, *et al.* (2014) for how the currents I_{RMP}, in such a set of external coils, changes the plasma transport in the vicinity of the rational surface defined by $q(r) = 3$.

The external coils mounted on the plasma wall create a radial magnetic pertur-bation $\delta B_r(\mathbf{x})$ inside the plasma that exerts a torque on the plasma [Fitzpatrick (1995)]. More generally, error magnetic fields created from external coil misalign-ments and large openings in the metallic walls of the vessel chamber create magnetic perturbations $\delta B_r(\boldsymbol{x})$ that are amplified at the low-order rational magnetic surfaces and lead to rotational torques. The website http://farside.ph.utexas.edu has a collection of works of Fitzpatrick on the physics of the plasma structures created by error magnetic fields. The description in terms of an analogy with the induction electric motor in a conference proceedings titled *Bifurcated states of rotation in the presence of error-fields, 1997* is particularly useful. The error fields create torques on the plasma from the currents in the metallic chamber walls. The problem is com-plicated, so we restrict the discussion here to the magnetic perturbations designed specifically to control the ELMs in the tokamak.

Of course, the application of the RMP coils is such as to reduce the plasma confinement just inside the plasma edge, defined by the last close magnetic surface. Just outside this last closed magnetic surface the magnetic field lines spiral around the chamber ending up terminating on the plasma walls and the divertor plate installed, so as to collect most of the plasma crossing the magnetic separatrix, channeling or diverting this hot plasma into special chambers with extra cooling and tungsten, which has the highest melting-point temperature. By use of the RMPs and special divertor chambers, the tokamak may operate in a steady state with the maximum plasma pressure allowed by the plasma current I_p and the size

R and a of the machine. This limit on the pressure with ideal conditions is called the Troyon limit and given in Chapter 1.

Turbulence simulations of transport barrier relaxations at the tokamak plasma edge have revealed that the control of such relaxations by RMPs is attributed to a local erosion of the barrier [Leconte, *et al.* (2009); Leconte, *et al.* (2010); Beyer, *et al.* (2011)]. This erosion at the resonance position is known to be caused, at least partly, by the enhancement of the radial heat flux in presence of the RMP due to the strong parallel heat flow along perturbed field lines, producing the so-called magnetic flutter flux [Fitzpatrick (1995); Callen (1977)]. However, in the presence of magnetic curvature, an additional transport mechanism exists and is linked to stationary convection cells associated with the magnetic islands induced by RMPs [Leconte, *et al.* (2009); Leconte, *et al.* (2010)]. In the presence of a mean poloidal velocity shear, this additional convective thermal transport can be considerably high, higher than than the thermal flux from the magnetic flutter flux [Beyer, *et al.* (2011)].

Two different equilibrium plasma states exists in the presence of an RMP. The first regime is characterized by the absence of mean-poloidal flow and a low level of convective thermal transport. The second regime shows mean-poloidal rotation and large-convective transport. Marcus, *et al.* (2014) show that these two equilibria depend on magnetic curvature and RMP amplitude. The simple cylindrical equilibrium without mean-poloidal rotation is found to be unstable, such that the plasma evolves self-consistently to the poloidally-rotating state where large convective transport is present.

Fitzpatrick (1995) analyzes the effect of static, externally generated, multi-harmonic, helical magnetic perturbation on H-mode plasmas. The models show that non-resonant harmonics from external coils give rise to significant toroidal flow damping in the pedestal and modify the poloidal flow damping. The resulting neoclassical ion flow causes a helical phase-shift to develop between the locked island chain and the resonant harmonic from the external perturbation. When the phase-shift exceeds a critical value, the chain unlocks from the resonant harmonic and starts to rotate, after which it decays away and is replaced by a helical current sheet. Above a critical amplitude of the resonant harmonic from the external perturbation, the island chain either unlocks or becomes unstable increasing the cross-field transport.

Analysis of the new helical equilibrium with rotation and the change in the plasma transport is described further in Chapter 7.3.

An extensive diagnostic system will be installed on the ITER to provide the measurements necessary to control, evaluate and optimize plasma performance in ITER which will further the understanding of plasma physics. These diagnostics include measurements of temperature, density, impurity concentration, and particle and energy confinement times. The system will comprise about 50 individual measuring systems drawn from the full range of modern plasma diagnostic tech-

Table 6.1 Parameters of Referenced Tokamaks in Historical Order

Machine	Years	R(m)	a(m)	B(T)	I(MA)	Divertor	Auxiliary Power
TFR	1975	1.5	0.37	4.5	0.7	no	1 MW ECRH
T-10	1973	1.0	0.2	6.0	0.5	no	0.7 MW NBI
PLT	1975	1.3	0.4	3.5	0.7	no	3 NBI 5 ICRH 1 LH
Alcator C	1979	0.64	0.17	12	0.9	no	4 LH
ASDEX	1980	1.5	0.4	3.0	0.5	yes	4.5 NBI 3 ICRH 2 LH
DIII	1980	1.5	0.5	2.6	0.6	yes	7.0 NBI 2.0 ECRH
TFTR	1982-1990	2.4	0.8	5.0	2.2	no	40 NBI 11 ICRF
JET	1989	3.0	1.25	3.5	7.0	yes	20 NBI 20 ICRF 7 LH
DIII-D	1986	1.67	0.67	2.1	1.0	yes	20 NBI 8 ICRF 2.0 ECRH
Tore Supra	1988	2.4	0.8	4.5	2.0	ergodic	12 ICRH 8 LH
FT-U	1990	1.0	0.3	8.0	1.3	no	2.0 ICRH 9.0 LH 1.0 ECRH
ASEDEX-U	1991	1.7	0.5	3.9	1.4	yes	10 NBI 6 ICRH 0.5 ECRH
JT-60U	1991	3.4	1.1	4.2	5.0	yes	40 NBI 7.0 ICRH 8.0 LHCD
TCV	1992	0.9	0.24	1.4	1.2	no	4.5 ECR
Alcator C-Mod	1993	0.67	0.22	9	1.1	yes	4.0 ICRF X LHCD
EAST	2006	1.7	0.4	3.5	1.0	yes	3.0 ICRF 3.5 LHR 0.5 ECH
KSTAR	2009	1.8	0.5	3.5	2.0	yes	14 NBI 6.0 ICRF 3.0 LH 4.0 ECR
ITER	2020	6.2	2.0	5.3	15	yes	50 NBI 40 ICRH 40 LH 20 ECH

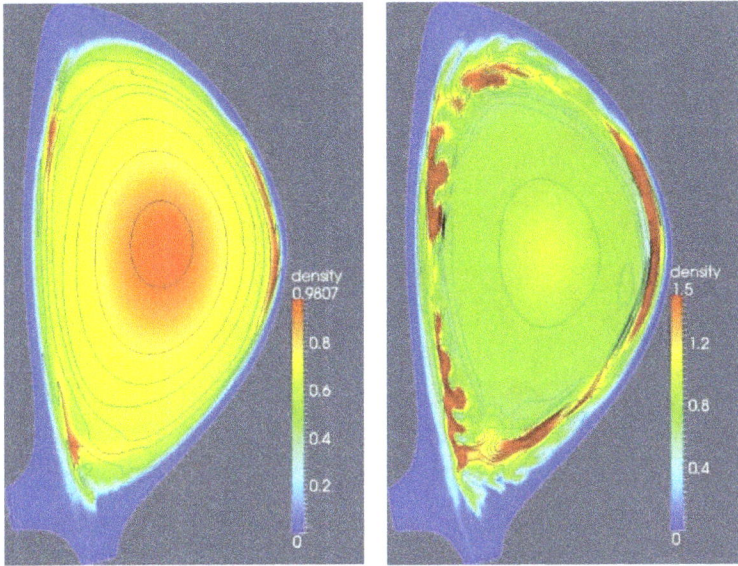

Fig. 6.1 A complex set of external coils described in Chapter 7 may be required for control of both the ELMs and the vertical displacement modes [VS] is shown in Loarte, *et al.* (2007).

niques, which includes lasers, X-rays, neutron cameras, impurity monitors, particle spectrometers, radiation bolometers, pressure and gas analysis, and optical fibers.

Because of the harsh environment inside the vacuum vessel, these systems will have to cope with a range of phenomena not previously encountered in diagnostic implementation, while all the while performing with accuracy and precision. The levels of neutral particle flux, neutron flux and fluence will be respectively about 5, 10 and 10,000 times higher than the harshest experienced in today's machines listed in Table 6.1 for comparison with ITER. The pulse length of the fusion reaction – or the amount of time the reaction is sustained – will be about 100 times longer than those of JET.

6.5 ELM Control with Pellet Injection

Edge localized modes (ELMs) may be controlled by using pellet injection to trigger the release of their energy at a predetermined level. Extensive simulations of the ability to use pellets to trigger the ELM release are reported in Futatani, *et al.* (2014). The simulations are benchmarked with data from DIII-D. The modeling shows that the key parameter for the triggering of ELMs by pellets is the value of the localized pressure perturbation, which leads to a threshold size for the pellets for a given injection velocity. The minimum pellet size for ELM triggering depends on injection geometry with the largest value being required for injection from the outer midplane, intermediate size pellets for injection near the X-point and the

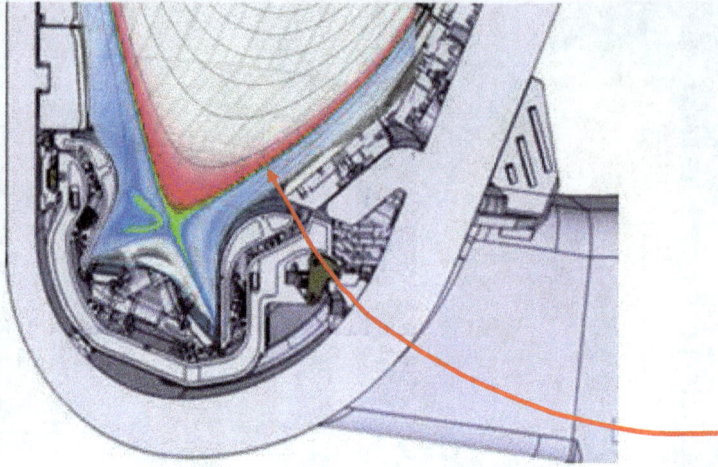

Fig. 6.2   Diagram showing the path of the pellet injection near the divertor chamber for triggering the early release of the ELM energy. This configuration appears as the most practical pellet injection configuration. This method of ELM control is called pellet pacing.

smallest pellets for injection from the high-field side. For ELM triggering in ITER at 15 MA with $Q_{DT} = 10$ the pellet size for ELM triggering from injection near the X-point and on the inner high magnetic field size of the plasma is shown to be a practical ELM control method. The pellet injection modeling was carried using an ITER-like discharge from DIII-D (shot no. 131498, $q_{95} = 3.5$, $\beta_N = 1.8$, $H_{98} = 1.1$). The edge transport barrier leads to the H-mode pedestal plasma that is reproduced in the JOREK simulations by a suitable choice of the radial dependence of the diffusion coefficients with reduced transport coefficients in the H-mode pedestal. As expected, the JOREK simulation shows that the edge plasma profiles are unstable to ballooning modes centered at $n = 10$ modes, leading to a nonlinear ELM as shown in Fig 6.3.

The JOREK simulations show that the ablation of the pellet leads to an earlier growth of the MHD activity for the $n = 1$ to 10 toroidal modes. The amplitude of magnetic energy increases with pellet size. In addition, for pellets above a critical pellet size (1.5 mm for these modeling conditions), a strong increase of the energy of the $n = 6$ harmonics is observed in the simulations. The results of the simulations show that pellet injection is an effective method to stop the growth of the large energy ELMs by causing them to go nonlinear earlier releasing their energy well before growing to their natural size.

Modeling of the triggering control of ELMs by pellet injection has identified the minimum pellet size required for stopping the growth of the ELMs before they reach the level of 5%-10% of the plasma energy as occurs in uncontrolled large ELMs when pushing the system to obtain the highest fusion Q values. The required fuel

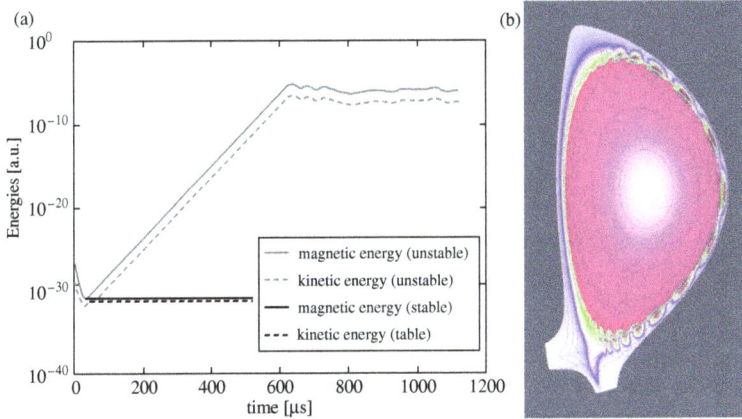

Fig. 6.3 (a) The energy of the $n = 10$ toroidal harmonic as a function of time for the pedestal at 100% of the ELM cycle and for the stable pedestal pressure at 70%. (b) Modeled plasma density and flow contours (lines) during an uncontrolled or natural ELM in the ITER-like DIII-D plasma with the plasma profiles.

throughput for an effective pellet control technique compatible with the installed fueling and tritium re-processing capabilities in ITER. The evaluation of the capabilities of the ELM control coil system in ITER for ELM suppression has a factor of ~ 2 margin in terms of coil current to achieve its design criteria. The consequences for the spatial distribution of the power fluxes at the divertor of ELM control with the RMP fields lead to substantial toroidal asymmetries in zones of the divertor target away from the separatrix. Therefore, specifications for the plasma rotation with the RMP coils may be required to control the magnitude of the localized erosion of the ITER divertor surfaces. Loarte, *et al.* (2014) argue that a plasma rotation frequency exceeding of 1 Hz is required in order to avoid unacceptable thermal cycling of the divertor target plates for the highest power fluxes. The possible use of the in-vessel vertical stability coils for ELM control as a back-up to the main ELM control systems in ITER is described in Loarte, *et al.* (2014).

The recycling of hydrogenic fuel and its retention in plasma facing materials considered for next-step devices is an active area of research in all fusion devices. For ITER this research is particularly important for determining the considerable amounts of tritium that potentially could be retained in the plasma facing components.

Past analysis has shown that the number of hydrogenic atoms in the surface of plasma facing components that interact with the plasma exceeds in general the total number of ions in the plasma. This leads to a complex coupled system between the hydrogenic species at the wall and the plasma fuel inventory. The recycling and retention behavior at the walls may strongly affect the fueling of the plasma while the plasma changes the composition of the wall surface [Loarte, *et al.* (2007); Loarte, *et al.* (2014)] describe the mechanisms that cause the retention of fuel in

Table 6.2   Parameters from the tritium experiments carried out in TFTR and JET extrapolated to ITER.

TFTR	JET	ITER
Parameter T-experiment	T-experiment and requirements	Tritium in-vessel inventory limit
2 g 20 g site inventory	350 g (guideline value) 1000 g	
T introduced in the torus		
5.2 g	35 g	50 g/pulse (400 s)
Discharge number and duration	708 pulses ≈ 33 min	500 pulses ≈ 250 min
Tritium retained in the torus before cleaning	1.7 g	11.5 g
Fraction of tritium removed by cleaning 50%	50%	100%
Tritium retained after cleaning	0.85 g	6 g

Table 6.3   Parameters of the balance of tritium for the experiments carried out in TFTR and JET and foreseen for ITER.

TFTR	JET	ITER projections
Parameter	T-experiment	T-experiment and requirements
Tritium in-vessel inventory limit	2 g 20 g site inventory	350 g (guideline value)
		1000 g (assumption for safety evaluation)
T-introduced in the torus	Discharge number and duration	~ 33 min, 500 pulses ~ 250 min
5.2 g, 35 g, 50 g/pulse (400 s)	708 pulses	

fusion devices are as follows.

(i) Direct implantation of ions and neutrals in a shallow surface layer and possible diffusion into the bulk. For this mechanism, the wall inventory is determined by material properties, diffusion and recombination coefficients and the concentration/strength of trapping sites. The fuel is retained temporarily ('dynamic retention') or quasi-permanently by solution and/or trapping. The amount of fuel retained by these mechanisms tends to saturate with time and/or plasma fluence, as the surface layer concentrations reach steady-state values and the trapping sites are filled.

(ii) The production of tritium by transmutation in the nuclear reactions. This occurs only for special wall materials (e.g. beryllium) and can cause the build up of a tritium inventory within the bulk material by trapping in microscopic defect sites and helium bubbles.

(iii) The co-deposition of hydrogen with eroded wall material forming hydrogen-rich co-deposits inside the fusion device. The rate of retention by this mechanism depends on: (a) the rate of erosion of the material and (b) the fuel retention capability of the re-deposited material. This retention mechanism is only relevant for few materials (e.g. carbon) in a certain temperature range. For example, diamond-like amorphous carbon films show retention (H/C ratio) of about 0.4 at room temperature, decreasing significantly only above $1000\,K$. On the other hand soft carbon films show a ratio $H/C > 1$ at room temperatures, which also decreases rapidly with increasing temperature. The amount of fuel retained by this mechanism does not appear to saturate in time. Thus, the mechanism raises more concern with respect to fuel retention in next-step devices.

For two D-T tokamaks the analysis of the tritium retention in walls is shown in Table 6.3. The last column gives the estimates for tritium retention in ITER.

6.5.1 *Database on fuel retention in present fusion devices*

The majority of the fuel retention data comes from fusion devices with carbon PFCs. The amount of fuel retention is determined either by comparing the fuel injected to the exhaust ('gas balance') or by ex situ analysis of wall tiles. The overall long-term deuterium fuel retention has been determined at JET by post-mortem surface analysis. These results show a large amount of hydrogen-rich carbon deposits at the inner divertor. The total amounts of carbon deposited at the inner divertor target are 900 and 500 g in the divertor S234.

References

Batista, A. M., Caldas, I. L., Lopes, S. R., Viana, R. L., Horton, W., and Morrison, P. J. (2006). Nonlinear three-mode interaction and drift-wave turbulence in a tokamak edge plasma, *Phys. Plasmas* **13**, p. 042510, http://dx.doi.org/10.1063/1.2184291.

Beklemishev, A. D. and Horton, W. (1992). *Phys. Fluids B* **4**, p. 200.

Belli, E. A., and Candy, J. (2008). Kinetic calculation of neoclassical transport including self-consistent electron and impurity dynamics, *Plasma Phys. Control. Fusion* **50**, p. 095010, doi:10.1088/0741-3335/50/9/095010.

Beyer, P., de Solminihac, F., Leconte, M., Garbet, X., Waelbroeck, F. L., Smolyakov, A. I., and Benkadda, S. (2011). Turbulence simulations of barrier relaxations and transport in the presence of magnetic islands at the tokamak edge, *Plasma Phys. Control. Fusion* **53**, p. 054003, doi:10.1088/0741-3335/53/5/054003.

Burrell, K. H., Evans, T. E., Doyle, E. J., Fenstermacher, M. E., Groebner, R. J., Leonard, A. W., Moyer, R. A., Osborne, T. H., Schaffer, M. J., Snyder, B. P., Thomas, P. R., West, W. P., Boedo, J. A., Garofalo, A. M., Gohil, P., Jackson, G. L., La Haye, R. J., Lasnier, C. J., Reimerdes, H., Rhodes, T. L., Scoville, J. T., Solomon, W. M., Thomas, D. M., Wang, G., Watkins, J. G., and Zeng, L. (2005). ELM suppression in low-edge collisionality H-mode discharges using $n = 3$ magnetic perturbations, *Plasma Phys. Control. Fusion* **47**, p. 12B, doi:10.1088/0741-3335/47/12B/S04.

Callen, J. D. (1977). Drift-wave turbulence effects on magnetic structure and plasma transport in tokamaks, *Phys. Rev. Lett.* **39**, pp. 1540-1543, http://link.aps.org/doi/10.1103/PhysRevLett.39.1540.

Connor, J. W., and Wilson, H. R. (2000). A review of theories of the L-H transition, *Plasma Phys. Control. Fusion* **42**, p. R1-R75, doi:10.1088/0741-3335/42/1/201.

Diamond, P. H., and Kim, Y-B. (1991). Theory of mean poloidal flow generation by turbulence, *Phys. Fluids B* **3**, p. 1626, http://dx.doi.org/10.1063/1.859681.

Dimits, A. M., Bateman, G., Beer, M. A., Cohen, B. I., Dorland, W., Hammett, G.W., Kim, C., Kinsey, J. E., Kotschenreuther, M., Kritz, A. H., Lao, L. L., Mandrekas, J., Nevins, W. M., Parker, S. E., Redd, A. J., Shumaker, D. E., Sydora, R., and Weiland, J. (2000). *Phys. Plasmas* **7**, p. 969, http://link.aip.org/link/doi/10.1063/1.873896.

Evans, T. E., Moyer, R. A., Thomas, P. R., Watkins, J. G., Osborne, T. H., Boedo, J. A., Doyle, E. J., Fenstermacher, M. E., Finken, K. H., Groebner, R. J., Groth, M., Harris, J. H., La Haye, R. J., Lasnier, C. J., Masuzaki, S., Ohyabu, N., Pretty, D. G., Rhodes, T. L., Reimerdes, H., Rudakov, D. L., Schaffer, M. J., Wang, G., and Zeng, L. (2004). Suppression of Large Edge-Localized Modes in High-Confinement DIII-

D Plasmas with a Stochastic Magnetic Boundary, *Phys. Rev. Lett.* **92**, p. 235003, http://dx.doi.org/10.1103/PhysRevLett.92.235003.

Evstatiev, E. G., Horton, W., and Morrison, P. J. (2003). Multi-wave model for plasma-wave interaction, *Phys. Plasmas* **10**, p. 4090, http://dx.doi.org/10.1063/1.1609989.

Finken, K. H., Unterberg, B., Xu, Y., Abdullaev, S. S., Jakubowski, M., Lehnen, M., de Bock, M. F. M., Bozhenkov, S., Brezinsek, S., Busch, C., Classen, I. G. J., Coenen, J. W., Harting, D., von Hellermann, M., Jachmich, S., Jaspers, R. J. E., Kikuchi, Y., Krämer-Flecken, A., Liang, Y., Mitri, M., Peleman, P., Pospieszczyk, A., Reiser, D., Reiter, D., Samm, U., Schega, D., Schmitz. O. (2007). Influence of the dynamic ergodic divertor on transport properties in TEXTOR, *Nucl. Fusion* **47**, 7, p. 522, doi:10.1088/0029-5515/47/7/004.

Fitzpatrick, R. (1995). Helical temperature perturbations associated with tearing modes in tokamak plasmas, *Phys. Plasmas* **2**, p. 825, http://dx.doi.org/10.1063/1.871434.

Garbet, X., Idomura, Y., Villard, L., and Watanabe, T. H. (2010). Gyrokinetic simulations of turbulent transport, *Nucl. Fusion* **50**, p. 043002, doi:10.1088/0029-5515/50/4/043002, stacks.iop.org/NF/50/043002.

Grierson, B. A., Burrell, K. H., Solomon, W. M., Budny, R. V., and Candy, J. (2013). Collisionality scaling of main-ion toroidal and poloidal rotation in low-torque DIII-D plasmas, *Nucl. Fusion* **53**, p. 063010, doi:10.1088/0029-5515/53/6/063010.

Hamaguchi, S., and Horton, W. (1992). Effects of sheared flows on ion-temperature-gradient-driven turbulent transport, *Phys. Fluids B* **4**, p. 319, http://dx.doi.org/10.1063/1.860280.

Horton, W., Park, H-B., Kwon, J-M., Strozzi, D., Morrison, P. J., and Choi, D-I. (1998). Drift wave test particle transport in reversed shear profile, *Phys. Plasmas* **5**, p. 3910, http://dx.doi.org/10.1063/1.873110.

Howard, L. N., and Krishnamurti, R. (1986). Large-scale flow in turbulent convection: a mathematical model, *J. Fluid Mech.* **170**, pp. 385-410, http://dx.doi.org/10.1017/S0022112086000940.

Kim, Y. B., Diamond, P. H., and Groebner, R. J. (1991). Neoclassical poloidal and toroidal rotation in tokamaks, *Phys. Fluids B* **2050**, http://dx.doi.org/10.1063/1.859671.

Leconte, M., Beyer, P., Garbet, X., and Benkadda, S. (2010). Effects of resonant magnetic perturbations on the dynamics of transport barrier relaxations in fusion plasmas, *Nucl. Fusion* **50**, p. 054008, doi:10.1088/0029-5515/50/5/054008.

Leconte, M., Beyer, P., Garbet, X., and Benkadda, S. (2009). Control of transport-barrier relaxations by resonant magnetic perturbations, *Phys. Rev. Lett.* **102**, 045006, doi:http://dx.doi.org/10.1103/PhysRevLett.102.045006.

Liang, Y., Koslowski, H. R, Thomas, P. R., Nardon, E., *et al.* (2007). Active control of Type-I edge-localized modes with $n = 1$ perturbation fields in the JET Tokamak, *Phys. Rev. Lett.* **98**, p. 265004, http://dx.doi.org/10.1103/PhysRevLett.98.265004.

Loarte, A., Lipschultz, B., Kukushkin, A. S., Matthews, G. F., Stangeby, P. C., Asakura, N., Counsell, G. F., Federici, G., Kallenbach, A., Krieger, K., Mahdavi, A., Philipps, V., Reiter, D., Roth, J., Strachan, J., Whyte, D., Doerner, R., Eich, T., Fundamenski, W., Herrmann, A., Fenstermacher, M., Ghendrih, P., Groth, M., Kirschner, A., Konoshima, S., LaBombard, B., Lang, P., Leonard, A. W., Monier-Garbet, P., Neu, R., Pacher, H., Pegourie, B., Pitts, R. A., Takamura, S., Terry, J., Tsitrone, E., and the ITPA Scrape-off Layer and Divertor Physics Topical Group (2007). Chapter 4: Power and particle control, *Nucl. Fusion* **47**, pp. S203-S263, doi:10.1088/0029-

5515/47/6/S04.

Loarte, A., Huysmans, G., Futatani, S., Baylor, L. R., Evans, T. E., Orlov, D. M., Schmitz, O., Becoulet, M., Cahyna, P., Gribov, Y., Kavin, A., Sashala Naik, A., Campbell, D. J., Casper, T., Daly, E., Frerichs, H., Kischner, A., Laengner, R., Lisgo, S., Pitts, R. A., Saibene, G., and Wingen, A. (2014). Progress on the application of ELM control schemes to ITER scenarios from the non-active phase to DT operation, *Nucl. Fusion* **54** p. 033007 (18pp), doi:10.1088/0029-5515/54/3/033007.

Marcus, F. A., Beyer, P., Fuhr, G., Monnier, A., and Benkadda, S. (2014). Convective radial energy flux due to resonant magnetic perturbations and magnetic curvature at the tokamak plasma edge, *Phys. Plasmas* **21**, p. 082502, doi:10.1063/1.4891437, http://dx.doi.org/10.1063/1.4891437.

Nunami, M., Watanabe, T.-H., Sugama, H., and Tanaka, K. (2012). Gyrokinetic turbulent transport simulation of a high ion temperature plasma in large helical device experiment, *Phys. Plasmas* **19**, p. 042504, http://dx.doi.org/10.1063/1.4704568.

Peeters, A. G., Angioni, C., Bortolon, A., Camenen, Y., Casson, F. J., Duval, B., Fiederspiel, L., Hornsby, W. A., Idomura, Y., Hein, T., Kluy, N., Mantica, P., Parra, F. I., Snodin, A. P., Szepesi, G., Strintzi, D., Tala, T., Tardini, G., de Vries, P., and Weiland, J. (2011). Overview of toroidal momentum transport, *Nucl. Fusion* **51**, p. 094027, doi:10.1088/0029-5515/51/9/094027.

Pütterich, T., Wolfrum, E., Dux, R., Maggi, C. F., and ASDEX Upgrade Team (2009). Evidence for strong inversed shear of toroidal rotation at the edge-transport barrier in the ASDEX Upgrade, *Phys. Rev. Lett.* **102**, p. 025001 and *J. Nucl. Mater.* **415**, p. S334-9, http://dx.doi.org/10.1103/PhysRevLett.102.025001.

Rhodes, T. L., Holland, C., Smith, S. P., White, A. E., Burrell, K. H., Candy, J., DeBoo, J. C., Doyle, E. J., Hillesheim, J. C., E. Kinsey, J., McKee, G. R., Mikkelsen, D., Peebles, W. A., Petty, C. C., Prater, R., Parker, S., Chen, Y., Schmitz, L., Staebler, G. M., Waltz, R. E., Wang, G., Yan, Z., and Zeng, L. (2011). L-mode validation studies of gyrokinetic turbulence simulations via multiscale and multifield turbulence measurements on the DIII-D tokamak, *Nucl. Fusion* **51**, 6, p. 063022, doi:10.1088/0029-5515/51/6/063022.

Rogister, A. (1994). Revisited neoclassical transport theory for steep, collisional plasma edge profiles, *Phys. Plasmas* **1**, 3, p. 619, http://dx.doi.org/10.1063/1.870807.

Tanaka, K., Michael, C., Vyacheslavov, L., Funaba, H.,Yokoyama, M., Ida, K., Yoshinuma, M., Nagaoka, K., Murakami, S., Wakasa, A., Ido, T., Shimizu, A., Nishiura, M., Takeiri, Y., Kaneko, O., Tsumori, K., Ikeda, K., Osakabe, M., Kawahata, K., and the LHD Experiment Group. (2010). Turbulence Response in the High Ti Discharge of the LHD, *Plasma Fusion Res.* **5**, p. S2053, http://dx.doi.org/10.1585/pfr.5.S2053.

Viezzer, E., Pütterich, T., Angioni, C., Bergmann, A., Dux, R., Fable, E., McDermott, R. M., Stroth, U., Wolfrum, E., and the ASDEX Upgrade Team (2014). Evidence for the neoclassical nature of the radial electric field in the edge transport barrier of ASDEX Upgrade, *Nucl. Fusion* **54**, p. 012003, doi: 10.1088/0029-5515/54/1/012003.

Watanabe,T.-H., Sugama, H., and Ferrando-Margalet, S. (2007). Gyrokinetic simulation of zonal flows and ion temperature gradient turbulence in helical systems, *Nucl. Fusion*, **47**, 9, p. 1383, doi:10.1088/0029-5515/47/9/041.

Wolfrum, E., Burckhart, A., Fischer, R., Hicks, N., Konz, C., Kurzan, B., Langer, B., Pütterich, T., Zohm, H., and the ASDEX Upgrade Team. (2009). Investigation of inter-ELM pedestal profiles in ASDEX Upgrade, *Plasma Phys. Control. Fusion* **51**, p. 124057, doi:10.1088/0741-3335/51/12/124057.

Chapter 7

Steady-State Operation

The tokamak is a current-confinement device in that the magnetic field produced by the large toroidal current carried by the plasma produces the magnetic field component that confines the particle orbits in a toroidal vessel. The toroidal magnetic field is required to allow the plasma with a large plasma current I_p to remain stable. Once the plasma is sufficiently hot a fraction of the toroidal current – called the bootstrap fraction – is generated by the pressure gradient of the plasma itself. The forming of the plasma and external control of the plasma current is carried out by using radio frequency waves called RF heating and current drive.

The most efficient RF for current drive has proven to be with lower-hybrid waves in the GHz frequency range launched from antennas as shown in Fig. 7.1. These types of RF antennas are called grills. The grill shown in Fig. 7.1 is compose of an array of 16×6 waveguides powered by Klystrons with the power phased between the rows of neighboring waveguides such as to direct the waves so that their toroidal momentum is transferred to the electrons carrying the toroidal plasma current. Thus, the current can be controlled and maintained by controlling the RF wave antenna. There are currently several ongoing Exploratory Plasma Research programs developing other methods to maintain the plasma confinement current but these exploratory methods – such as the helicity injection experiments at the University of Washington and the University of Wisconsin – require further development. These exploratory methods use external coils to inject plasma with magnetic helicity that combines through reconnection dynamics with the toroidal plasma current. The helicity inject systems are complex but show efficiency in the power required to drive the toroidal plasma current. These research programs are called exploratory and discovery plasma research in the US Department of Energy Office of Plasma Science. The Wisconsin University experiment is related to the reversed field pinch plasma and the University of Washington experiment is called helicity injection into a tokamak (HIT-SI). NSTX has a helicity injection research program. Since these methods are not in the ITER project, we do not describe the physics here.

The RF current drive method has been successfully demonstrated on Tore Supra, Alcator C-Mod, and recently on EAST to show that the plasma current can be

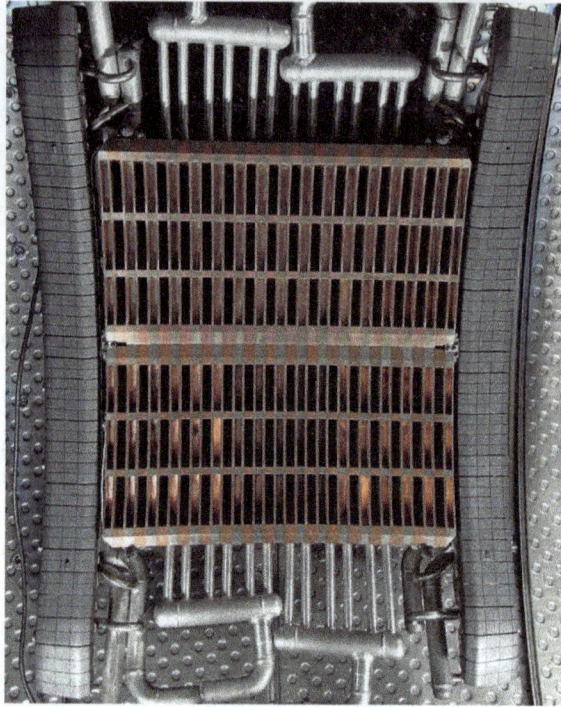

Fig. 7.1   The LHCD phased array antenna designed for Tore Supra with one $TE_{1,0}$ mode and two TM modes launching 3.7 GHz traveling waves. The spectral distribution is determined by the 16×6 phased array wave-guides shown in the matrix. The antenna runs steady state at megawatt power levels with cooling provided by the water tubes seen at the top and bottom of the grill with the vertical tubes running from the four horizontal chambers – two above and two below the antenna grill. The individual wave-guide openings are 14.6 mm wide in the toroidal direction and 76 mm high in the poloidal direction. The spectrum of the emitted waves is centered at $k_\theta = 0$ and peaked in $k_\parallel$ with the $n_\parallel$ spectrum centered at -1.7. The minus sign denotes that the waves propagate opposite to the toroidal plasma current and thus parallel to the electron fluid velocity $u_\parallel(r, t)$ producing the toroidal plasma current.

driven and controlled for long periods. The transformer action of the pulsed toka-maks has current driving time limited by the flux in Webers [1 Wb = 1 Volt × 1 sec] that the central solenoid can provide. After this time the large toroidal current I_p is driven by the RF waves.

In the EAST tokamak the ELMy H-mode regime was obtained with 1 MW of LHW power. For the H-mode at the 1 MW of LHW level required extensive Li wall coating by evaporation and Li powder injection. The threshold power for H-mode access follows the international tokamak scaling even in the low density range and a threshold in density was identified in EAST. This new result has a favorable implication for ITER since ITER will operate at marginal power levels at the beginning of its operation. EAST H-mode results demonstrate that LHW is a promising heating alternative for H-mode access at low-threshold power, providing steady-state capability and compatibility with advanced tokamak scenarios.

In this chapter the theory and experiments for driving and controlling the toroidal plasma current is reviewed. The plasma physics issues are discussed in terms of the closed-loop dynamics of driving the current and maintaining optimal current profiles for the reactor operation. Without a feedback loop on the RF current drive, the plasma can enter into unstable regimes or the gradients of the plasma can drive high-frequency waves, increasing the small space-scale electron driven drift wave plasma turbulence, producing radial diffusion of the thermal electrons, and spreading of the current profiles. With sufficiently strong LHCD there are bifurcations in the coupled RF driven waves and the turbulent transport dynamics resulting in bifurcations to an improved confinement state from reversed magnetic shear. The natural plasma turbulence produces stochastic scattering of the high-frequency rays which makes the prediction of the distribution of the rays and the associated wave-heating a statistical problem. Thus, one introduces a Fokker-Planck equation for the probably distributions of the RF rays of the lower-hybrid waves. The general frame work of the coupled system of coupled high-frequency current driving rays with the low-frequency turbulent transport determines the profiles of density and temperatures. This complex dynamics is discussed and modeled with a low-order dynamical system of ordinary differential equations for real-time control.

The use of high-frequency electromagnetic waves launched from antennas outside the plasma is a mature, well-developed discipline in terms of the design and the application to control high-temperature laboratory plasmas. Extensive theoretical-simulation modeling have been carried out and used to compute the expected changes in the ion and electron distributions functions. For fusion plasmas the most efficient and thoroughly-explored, high-frequency electromagnetic control methods are (1) the lower-hybrid (LH) waves in the centimeter wavelength range and (2) the 2nd and 3rd harmonic electron cyclotron (EC) waves in the millimeter wavelength range. Both methods have shown the ability to heat and shape the electron velocity distribution functions so as to control the toroidal plasma current. Proof-of-Principle experiments were carried out in the 1980s showing that externally-injected lower-hybrid waves in the gigahertz frequency range could maintain the tokamak current without the induction electric field. The recent results in the period from 2000 to 2015 are described. The issues pertaining to the use of these two plasma current control and heating methods for driving a steady state tokamak reactor with the parameters of ITER are described. The extrapolation to the larger DEMO system under design is an important open research problem.

7.1 Neoclassical Bootstrap Current

Once the plasma is sufficiently hot and collisionless, in the sense that a substantial fraction of the electrons are mirrored by the increase of the magnetic field strength as they travel from the outside of magnetic surface, where $B = B_{min}$ to the inside of the torus, where $B = B_{\mathrm{max}}$, then collisional friction between these mirror trapped electrons and the passing electrons gives rise to a pressure gradient driven toroidal current called the bootstrap current. The name bootstrap describes the fact that the plasma must first obtain the high-temperature state for the collisionality measure given by ν_e^* is to be small. The ν_e^* defined in Chapter 1, and repeated below in more detail, measures the amount of collisional scattering of a thermal electron in one bounce period as it is mirrored by the $R_M = B_{\mathrm{max}}/B_{\mathrm{min}}$. Once the plasma is into the regime of low collisionality as measured by $\nu_e^* \ll 1$, then a substantial toroidal plasma current is generated by the electron density and pressure gradients. This boot strap current vanishes on the magnetic axis where there is no mirroring of trapped particles.

One measure of the success of a tokamak regime is the degree to which the plasma is into this low collisionality "banana regime". The name is used since the guiding center motion of the charged particles traces out the shape of a banana in the regime where the particles are mirrored by the $R_M = B_{\mathrm{max}}/B_{\mathrm{min}} > 1$ increasing magnetic field.

To what extent the quasi-steady-state tokamak can be maintained by the bootstrap current is unknown and will be a major contribution from the ITER experiments to the design of the DEMO tokamak. Clearly, a degree of external toroidal current control will be required from RF waves. There are a series of bi-annual international meetings on the RF heating and control of plasmas with the proceedings of the 2013 meeting [Radio-Frequency (2013)]. The current drive methods were first demonstrated in the 1980s as described in the 1985 Proceedings of Radio Frequency Plasma Heating (AIP) [Swanson (1985)]. The 1985 Proceedings contains nineteen reports on experiments, theory and simulations of lower-hybrid heating and current drive. Those reports demonstrated that lower-hybrid antennas fed by waveguides could efficiently deliver 30 GHz, 100 kW power RF waves propagating approximately along the magnetic field that would shape the electron velocity distribution functions so as to provide high-efficiency heating and plasma current control. The experiments on Petula-B and Alcator both produced steady state periods where the toroidal plasma current was maintained by the LH waves and the Ohmic heating induction voltage dropped to negligible levels. Similar reports on the PLT and FT tokamak experiments also showed the feasibility of maintaining the tokamak plasma with lower-hybrid current drive (LHCD). In addition, the 1985 AIP Conference Proceedings contains nine contributions on electron cyclotron heating (ECH) experiments and theory with antennas launching millimeter wavelength (100-170 GHz) electromagnetic waves driving up the perpendicular electron energies for a wide range of plasmas including PLT and plans for TFTR [Mazzucato, *et al.* (2008); Mazzucato, *et al.* (2009); Mazzucato (2010)]. The effects of the high-power heating on confinement and transport are analyzed by Kritz, *et al.* (2002).

Now, we know from the current drive experiments in the 2000s, as described for the experiments in Tore Supra [Peysson, *et al.* (2011); Goniche, *et al.* (2010); Becker

(1990)], Alcator C-MOD [Bonoli, *et al.* (1988)], JET, and JT-60U how to apply high-frequency current drive and heating with high-efficiency in large, fusion-grade laboratory plasmas. Still there remain issues, discussed in this chapter, regarding the optimal design choices for the heating and current drive mechanisms for the large ($R = 6$ m, $a \simeq 2$ m, $B_T = 5$ T, $I_p = 15$ MA) scale ITER and DEMO plasmas.

An ultra-high field extension of the Alcator experiment named IGNITOR was proposed to produce a short – less than ten-second pulse – that would achieve ignition at low plasma in the Alcator density scaling law regime of Ohmic Heating. The merits of the IGNITOR versus the ITER approach were debated in 2000 at the CEA headquarters in Paris with B. Coppi and K. Lackner presenting the arguments for the high-field short pulse versus the lower superconducting field ITER machine. The committee decided that the ITER approach was closer to the next step towards a fusion power station of the DEMO type. The short-pulse IGNITOR experiment is being pursued in an agreement between Italy and Russia.

In 2007 the neutral beam injection (NBI) system power supplies were modified in JT-60U to extend the injection periods to 30 s for three perpendicular injectors and negative ion injectors (N-NBI) with 3 MW for 30 seconds giving 90 MJ. The total NBI and N-NBI achieved was 450 MJ. The direct electron cyclotron heating (ECH) power from four gyrotrons was 3 MW for 5 s for 15 MJ of injected energy. These additions allowed new experiments for long pulse, quasi-steady-state operation. In the long pulse operation high bootstrap fraction from 50-80% was achieved in the reserved magnetic shear plasmas. These results have driven the plans for the development of the next generation long pulse Japanese tokamak called JT-60SA where SA is short for super advanced. The focus of the JT-60SA is to investigate issues for the DEMO fusion reactor and thus requires quasi-steady operation by optimizing the pressure gradient driven bootstrap current J_{bs} and the RF based current drive system. The DEMO design requires more research on the "no wall" dynamics of the system since steady-state operation time of minutes will exceed the resistive wall diffusion time by an order of magnitude.

The 2007 experiments in JT-60U produced long pulses with both weak magnetic shear (WS) and reversed magnetic shear (RS) which explored the beta limits for stable operation. These beta limits are expressed in terms of the dimensionless beta parameter β_N defined as the coefficient in the dimensional formula for MHD stability in Chapter 2. An alternative view of the success of the confinement is expressed by the dimensionless confinement time parameter H_{H98} which is the dimensionless ratio factor used to the measure energy confinement time τ_E divided by the dimensionless H-mode empirical formula derived from the 1998 tokamak database. The H-mode regime is reached when the injected power exceeds a threshold value $P_{L \rightarrow H}$. All plans for the future advanced tokamaks require auxiliary heating powers well above the H-mode threshold heating power described in Chapter 5.

In the 2007 experiments in JT-60U, the record results achieved were 25 s pulses with $\beta_N \sim 2.6$ corresponding to the beta-poloidal of 2.3 with confinement time scaling factor of $\tau_{H98} = 1.7$ and the dimensionless bootstrap fraction $f_{BS} = 0.7$. This pulse length is about $15\,\tau_R$ resistive wall diffusion times as in the no-wall stability regime of MHD theory. In addition, the density measured by the ratio to the Greenwald density limit formula is high so that these reversed shear discharges in JT-60U are perhaps the best models achieved at the time of writing (2014) for the next generation of tokamak reactors and for ITER.

In the large tokamak the millimeter to centimeter wavelength RF (few GHz LH and 170 GHz ECH) wave signals suffer from scattering of the electromagnetic waves from the intrinsic density fluctuations of the plasmas. The electron density fluctuations give rise to both coherent and incoherent scattering of the electromagnetic waves producing an unpredictable propagation path and distortions of the wave cycles through phase shifts and frequency drifting of the electromagnetic waves passing through the plasma turbulence. The problem is often called that of the scintillation and fading of the RF signals terms used for the same problem for visible light passing through the neutral atmosphere. For the RF frequencies with centimeter wavelengths scintillation from the plasma turbulence dominates the uncertainty in the performance of the system.

Let us briefly review some of the fundamental RF scattering processes taking place in the laboratory and the near-Earth propagation of high-frequency electromagnetic waves. Several simple models are introduced to describe the types of distortions and uncertainties in the propagation paths and phase distortions of the high-frequency signals. Figure 7.2 defines the geometry used in describing the scattering of high-frequency waves by a plasma fluctuation with wavenumber $\boldsymbol{q}$ and frequency Ω.

7.2 Scattering of Radio Frequency-RF Waves in Turbulent Plasmas

The random density fluctuations scatter the RF wavenumber vector and the frequency spectrum through the Doppler shifts from the motion of the electrons in the plasma turbulence. In the frame moving with the plasma gas velocity v_{pl}, the electron density fluctuations are static, or frozen, over the short-time interval for the RF wave to propagate through the fluctuation.

For a spherical wave $\boldsymbol{E} = \widehat{\epsilon}[A\exp(ikr)/r]$ the scattering function is given in terms of the electron density fluctuations through

$$H(k,\rho) = \frac{k^2}{4}\int_0^x [A(0) - A(\rho)]dx' = 4\pi k^2 \int_0^x dx' \int_0^\infty [1 - J_0(k\rho)]\Phi_n(k)k\,dk \quad (7.1)$$

where $\Phi_{\delta n_e}(k)$ is the isotropic electron density correlation function [Ishimaru (1978)].

One introduces and defines the two correlations lengths ℓ_0 and L_0, called the inner and outer scale-lengths of the electron density turbulence, and ρ the distance

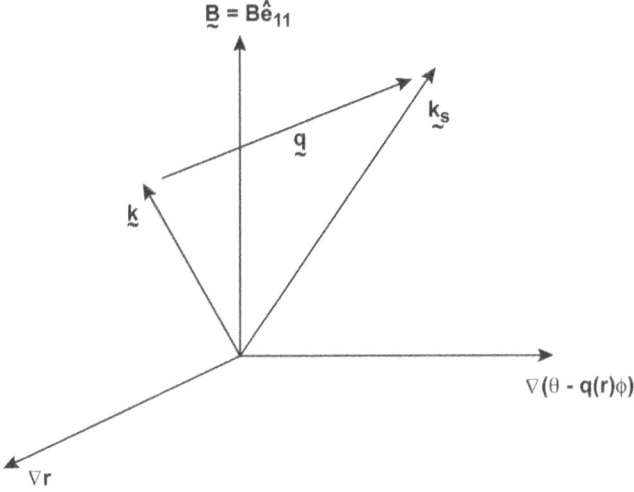

Fig. 7.2 Scattering of the incident RF wave **k** by the electron density fluctuation **q** to a new RF wave k_s. The scattering broadens the RF wave spectrum and the RF driven part of the electron phase space density.

from scattering volume to **x**. For example, change in the amplitude of the wave is computed with RF scattering theory for a Kolmogorov-like density spectrum of strength $C_n^2 \rho^{-3}/\ell$ for the two-point correlation function that the change it the amplitude A of the RF wave in the three regions of spatial scales is given approximately by

$$A(0) - A(\rho) = \begin{cases} 6.56 \; C_n^2 \rho^2 / \ell_0^{1/3} & \rho < \ell_0 \\[2mm] 5.83 \; C_n^2 \rho^{5/3} & \ell_0 < \rho < L_0 \\[2mm] 3.127 \; C_n^2 L_0^{5/3} & \rho > L_0 \; . \end{cases} \tag{7.2}$$

For example, in atmospheric turbulence the reference value is $C_n^2 = 10^{-14}$ and the scattering is seen as scintillation of the stars in the night sky. In turbulent plasmas the value of C_n^2 is orders of magnitude larger and arises from the drift wave turbulence of the electron density fluctuations. The Fokker-Planck equation for the scattering of the LHCD rays by the drift wave turbulence is analyzed in [Horton, *et al.* (2013)].

7.3 ELM Control for Steady-State Plasma Operation with Resonant Magnetic Perturbations

As discussed in Chapter 5, when the tokamak plasma is driven by auxiliary heating and current drive to its highest performance in fusion power, the core plasma emits large structures called edge localized modes (ELMs) into the scrape-off layer.

The strongest of these bursts will occur in quasi-periodic sequence emitting structures with 5 to 10% of the core plasma energy. These events must be reduced to smaller controlled bursts and the leading method for doing so is to introduce external coils for driving magnetic perturbations that allow the plasma to emit the thermal energy in smaller and more frequent bursts, similar to the Type III ELMs. The external coils are designed with a specific winding number to create helical magnetic perturbations at the low-order rational surface near the magnetic separatrix, where the steep plasma pressure gradient buildup exceeds the onset value for the ballooning-peeling modes discussed in Chapters 4–6. Resonant magnetic perturbations are created by coils with a specific toroidal winding number ℓ located outside the vacuum vessel wall

$$J_{\text{RMP}} = J_{\text{RMP}}(r) \cos \zeta = J_0 \left(r - r_{\text{RMP}}\right) \sum_m \cos(m\theta - \ell\phi). \qquad (7.3)$$

A sum over poloidal currents is used to spread the perturbation over a close set of rational surfaces typically close to the $q(r_{\text{res}}) = 3$ mode rational surface, where the large, dangerous ELMs are typically observed as the plasma is pushed to its highest performance.

To describe the complex but critical ELM control mechanisms we follow Beyer, *et al.* (2007), Fuhr, *et al.* (2008), and Monnier, *et al.* (2014) who use three reduced and normalized partial differential equations 3PDE model with an external antenna coil currents J_{RMP} for driving the magnetic islands at a specific rational surface. Here we simplify the analysis by leaving out the D-shaping of the magnetic surfaces which introduces further complex poloidal mode coupling to the problem.

In the following driven-damped PDEs for the plasma vorticity from the $E \times B$ convection velocity, the pressure equation driven by the plasma energy source $S(x)$ in the inner domain well inside $q = 3$ resonant magnetic surface, and the PDE for the magnetic flux function ψ, from which the equations are used to simulate the spatio-temporal behavior of the RMP-controlled plasma pressure p, electrostatic potential ϕ, and magnetic flux ψ, which is driven by the external magnetic coils with current density $J_{\text{RMP}}(r, t)$, placed at the outer edge of the simulation domain. The current density $J_{\text{RMP}}(r, t)$ is modeled with toroidal mode number $\ell = 4$ and the series of equal amplitude poloidal mode numbers $m = 10, 11, 12, 13, 14$, so that the steady-state external antenna is driving a magnetic perturbation resonant with a ballooning interchange mode on the $q = 12/4$ rational magnetic surface. Thus, the model is simple enough to allow analytic progress for comparison with the 3D numerical simulations. Two cases are considered: (1) the limit of cylindrical plasma with a magnetic curvature term modeled with $\mathbf{G} = g\delta/\delta\theta$ and (2) a model with more complete operator for the horizontal direction of the gradient and curvature vector for the toroidal geometry where $\mathbf{G} = \sin\theta\, \partial_x + \cos\theta\, \partial_y$.

The system of three driven-damped coupled nonlinear partial differential equations is given by

$$\partial_t \nabla_\perp^2 \phi + \{\phi, \nabla_\perp^2 \phi\} = -\alpha^{-1} \nabla_\parallel \nabla_\perp^2 \psi + \mathbf{G}p + \nu \nabla_\perp^4 \phi, \tag{7.4}$$

$$\partial_t p + \{\phi, p\} = -\delta_c \mathbf{G}\phi + \chi_\parallel \nabla_\parallel^2 p + \nabla_\perp \cdot [\chi_\perp(x)\nabla_\perp p] + S(x), \tag{7.5}$$

$$\partial_t \psi = -\nabla_\parallel \phi + \alpha^{-1} \nabla_\perp^2 \psi - \alpha^{-1} J_{\mathrm{RMP}}(\mathbf{r}, t). \tag{7.6}$$

In toroidal coordinates (r, θ, φ) and in a slab geometry (x, y, z) in the vicinity of a reference surface $r = r_0$ at the plasma edge, i.e. $x = (r - r_0)/\xi_{\mathrm{bal}}$, $y = r_0\theta/\xi_{\mathrm{bal}}$, $z = R_0\varphi/L_s$, the normalized operators are

$$\nabla_\parallel = \partial_z + (\kappa/q_0 - x)\,\partial_y - \{\psi, \cdot\} \quad \text{with} \quad \kappa = \frac{L_s r_0}{R_0 \xi_{\mathrm{bal}}}, \tag{7.7}$$

$$\nabla_\perp^2 = \partial_x^2 + \partial_y^2 \quad \text{and} \quad \{\phi, \cdot\} = \partial_x \phi \partial_y - \partial_y \phi \partial_x,$$

$$\mathbf{G} = \sin\theta\,\partial_x + \cos\theta\,\partial_y, \tag{7.8}$$

in case of toroidal curvature. Here, $q_0 = q(r_0)$ is the safety factor at the reference surface; R_0 is the major radius of the magnetic axis and L_s is the magnetic shear length. For the cylindrical model the operator $\mathbf{G}$ reduces to $\mathbf{G} = \partial_y$.

7.3.1 *Resistive MHD normalization*

For controlled magnetic fusion experiments one is interested in systems that are stable to the fast MHD instabilities. Thus, we introduce the normalization that arises from the dynamics of the resistive MHD instabilities. These dimensionless units in resistive MHD models are different from those used for the drift wave instabilities that have higher frequencies and smaller scale length associated with ion gyroradius. Thus, one must alter the different normalized scales used in the MHD, the resistive MHD and the drift wave theory and simulations. The drift wave models use space and time variables normalized to the ion gyroradius taken with the electron temperature called ρ_s, where s is for the sound speed $c_s = \sqrt{(T_e/m_i)}$, using the working gas ion mass. For the RMP simulations the resistive MHD equations are used. RMP and magnetic reconnection simulations are concerned with MHD toroidal plasma gradients below the threshold for pressure gradient producing fast ideal MHD instabilities. A small resistivity breaks the frozen influx condition and allows a slower resistivity-controlled pressure gradient driven instability and the still lower magnetic reconnection instability. The following equations are written in variable for dynamics in resistive instabilities in an MHD stable plasma. For this slow dynamics the instabilities driven by radial gradients of the pressure $p(r)$ and the toroidal current density $j_\phi(r)$ are of critical importance

For resistive MHD instabilities the parallel ($\parallel$) lengths and perpendicular ($\perp$) to the unperturbed magnetic field are normalized by L_s and ξ_{bal}, respectively, and time is normalized by τ_{int}, where the resistive ballooning radial correlation length ξ_{bal} and the interchange time τ_{int} are given by

$$\xi_{\mathrm{bal}} = \left(\frac{L_p}{\tau_e c_s}\frac{m_e}{m_i}\right)^{\frac{1}{2}} \frac{L_s}{L_p}\left(\frac{L_p}{R_0}\right)^{\frac{1}{4}}\rho_s \quad \text{and} \quad \tau_{\mathrm{int}} = \frac{(L_p R_0)^{\frac{1}{2}}}{\sqrt{2}c_s}, \tag{7.9}$$

where τ_e is the reference values of the electron collision time, c_s is the sound speed, ρ_s is the ion Larmor radius at the electron temperature T_e. The pressure-gradient length L_p is controlled by the external plasma heating and is key parameter for the strength of the resistive MHD instability in a given plasma discharge. For a typical weakly collisional tokamak plasma edge, one typically finds $\xi_{\rm bal} \sim \rho_s$ and $\tau_{\rm int} \sim 10\text{-}100\, L_p/c_s$.

In the reduced, dimensionless ballooning-pressure equations gradient the dimensionless pressure gradient parameter is

$$\alpha = \frac{\beta L_s^2}{L_p R_0}$$

which is similar to the normalized pressure gradient $\alpha_{\rm MHD}$ used for collisionless MHD tokamak stability theory

$$\alpha_{\rm MHD} = \alpha \left(\frac{q_0 R_0}{L_s} \right)^2, \tag{7.10}$$

where β is the ratio of plasma pressure to magnetic pressure. The parallel and perpendicular heat-conductivity coefficients $\chi_\parallel$ and $\chi_\perp$ are normalized by $L_s^2/\tau_{\rm int}$ and $\xi_{\rm bal}^2/\tau_{\rm int}$, respectively. Therefore, a ratio of the normalized coefficients of $\chi_\parallel/\chi_\perp \sim 1$, corresponds to a ratio of the dimensional coefficients of $L_s^2/\xi_{\rm bal}^2 \sim 10^7 - 10^8$. In the present simulations, we use $\alpha = 0.1$, $\nu = \chi_\perp = 0.93$, $\chi_\parallel = 1$ and the curvature parameter $\delta_c = \frac{5}{3}2L_p/R_0$ is set to $\delta_c = 0.01$ for the steep pressure gradients close to the scrape off layer just inside the magnetic separatrix.

7.3.2   *RMP simulations with the 3D resistive MHD model*

The main computational domain corresponds to the volume between the toroidal magnetic surfaces defined by $q = 2.5$ and $q = 3.5$ that surround the rational surface with $q = q_0 = 3$. Here, a linear q-profile is assumed to be monotonically increasing with radius. The local radial position is specified with the dimensionless variable x in a reduced domain from with $x_{\rm min} < x < x_{\rm max}$ that includes the location of the RMP coils. The dimensionless simulation described here uses the parameters $\xi_{\rm bal}/r_0 = 1/500$, $L_s/R_0 = 1$. The complete computational domain is taken as delimited by $x_{\rm min} < x_{q=2.5}$ and $x_{\rm max} > x_{q=3.5}$ to include the interior high pressure plasma source $S(x)$ and the exterior wall region. The thermal source S is located in the inner buffer zone $x_{\rm min} < x < x_{q=2.5}$ and gives rise to a constant incoming energy flux, $Q_{\rm tot} = \int_{x_{\rm min}}^{x_{q=2.5}} S\, dx$ from the plasma center into the main computational domain. The external RMP antenna current $J_{\rm RMP}$ is located in the outer buffer zone $x_{q=3.5} < x < x_{\rm max}$ (Fig. 7.3a) where the current density in the external RMP coils is taken as

$$J_{\rm RMP} = J_0(x) \cos \zeta = J_0(x) \sum_m \cos(m\theta - \ell\phi) \tag{7.11}$$

in Marcus, *et al.* (2014). The external current density from the RMP coils is modeled with the central mode $(m_0, n_0) = (12, 4)$ and a set of neighboring azimuthal mode

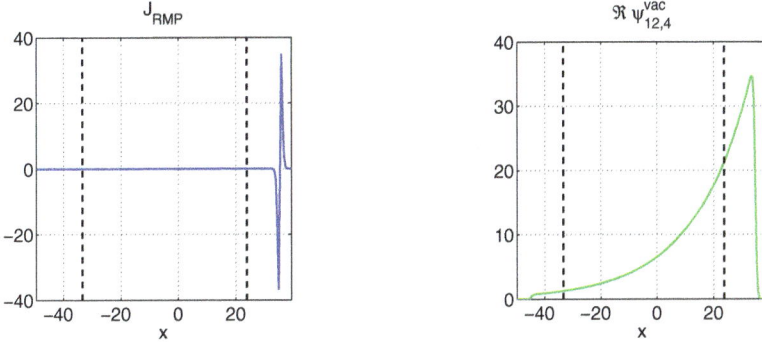

Fig. 7.3 The location and structure of the RMP antenna on the left is shown in the left frame. The magnetic flux function driven by the antenna at the $m = 12$, $n = 4$ response of the magnetic flux function $\psi(r)$ driven in the plasma domain is shown in on the right.

numbers $m = 10$ to 14. These external current winding induce a resonant magnetic perturbation in the plasma at the $q = q_0 = 3$, as shown in Fig. 7.3.

The surface averaged radial pressure profile $\bar{p}(x,t) = \langle p \rangle_{y,z}$ evolves self-consistently accordingly from the three thermal fluxes and the interior heating term $S(x)$ through the energy transport equation (the toroidal and poloidal average $\langle \cdot \rangle_{yz}$ of Eq. (7.5)),

$$\partial_t \bar{p} = -\partial_x \left(Q_{\mathrm{conv}} + Q_{\mathrm{coll}} + Q_{\delta B} \right) + S(x) \tag{7.12}$$

with the convective, collisional and magnetic fluctuation thermal fluxes given by

$$Q_{\mathrm{conv}} = \langle p\, v_r \rangle, \quad Q_{\mathrm{coll}} = -\chi_\perp \partial_x \bar{p}, \quad Q_{\delta B} = -\chi_\parallel \left\langle \partial_y \psi \nabla_\parallel p_{y,z} \right\rangle. \tag{7.13}$$

In a stationary state, integration of Eq. (7.12) across the radial leads to the energy-flux balance

$$Q_{\mathrm{conv}}(x) + Q_{\mathrm{coll}}(x) + Q_{\delta B}(x) = Q_{\mathrm{tot}} \tag{7.14}$$

for a steady state plasma.

7.4 Helical Equilibrium Plasma States Created by the External RMP Currents

The RMP coils change the equilibrium plasma to a helical equilibrium. In the limit $r/R \to 0$ there is no linear coupling between neighboring poloidal wavenumbers m, $m - 1$, $m + 1$ modes and the equilibrium is of the form

$$\begin{pmatrix} \phi_{\mathrm{eq}} \\ p_{\mathrm{eq}} \\ \psi_{\mathrm{eq}} \end{pmatrix} = \begin{pmatrix} \bar{\phi}(x) \\ \bar{p}(x) \\ 0 \end{pmatrix} + \begin{pmatrix} \phi_1(x) \\ p_1(x) \\ \psi_1(x) \end{pmatrix} \exp(i\zeta) + \mathrm{c.c.}, \tag{7.15}$$

where the bar describes the axisymmetric part of the field and the index "1" designates the $(m, n) = (m_0, n_0) = (12, 4)$ helical components. Neglecting the toroidal curvature there is the single driven helical component given by

$$(\phi_1, p_1, \psi_1) \equiv (\phi_{m_0,n_0}, p_{m_0,n_0}, \psi_{m_0,n_0}) \equiv (\phi_{12,4}, p_{12,4}, \psi_{12,4})$$

varying with the helical coordinate $\zeta = m\theta - n\varphi$ driven by the RMP antenna.

The magnetic perturbations $\delta B_r^{\mathrm{RMP}}$ create a rotational torque from $j_\phi \delta B_r^{\mathrm{RMP}}$ in the plasma creating the fields shown in Fig. 7.4. The perturbation equations

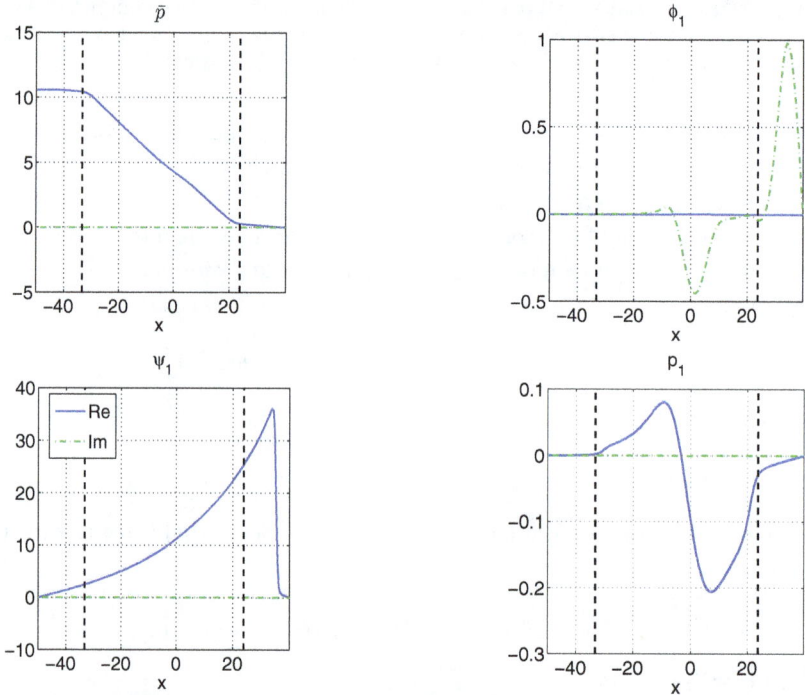

Fig. 7.4   Radial profiles of the fields $\bar{p}$, ϕ_1, p_1 and ψ_1 corresponding to the helical equilibrium described in Eq. (7.16). The magnetic perturbation is in phase with the external current and the plasma is not rotating ($\bar{v}_E = \partial_y \bar{\phi} = 0$), but this equilibrium is unstable to the onset of rotation according to the results obtained with numerical simulations. The dashed vertical lines mark the location of the rational magnetic surfaces with $q = 5/2$ and $q = 7/2$.

give symmetric solutions corresponding to a helical equilibrium where the induced magnetic perturbation and the associated pressure perturbation are in phase with the external current (7.11), i.e. $\mathrm{Im}\,\psi_1 = 0$, $\mathrm{Im}\,p_1 = 0$, and the potential variation is in phase quadrature, $\mathrm{Re}\,\phi_1 = 0$. In this equilibrium, the poloidal rotation is zero, $\bar{v}_E = -\partial_y \bar{\phi} = 0$. In summary, the driven helical solutions of Eqs. (7.4), (7.5), and

(7.6) reduce to

$$
\begin{pmatrix} \phi_{\text{eq}}^{(1)} \\ p_{\text{eq}}^{(1)} \\ \psi_{\text{eq}}^{(1)} \end{pmatrix} = \begin{pmatrix} 0 \\ \bar{p}(x) \\ 0 \end{pmatrix} + \begin{pmatrix} -2\phi_1^I(x)\sin\zeta \\ 2p_1^R(x)\cos\zeta \\ 2\psi_1^R(x)\cos\zeta \end{pmatrix}, \tag{7.16}
$$

where ϕ_1^I, p_1^R and ψ_1^R are real fields corresponding to the imaginary (I) and real (R) parts of ϕ_1, p_1 and ψ_1, respectively. Typical radial profiles of $\bar{p}$, ϕ^I, p^R and ψ^R are shown in Fig. 7.4. Note that there is now a non-vanishing convective flux Q_{conv} induced by RMP antenna in the equilibrium (7.16), as the pressure and potential perturbations are phase shifted by $\pi/2$. In the cylindrical model this convective flux Q_{conv} is much smaller than the magnetic flutter flux $Q_{\delta B}$.

The RMP perturbation drives the plasma to a new helical-stationary state with non-vanishing sheared-plasma rotation from the torque created by δB_r^{RMP}. The new rotating equilibrium state is characterized by: (1) a phase shift between the external current and the magnetic perturbation induced in the plasma and (2) a large convective flux $Q_{\text{conv}} \gg Q_{\delta B}$. Typical radial profiles of the axisymmetric components as well as the helical amplitudes and the fluxes are shown in Fig. 7.10. Note that the rotation velocity vanishes close to the resonant surface, which is consistent with a nearly complete penetration of the magnetic perturbation [Monnier, *et al.* (2014)], i.e. the amplitudes of the $\psi_{12,4}(x)$ at the resonance surface $x = 0$ are similar in both equilibria and close to the maximum value corresponding to the vacuum case.

In the rotating steady state, without the toroidal $\cos\theta$-mode coupling, the fields are $\widehat{\psi}_{12,4}^{\text{rotationstate}} \approx \widehat{\psi}_{12,4}^{\text{symmetricstate}} \approx \widehat{\psi}_{12,4}^{\text{vacuum}}$,

$$
\text{where} \quad \widehat{\psi}_{12,4} = |\psi_{12,4}(x=0)|. \tag{7.17}
$$

The RMP creates a strong radial convective flux $\langle v_x \delta p \rangle$ that flattens the pressure profile on the resonant surface as designed to eliminate or control the ELMs. Two-dimensional maps of the pressure p and the non-axisymmetric part of potential $\phi - \bar{\phi}$ corresponding to the two different equilibrium states are shown in Fig. 7.6. The periodic structure that corresponds to the flattening of the pressure on the magnetic islands is symmetric in the symmetric state (Fig. 7.6a) and distorted in the rotation state (Fig. 7.6b) from the shear flow. The potential structure together with the pressure structure, as illustrated in Fig. 7.9d, is responsible for the strong convective flux in the rotation state. With increasing amplitude of the external RMP current, both the magnetic flutter flux and the convective flux increase in the stable rotating equilibrium, but the convective thermal flux is always larger than the thermal flux produced by the magnetic flutters as shown in Fig. 7.7.

7.4.1 *Transition to the rotating state with strong convective flux*

The broken symmetry introduced by the RMP coils produces a transition to a rotating helical plasma state, as illustrated by performing time integrations in two

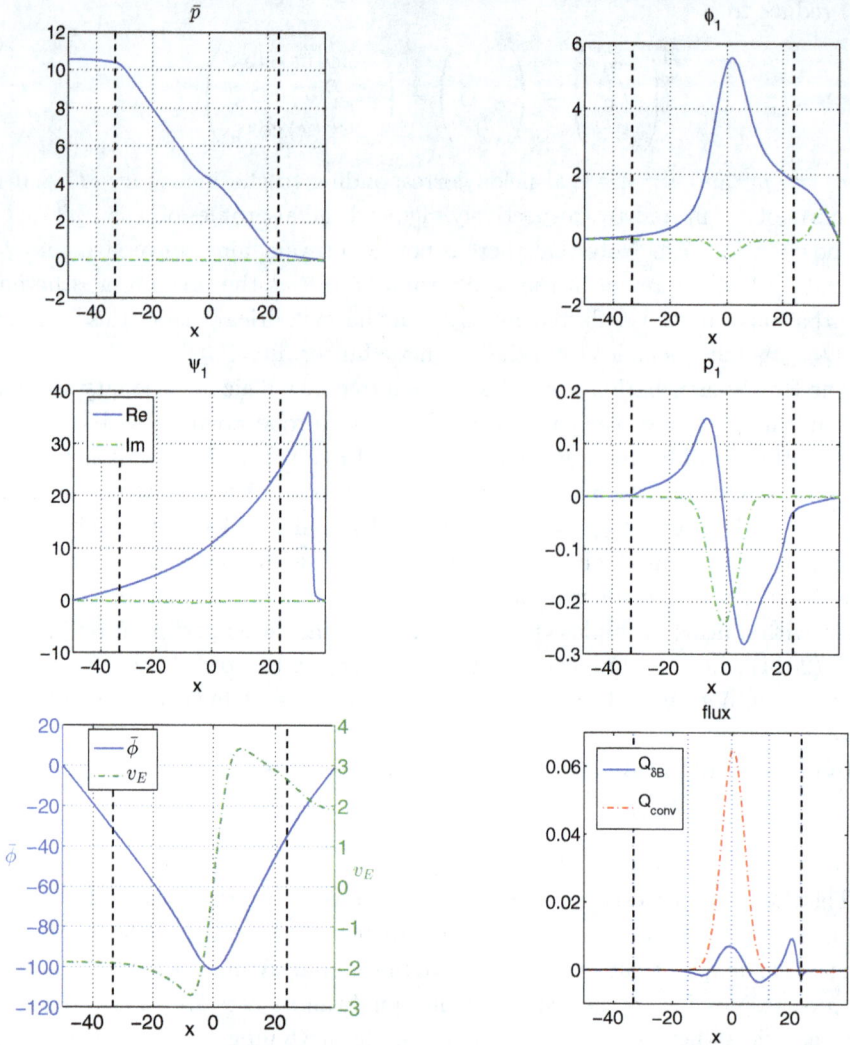

Fig. 7.5   Radial profiles of the fields $\bar{p}$, ψ_1, p_1, ϕ_1, the poloidal rotation $\bar{v}_E = \partial_y \bar{\phi}$ and the fluxes Q_{conv}, $Q_{\delta B}$ driven by the helical equilibrium are shown in the series of six frames. The magnetic perturbation is phase shifted with respect to the external RMP antenna current.

successive phases of the evolution. The evolution of the convective and magnetic flutter fluxes at the resonant surface is shown in Fig. 7.8a. In an early phase of the integration (from $t = 500$), the rotation is forced to zero and the system is rapidly evolving to the stationary state obtained in Eq. (7.16), where the convective flux Q_{conv} is much smaller than the flutter flux $Q_{\delta B}$. Then, in the late phase, at $t = 8000$ and onward, the constraint on the poloidal rotation is released and the system evolves self-consistently to the rotating state characterized by $Q_{\mathrm{conv}} \gg Q_{\delta B}$.

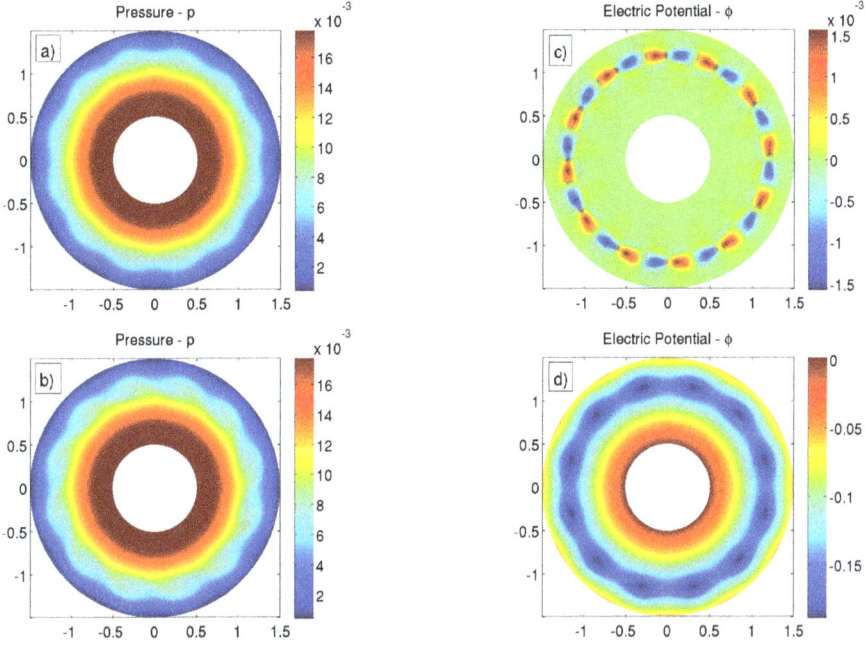

Fig. 7.6 Two-dimensional maps of the pressure p and the non-axisymmetric part of potential $\phi - \bar{\phi}$ corresponding to the two different equilibrium states in cylindrical geometry. Pictures a) and c) depict the pressure and electric potential profiles for the case described by system in equilibrium in Eq. (7.17) with the profiles shown on Fig. 7.4. This system evolves to the state depicted on frames b) and d) with poloidal rotation, which the phase-shift between p and $\partial \phi$ determines the high-convective flux and the poloidal rotation.

The instability of the symmetric equilibrium is characterized by the growth rate of a small perturbation to the initial equilibrium. A series of such numerical simulations for different amplitudes of the external magnetic perturbation shows that the growth rate is approximately constant up to a critical value of the RMP perturbation amplitude. Above this critical amplitude, the growth rate is strongly increasing with the amplitude of the external RMP perturbation. This is illustrated in Fig. 7.8b where the growth rate is plotted against the magnetic island width W linked to the RMP perturbation amplitude by

$$W = 4\sqrt{2\widehat{\psi}_{12,4}}. \tag{7.18}$$

The time growth rate of Q_{conv} strongly increases for island widths $W > 34$. In that case, the half-width of the island $W/2 > 17$ approaches the distance between the resonant surface and the external boundary of the main computational domain $x_{q=3.5} - x_{q=3} = 23$ and the island likely is influenced by the boundary. Also, higher-order harmonics become significant for $W > W_c \approx 22$, where the critical island width is given by Fitzpatrick (1995) $W_c = [(8/m_0)(r_0/\xi)]^{1/2}(\chi_\perp/\chi_\parallel)^{1/4}$. For magnetic island $W = 18$ width, the amplitude of the second order $(m, n) = (24, 8)$

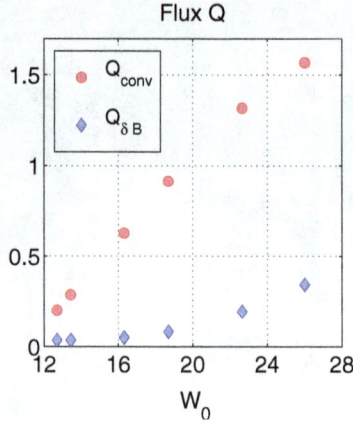

Fig. 7.7   Radially-integrated convective and flutter fluxes for the cylindrical case, $Q_{\text{conv}} = \int_{q=2.5}^{q=3.5} Q_{\text{conv}}(x)\,dx$ and $Q_{\delta B} = \int_{q=2.5}^{q=3.5} Q_{\delta B}(x)\,dx$, as a function of the magnetic perturbation amplitude expressed in terms of the vacuum island width W_0. Even for low values of external resonant magnetic perturbation ψ_0 (small island width W_0), Q_{conv} is higher than $Q_{\delta B}$.

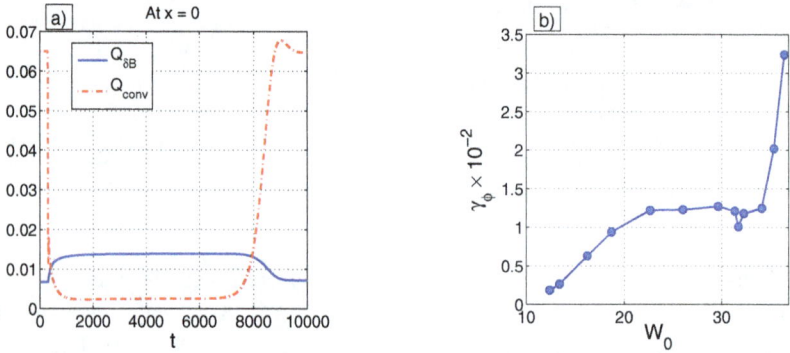

Fig. 7.8   (a) Evolution of the convective (dot-dashed line) and magnetic flutter (solid line) fluxes. Forcing the rotation to zero, the convective flux decreases and the system responds increasing the flux of the magnetic flutter. Releasing the constrain, the system evolves to a stable condition with Q_{conv} bigger than $Q_{\delta B}$. (b) The growth rate of the convective flux plotted against the width of the magnetic island W_0.

mode is one order of magnitude lower than the amplitude of the main harmonics, but when the magnetic island width reaches $W = 36.5$ (not shown) the second harmonic is only lower by a factor of 0.3.

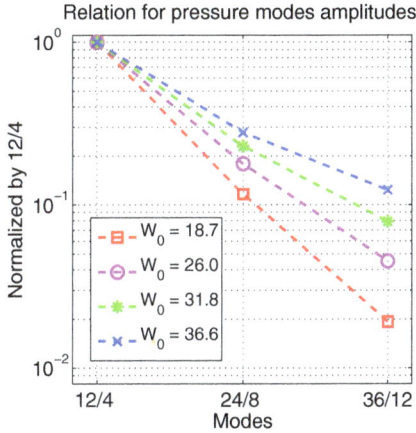

Fig. 7.9 Amplitude of the higher-order resonant pressure modes with increasing mode numbers. Larger magnetic island widths W_0 increases the amplitude of the higher order resonant magnetic perturbations which contributes to the increase of Q_{conv} and poloidal rotation effects.

7.5 Rotating States in Toroidal Geometry with Multiple RMPs

The outward toroidal curvature from $r\cos(\theta)/R$ gives rise to linear coupling between $m-1$, m, and $m+1$ modes. In particular, in the vorticity equation (7.4), a (m, n) harmonic of the pressure p, couples to $(m-1, n)$ and $(m+1, n)$ harmonics of the electrostatic potential ϕ. If a self-consistent state with rotation and convective transport similar to the one shown in Fig. 7.9 exists in toroidal geometry, then the state involves multiple harmonics. Marcus, *et al.* (2014) show that there are multiple harmonic RMP with poloidal wavenumbers $m = 10, 11, 12, 13, 14$ and toroidal wavenumber $n = 4$. In order to understand these toroidal effects clearly, we first apply the multiple harmonics of the RMP in the cylindrical curvature model with $r/R \approx 0$.

A stable equilibrium with rotation and important convective flux is recovered in this case. The plasma self-organizes such that the rotation velocity vanishes close to the resonant surfaces $q = 10/4, 11/4, \ldots, 14/4$. The convective and magnetic flutter fluxes show local maxima close to these resonant surfaces. The width of the $(12,4)$ island is $W_{12,4} = 18.5$ and the width of the closest neighboring island $(13,4)$ is $W_{13,4} = 18.4$, so the corresponding Chirikov overlapping parameter [Chirikov (1983)] is

$$\sigma = \frac{W_{12,4} + W_{13,4}}{2x_{q=3.25}} \approx 1.4.$$

The islands therefore are overlapping, but this overlapping is sufficiently weak that distinct local maxima are observed for the Q_{conv} and $Q_{\delta B}$ fluxes.

Repeating the calculation with finite $r\cos(\theta)/R$ gives rise to multiple harmonics RMP in the toroidal curvature case, the plasma also evolves to a new steady-state

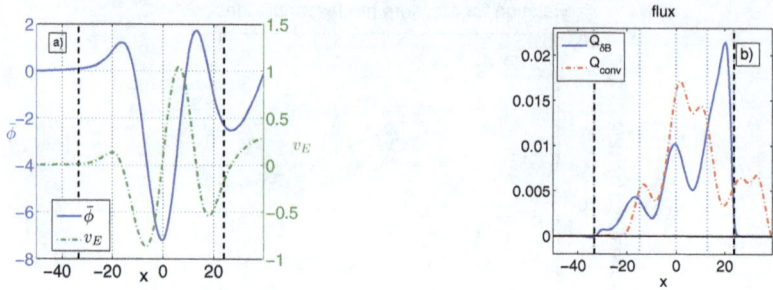

Fig. 7.10   Multiple harmonics RMP for toroidal curvature island width $W_0 = 18.7$. a) depicts the equilibrium electric potential and poloidal rotation $v_E(x)$ and b) the difference in the flux intensities for Q_{conv} and $Q_{\delta B}$.

equilibrium with fast poloidal rotation and a strong convective thermal flux. These states are illustrated in Fig. 7.10.

7.6   Issues with RMP for Controlling ELMs

The helical magnetic perturbation driven by external coils produces a more complex plasma state where the perturbations become phase shifted and the plasma rotates. This is due to the coupling between pressure and electrostatic potential perturbations induced by the external coils. In the rotating RMP state, the phase between pressure and electrostatic potential is such that there is a significant convective thermal flux $\langle v_r \delta p \rangle$ that exceeds the thermal flux from the magnetic flutter $\langle \delta p \delta B_r \rangle$ by a factor of 3 to 10. Thus, the plasma confinement regime is changed across a critical value of the RMP coil currents $I_{\text{RMP}}^{\text{crit}}$.

In the toroidal geometry the single harmonic RMP perturbation is replaced by multiple RMP-induced modes that drives poloidal plasma rotation. The system reaches a new final steady state with a high-convective thermal flux. The RMP coupling mechanism between pressure and the electric potential drives the plasma rotation and the associated strong convective thermal flux in the multiple RMP toroidal geometry. The combined toroidal $[\epsilon \cos \theta]$ and RMP $[\cos(m\theta - n\phi)]$ perturbations produce a complex plasma dynamics with significant quasi-steady-state thermal losses and plasma rotation due to the broken symmetry of system. The new system shows some properties similar to those of the helical toroidal plasma confinement system – such as LHD machine at NIFS – which has an outer layer of stochastic magnetic field lines.

Thus, there are advantages and disadvantages for using RMP for controlling ELMs. The advantage is the thermal flux is smoothed out in time avoiding the giant ELM heat loads into the SOL and on to the vessel walls. The disadvantage is that the toroidal symmetry is broken which introduces a complex set of new plasma dynamics with an increased convective thermal flux. The trade-off is complicated and not resolved at the time of writing (2015). Other methods of controlling the giant ELMs in ITER need to be explored.

The EMEDGE 3D simulations show that the simple static equilibrium in which the magnetic perturbation inside the plasma is in phase with the external perturbation and the plasma is not rotating is unstable to a transition to a rotating complex 3D plasma state. The RMP driven plasma self-organizes into a more complex state where the perturbations becoming phase shifted producing the plasma rotation. This is due to the nonlinear coupling between pressure and electrostatic potential perturbations induced by the $\cos(\theta)$ variation of the toroidal magnetic curvature and gradient of the magnetic field. In the stable equilibrium state, the phase between pressure and electrostatic potential is such that a rotating plasma with a significant outward thermal convective flux is produced. The convective thermal flux then exceeds the thermal flux from the magnetic flutter by a factor of 3-10 times in the simulations. Beyer, *et al.* (2007) and Marcus, *et al.* (2014) show that with the symmetry breaking RMP, the plasma reaches a final steady state with a high-convective flux and an induced poloidal rotation. The onset of the large convective thermal flux occurs when the RMP perturbation amplitude is above a critical value for which the Chirikov overlapping parameter is greater than unity at the principle resonant magnetic surface driven by the RMP coil.

In summary, the RMP coils provide an external control tool for interrupting the growth of Type-I ELM structures. The RMP is being evaluated for installation on the ITER machine for limiting the growth of dangerously large Type I ELMs that could produce unacceptable transient thermal loads on the confinement vessel walls. The RMP coils, however, produce large helical perturbations on the plasma breaking the toroidal symmetry of tokamak plasma and lead to enhance δB_r with associated stochastic magnetic field transport and enhanced convective $\langle v_r \delta p \rangle$ transport. Owing to the complexity of the 3D nonlinear dynamics, finding the trade-off of these effects from the RMP coils is currently requiring extensive high performance computer simulations for their design in the ITER machine.

7.7 RF Driven Anisotropic High Energy Electron Phase-Space Distribution Functions

High-power radio frequency (RF) are launched from external antennas and in Fig. 7.1 are used to manipulate the electron phase space distribution function. The most important example is the application of the RF waves to maintain the electron current that produces the confining magnetic field. The method is called the lower hybrid current drive (LHCD) and has been successfully used in several tokamaks to maintain a steady state – after the ohmic transformer has ended its period of operation.

Lower-hybrid current drive suggested by Fisch (1978) and demonstrated on PLT tokamak [Bernabei, *et al.* (1982)] is to be the most robust and efficient method of driving the tokamak current with external radio frequency waves in steady-state tokamak operation [Litaudon (2011)]. The method has been tested in Tore Supra for discharges at megawatts of RF power over periods up to six minutes [Hoang, *et al.* (2003); Bucalossi (2011)].

The Eighth IAEA Technical Meeting on Steady-State Operation of Magnetic Fusion Devices; 26-29 May 2015, Nara, Japan, identified the following key issues for future research for ITER and the following DEMO fusion reactors: (1) super-conducting devices, (2) long-pluse operation, (3) steady-state fusion-plasma wall interactions, and (4) particle control and power exhaust.

The Tore Supra power deposition code is ALCYON and the transport code is CRONOS, given in Artaud, *et al.* (2010). Simulations have been performed with CRONOS using ALCYON for the FW power deposition calculation and under way for current drive in ITER.

The WEST machine project at Cadarache is a model of ITER tokamak that was approved by the CEA General Administrator, Bernard Bigot in 2012. The WEST (Tungsten (W) Environment in Steady-State Tokamak) machine follows the 20 years of innovation and discovery in long-pulse RF controlled plasmas in Tore Supra that is described in detail in the special issue of Fusion Science and Technology [Ekedahl, *et al.* (2009)].

Intense electron heating from electron cyclotron resonant waves (ECRH) and from lower-hybrid waves (LHRH and LHCD) distort the electron phase space distribution functions producing significant changes in the unstable parameter space for the ETG turbulence. Heavy parallel velocity tails from LHCD and approach the "runaway" distributions [Knoepfel and Spong (1979)]. These RF driven electron phase space distribution functions have lower rates of electron Landau damping and thus higher rates of radial gradient driven growth rates for ETG turbulence.

The electron cyclotron resonance heating in TCV is shown through tomographic reconstruction of the ECE emission on the high-field side of the plasma to create a two-component energy distribution with the high-energy part with components having an effective temperature of 10 to 50 keV. The high-energy part is essential in creating an increased absorption of the second 2X and third 3X harmonic heating with the extraordinary waves (polarization perpendicular to B_0) polarization. The extraordinary wave has the wave electric field vector pointing perpendicular to the magnetic field, which then gives much stronger coupling to the plasma through mode conversions to electrostatic waves than the ordinary wave (polarization parallel to B_0) with the wave electric field parallel to the toroidal magnetic field [Blanchard, *et al.* (2002)].

The electron distributions functions computed from LHCD wave codes produce skewed parallel velocity distributions in contrast to drifting Maxwellians. These skewed distributions have large parallel thermal fluxes $q_\parallel$ with steep radial gradients. They maintain the toroidal plasma current in the steady state for many energy confinement times in Tore Supra.

A velocity distribution with a large parallel heat flux $q_\parallel(r)$ has a velocity skewness that makes the distribution intrinsically different from that of a shifted, bi-Maxwellian. Thus, it is important to compute the kinetic Vlasov response functions for these skewed distribution functions. Methods for driving an electron drift velocity to maintain a steady-state tokamak plasma current I_p, will generically create a large parallel thermal flux $q_\parallel(r)$ with radial gradients. The best understood method for driving a steady toroidal plasma current is with microwaves focused by the antennas to carry the required unidirectional RF wave (or photon momentum) to be absorbed by the electrons. This involves the use of high-power levels of microwaves launched from multiple Klystrons from carefully designed antenna structures as shown in Fig. 7.1. The RF waves become lower-hybrid plasma waves in the plasma. This method of current drive is designated as LHCD for lower-hybrid current drive. Contours of the constant levels of a the electron distribution functions computed for the simulations of $P_{\text{LHCD}} = 8\,\text{MW}$ apply to the Tore Supra plasma using the CIEL2F3 simulation code. Essentially, all the electric current $j_\parallel = -en_e u_\parallel$ and parallel thermal flux $q_\parallel$ is carried by the passing electrons. The low-frequency plasma drift wave response function for the LHCD electron phase space distribution is defined as P_{LHCD} and given in Eq. (7.19).

In Table 5.2 the drift wave stability parameters for the LHCD driven plasmas are given. The dimensionless parameter $\alpha_q = q_\parallel/n_e T_e v_e$, defined by the generalized drift wave driving term is completed. For example, the LHCD distribution of f_e and $q_\parallel$ have a large gradient of $q_\parallel(r)$, giving $L_{T_e}/L_{q_\parallel} \gtrsim 5$. The combined ray tracing/Fokker-Planck codes DELPHINE of Peysson [Peysson, *et al.* (2011); Saoutic, *et al.* (1994)] and the newer package called LUKE and CP3O are used to compute the electron phase space distribution functions and the propagation of the RF waves in the in homogeneous plasma with the ray equations.

The $P^{\text{LHCD}}(\omega, k_\parallel, k_\perp)$ dispersion function reduces to the well-known $Z(\omega/k_\parallel v_T) I_0 e^{-b}$ response function when the electron distribution function is replaced by a non-relativistic bi-Maxwellian with $v_T = (T_\parallel/m_e)^{1/2}$ and $b = k_\perp^2 T_\perp/m\omega_c^2$. The relativistic electrostatic plasma response function is computed in momentum space as

$$P(\omega, k_\parallel, k_\perp) = -\int_0^\infty \int_{-\infty}^{+\infty} \frac{\left(\frac{\omega}{v_\parallel}\frac{\partial f}{\partial p_\parallel} + \frac{k_y}{eB}\frac{\partial f}{\partial x}\right) J_0^2\left(\frac{k_\perp p_\perp}{eB}\right) dp_\parallel dp_\perp^2}{\omega - k_\parallel v_\parallel \omega_D + i\varepsilon} \tag{7.19}$$

with the Landau initial value rule given by $\varepsilon \to 0^+$ for real ω. For $\text{Im}(\omega) < 0$ the function $P(\omega)$ is the analytic continuation of the polarization function $P(\omega, k_\parallel, k_\perp)$ defined in Eq. (7.19). The phase distribution function $f(p_\parallel, p_\perp, x)$ has relativistic electrons with $p/mc = \gamma > 1$ as verified from the X-ray spectrum in the LHCD experiments.

Theoretical and experimental studies show that sufficiently strong LH heating and current drive excites drift waves in a tokamak plasma. Benkadda, *et al.* (1996)

derive the conditions for the LH pump wave amplitudes to be sufficient to excite drift waves by a modulational instability. The modulational excitation occurs for long-wavelength drift oscillations, with the wavelengths exceeding the length of the LH pump wave (a few centimeters in the plasma). The instability is described by a dispersion equation similar to the equation for the usual hydrodynamical beam-plasma instability. The authors treat the excitation of the long-wavelength drift waves as a modulational excitation by a beam of the LH waves. The results are compared with data from the Tore Supra LHCD experiments which showed an increase of turbulence under conditions with intense lower hybrid heating and current drive.

More recent LHCD current drive experiments in Tore Supra have produced high-quality steady-state plasmas extending over six minutes with more than one gigajoule of energy passing through the plasma. The method LHCD is the best experimentally proven method to maintain a steady-state current drive in fusion grade tokamak plasmas.

References

Artaud, J. F., Basiuk, V., Imbeaux, F., Schneider, M., Garcia, J., Giruzzi, G., Huynh, P., Aniel, T., Albajar, F., Ané, J. M., Bécoulet, A., Bourdelle, C., Casati, A., Colas, L., Decker, J., Dumont, R., Eriksson, L. G., Garbet, X., Guirlet, R., Hertout, P., Hoang, G. T., Houlberg, W., Huysmans, G., Joffrin, E., Kim, S. H., Köhl, F., Lister, J., Litaudon, X., Maget, P., Masset, R., Pégourié, B., Peysson, Y., Thomas, P., Tsitrone, E., and Turco, F. (2010). The CRONOS suite of codes for integrated tokamak modeling, *Nucl. Fusion* **50**, p. 043001, doi:10.1088/0029-5515/50/4/043001.

Becker, G. (1990). Analysis of energy and particle transport and density profile peaking in the improved ohmic confinement regime, *Nucl. Fusion* **30**, 11, p. 2285, doi:10.1088/0029-5515/30/11/006.

Benkadda, S., Popel, S. I., Tsytovich, V. N., Devynck, P., and Laviron, C. (1996).Modulational excitation of drift waves by a beam of lowerhybrid waves, *Phys. Plasmas* **3**, p. 571, http://dx.doi.org/10.1063/1.871884.

Bernabei, S., Daughney, C., Efthimion, P., Hooke, W., Hosea, J., Jobes, F., Martin, A., Mazzucato, E., Meservey, E., Motley, R., Stevens, J., Von Goeler, S., and Wilson, R. (1982). Lower-hybrid current drive in the PLT tokamak, *Phys. Rev. Lett.* **49**, pp. 1255-1258, http://link.aps.org/doi/10.1103/PhysRevLett.49.1255.

Beyer, P., Benkadda, S., Führ-Chaudier, G., Garbet, X., Ghendrih, Ph., and Sarazin, Y. (2007). Turbulence simulations of transport barrier relaxations in tokamak edge plasmas, *Plasma Phys. Control. Fusion* **49**, p. 507, doi:10.1088/0741-3335/49/4/013.

Blanchard, P., Alberti, S., Coda, S., Weisen, H., Nikkola, P., and Klimanov, I. (2002). High-field side measurements of non-thermal electron cyclotron emission on TCV plasmas with ECH and ECCD, *Plasma Phys. Control. Fusion* **44**, p. 2231, doi:10.1088/0741-3335/44/10/310.

Bonoli, P. T., Porkolab, M., Takase, Y., and Knowlton, S. F. (1988). Numerical modeling of lower-hybrid RF heating and current drive experiments in the Alcator-C tokamak, *Nucl. Fusion* **28**, p. 991, doi:10.1088/0029-5515/28/6/004.

Bucalossi, J., Argouarch, A., Basiuk, V., Baulaigue, O., Bayetti, P., Bécoulet, M., Bertrand, B., Brémond, S., Cara, P., Chantant, M., Corre, Y., Courtois, X., Doceul, L., Ekedahl, A., Faisse, F., Firdaouss, M., Garcia, J., Gargiulo, L., Gil, C., Grisolia, C., Gunn, J., Hacquin, S. Hertout, P., Huysmans, G., Imbeaux, F., Jiolat, G., Joanny, M., Jourd'heuil, L., Jouve, M., Kukushkin, A., Lipa, M., Lisgo, S., Loarer, T., Maget, P., Magne, R., Marandet, Y., Martinez, A., Mazon, D., Meyer, O., Missirlian, M., Monier-Garbet, P., Moreau, P., Nardon, E., Panayotis, S., Pégourié, B., Pitts, R. A., Portafaix, C., Richou, M., Sabot, R., Saille, A., Saint-Laurent, F., Samaille, F., Simonin, A., and Tsitrone, E. (2011). Feasibility study of an actively cooled tungsten divertor in Tore Supra for ITER technology testing, *Fusion Eng. Design* **86**, pp. 684-688, doi:10.1016/j.fusengdes.2011.01.114.

Chirikov, B. V. (1983). Chaotic dynamics in Hamiltonian systems with divided phase space, *Lecture Notes in Physics***179**, pp. 29-46.

Fisch, N. (1997). Confining a tokamak plasma with rf-driven currents, *Phys. Rev. Lett.* **41**, p. 873, Erratum *Phys. Rev. Lett.* **42**, p. 410 (1979), doi:http://dx.doi.org/10.1103/PhysRevLett.41.873.

Fitzpatrick, R. (1995). Helical temperature perturbations associated with tearing modes in tokamak plasmas, *Phys. Plasmas* **2**, 825, http://dx.doi.org/10.1063/1.871434.

Goniche, M., Amicucci, L., Baranov, Y., *et al.* (2010). Lower-hybrid current drive for the steady-state scenario, *Plasma Phys. Control. Fusion* **52**, p. 124031, doi:10.1088/0741-3335/52/12/124031.

Hoang, G. T., Horton, W. Bourdelle, C., Hu, B., Garbet, X., and Ottaviani, M. (2003). Analysis of the critical electron temperature gradient in Tore Supra, *Phys. Plasmas* **10**, pp. 405-412, doi:10.1063/1.1534113.

Horton, W., Hu, B., Dong, J. Q., and Zhu, P. (2003). Turbulent electrons transport in tokamaks, *New J. Phys.* **5**, pp. 14.1-14.34, Jttp://www.njp.org.

Horton, W., Goniche, M., Peysson, Y., Decker, J., Ekedahl, A., and Litaudo, X. (2013). Penetration of lower-hybrid current drive waves in tokamaks, *Phys. Plasmas* **20**, p. 112508, doi:10.1063/1.4831981, http://dx.doi.org/10.1063/1.4831981.

Imbeaux, F., and Peysson, Y. (2005). Ray-tracing and Fokker-Planck modeling of the effect of plasma current on the propagation and absorption of lower-hybrid waves, *Plasma Phys. Control. Fusion* **47**, p. 204, doi:10.1088/0741-3335/47/11/012.

Ishimaru, A. (1978) Wave Propagation and Scattering in Random Media, **2** (Academic Press, New York) ISBN:0-12-374702-3.

Knoepfel, H., and Spong, D. A. (1979). Runaway electrons in toroidal discharges, *Nucl. Fusion* **19**, 785, doi:10.1088/0029-5515/19/6/008.

Kritz, A. H., Bateman, G., Onjun, T., Pankin, A., and Nguyen, C. (2002). Testing H-mode pedestal and core transport models using predictive integrated modeling simulations, *29th EPS Conf. Control. Fusion and Plasma Phys.* **26B** (ECA), (Montreux, 2002, Geneva: European Physical Society) p. D-5.001.

Litaudon, X. (2011). Real-time control of advanced scenarios for steady-state tokamak operation, *Fusion Sci. Tech.* **59**, 3, p. 469-485.

Marcus, F. A., Beyer, P., Fuhr, G., Monnier, A., and Benkadda, S. (2014). Convective radial energy flux due to resonant magnetic perturbations and magnetic curvature at the tokamak plasma edge, *Phys. Plasmas* **21**, p. 082502, doi:10.1063/1.4891437, http://dx.doi.org/10.1063/1.4891437.

Mazzucato, E. (2010). Study of turbulent fluctuations driven by the electron temperature gradient in the National Spherical Torus Experiment, *Nucl. Fusion* **50**, p. 029801, http://iopscience.iop.org/0029-5515/50/2/029801.

Mazzucato, E., Bell, R. E., Ethier, S., Hosea, J. C., Kaye, S. M., LeBlanc, B. P., Lee, W. W., Ryan, P. M., Smith, D. R., Wang, W. X., Wilson, J. R., and Yuh, H. (2009). Study of turbulent fluctuations driven by the electron temperature gradient in the National Spherical Torus Experiment, *Nucl. Fusion* **49**, 055001, doi:10.1088/0029-5515/49/5/055001.

Mazzucato, E., Smith, D. R., Bell, R. E., Kaye, S. M., Hosea, J. C., LeBlanc, B. P., Wilson, J. R., Ryan, P. M., Domier, C. W., Luhmann, N. C., Jr., Yuh, H., Lee, W., and Park, H. (2008). Short-scale turbulent fluctuations driven by the electron-temperature gradient in the National Spherical Torus Experiment, *Phys. Rev. Lett.* **101**, p. 075001-1, doi:10.1103/PhysRevLett.101.075001, http://link.

aps.org/doi/10.1103/PhysRevLett.101.075001.

Monnier, A., Fuhr, G., Beyer, P., Marcus, F. A., Benkadda, S., and Garbet, X. (2014). Penetration of resonant magnetic perturbations in turbulent edge plasmas, *Nucl. Fusion* **54**, p. 064018, doi:10.1088/0029-5515/54/6/064018.

Peysson, Y., Decker, J., Morini, L., and Coda, S. (2011). RF current drive and plasma fluctuations, *Plasma Phys. Control. Fusion* **53**, p. 124028, doi:10.1088/0741-3335/53/12/124028.

Angelo A. Tuccillo, and Ceccuzzi, Silvio. (2013). Radio Frequency Power in Plasmas: Proceedings of the 20th Topical Conference (Sorrento, Italy, June 25-28, 2013), Vol. 1580, ISBN:978-0-7354-1210-1.

Saoutic, B., Beaumont, B., Becoulet, A., Bizaro, J. P., Fraboulet, D., Garbet, X., Goniche, M., Guiziou, L., Hoang, G. T., Hutter, T., Joffrin, E., Kuus, H., Litaudon, X., Mollard, P., Moreau, D., Nguyen, F., Pecquet, A. L., Peysson, Y., Rey, G., van Houtte, D., and Zabiego, M. (1994). High-power ICRF and LHCD experiments on Tore Supra, *American Institute of Physics* **289**, pp. 24-31, http://link.aip.org/link/?APC/289/24/1.

Swanson, D. G. (editor). (1985). Radio frequency plasma heating: Sixth Topical Conference (Callaway Gardens, GA), ISBN:0883183285.

Chapter 8

Plasma Diagnostics

There is a wide range of diagnostics designed to measure the properties of the plasmas in ITER. The extensive array of over fifty diagnostics will provide the measurements necessary to control, evaluate and optimize plasma fusion performance in ITER. These systems include measurements of temperature, density, impurity concentration, and particle and energy confinement times. As an example we first consider the electromagnetic plasma radiation. The electrons emit electromagnetic waves from their acceleration both during their close encounters with other charged particles and from their cyclotron acceleration in the strong magnetic field [Jackson (1999)]. The first type of radiation called Bremsstrahlung radiation results from the acceleration of the electrons as they respond to the Coulomb force between the ions and electrons producing a stronger (by three orders of magnitude) acceleration of the electrons than of the ions. The presence of impurity ions such as Be^{+4} or Ni^{+Z} with their respective densities n_Z in charge state Z produces the broad spectrum radiation P_{brem} proportional to the sum over all the ions, labeled subscript i, in various charged stages and with various masses. The plasma radiation is proportional to $n_Z Z^2 / m_z$ summed over all ions times the electron density n_e. The radiation is intrinsic to the hot plasma. The sum over all z of $n_Z Z^2 / n_e$ defines the effective ion charge Z_{eff} in the plasma. The electrons are also accelerated by their cyclotron frequency $\omega_{ce} = eB/m_e$ in their circular Larmor orbit about the magnetic field. This acceleration gives radiation that is proportional to the electron density and temperature.

Taken together these radiated electromagnetic waves give important, nonperturbative measures of the temperature, density and impurity concentration. There are many more features of radiation from the plasmas used to determine its properties as described in works on plasma diagnostics [Hutchinson (2002)]. Another key diagnostic from electromagnetic fields from accelerated elections is the Thomson scattering of an incident laser beam. The effective scattering cross section for the incident photons is given in Table 8.1 as σ_{Th}. The frequency spectrum of the scattered photons are used to infer both the electron density profile $n_e(r,t)$ and the electron temperature profile $T_e(r,t)$ in plasmas. This Thomson scattering diagnostic is highly developed and used in all high-temperature fusion experiments. Here the plasma electron is accelerated by the electric field in the incident laser field and the resulting radiation is measured in a series of photon detectors in directions with increasing angles away from the direction of the incident beam. This technique is described in detail in Jackson, Hutchinson, and numerous other texts on electromagnetic waves and plasma diagnostics. The dramatic verification of the announcement of the break through of key electrons temperatures in magnetic confinement came from the British team [Peacock, *et al.* (1969)] that took the newly developed laser Thomson scattering diagnostic to the Khurchatov Institute in Moscow measuring the electron temperature verifying the results announced by Artsimovich in 1967.

8.1 Plasma Spectroscopy

Spectroscopy is probably the first key diagnostic method for examining many materials and especially hot gases and plasmas. The element helium was discovered by Pierre Janssen in 1868 from the spectral analysis of solar radiation from the limb of the sun during a total solar eclipse. The yellow line at 487 nm was close to but distinct from the known lines of sodium so he postulated the existence of the new element helium named for Helios. The emission line was subsequently matched to the unknown gas from wells in Texas.

The electron temperature and the shifts in the electron velocity distribution functions are measured by Thomson scattering experiments. In Thomson scattering a laser or high-frequency RF beam, with the frequency much higher than the electron plasma frequency, passes through the plasma and is recorded in a series of detectors at various positions around the plasma. By the geometry of the transmitted beam and the received beam, and the shape of the frequency spectrum in the receivers, one derives the electron density fluctuation level at the vector wavenumber $\mathbf{k} = \mathbf{k}_{\text{out}} - \mathbf{k}_{\text{in}}$ that scattered the high-frequency input beam. The scattering

Table 8.1   Atomic and Nuclear Data.

c	velocity of light	3.00×10^{10} cm/s
e	electronic charge	4.80×10^{-10} esn
		1.6×10^{-19} C
m_e	electron mass	9.11×10^{-28} g
		511 KeV
h	Planck's constant	6.62×10^{-27} erg
$\frac{h}{m_e c}$	Compton wavelength	2.43×10^{-10} cm
$\mu = \frac{e\hbar}{2m_e}$	Bohr magneton	9.27×10^{-21} erg/gauss
$\frac{m_p}{m_e}$	proton to electron mass ratio	1836
$m_e c^2$	rest mass of electron	511 KeV
k_B	Boltzmann constant	1.38×10^{-16} erg/K
N_A	Avogradro's number	6.02×10^{23}
πa_0^2	Atomic cross-section	8.8×10^{-17} cm^2
1 barn	nuclear cross-section	10^{-24} cm^2
σ_{Th}	Thomson cross-section	6.66×10^{-26} cm^2

cross-section for the process is well known [Jackson (1999)] and is called the Thomson scattering cross-section given by $\sigma_{Th} = (8\pi/3)e^4/m_e^2 c^4 = 6.66 \times 10^{-26} \mathrm{cm}^2$ and the classical radius of the electron is $r_0 = e^2/m_e c^2$. Here, we deviate from the MKS units to the cgs units used in atomic and nuclear physics. The Bohr radius is $a_0 = 137^2 r_0$ where the constant $1/137$ arises from the dimensionless interaction parameter of quantum electrodynamics $\alpha = e^2/\hbar c \cong 1/137$. For quick reference Table 8.3 gives a number of atomic and nuclear parameters.

8.2 Beam induced Plasma Spectroscopy

The neutral atoms that remain in the laboratory plasma arise primarily from influxes from the walls and provide valuable information about the core plasma by their emissions from ion-impact ionization and charge exchange with high-temperature ions in the plasma core. In deuterium discharges the emission from electrons dropping to the $n = 1$ and $n = 2$ Bohr states called the Lyman and Balmer line emission series has been a key diagnostic and is the primary diagnostic used to monitor the L to H transitions in the tokamak. In addition, the line transitions from the charge exchange collisions of neutral atoms and plasma ions give important diagnostic information of the otherwise invisible hot plasma ions. In this atomic collision the hot plasma ions become neutral atoms and leave the plasma in a straight line with the same velocity it had at the time and position of the collision. Using appropriate external chambers (a stripping cell) to collect these high-energy neutrals (mass and energy spectrometers) gives a direct measure of the hot, fast ions in the core plasma.

The method of measuring the core ion distributions with charge exchange and recombination collisions is strongly enhanced by injecting well-defined beams of collimated neutral atoms with precise direction and energy. New diagnostic methods using the previously highly-developed method of forming energetic neutral beams for heating the plasma have been extended into methods for injecting precise monoenergetic neutral beams for probing the plasmas. This combination of a diagnostic neutral beam and the charge exchange recombination spectroscopy is probably the most important diagnostic available for assessing the success of an experiment for fusion power.

Here we will limit the discussion to a few of the more complex and important diagnostics for plasmas – not widely discussed in the text books – that requires a probing of the plasma with an injected beam of neutral atoms. We use the case of a beam of neutral hydrogen atoms injected with energies well above the temperature of the plasma. The technique is called Charge Recombination Spectroscopy. The hydrogen neutral beam accelerator is shown in Fig. 8.1 and the system is useful for a variety of diagnostic systems as described briefly in this chapter (Chapter 8, pp. 344-354 of Hutchinson). Further detail on the range the diagnostics is found at `https://www.iter.org/mach/diagnostics`. The plasma measuring systems is

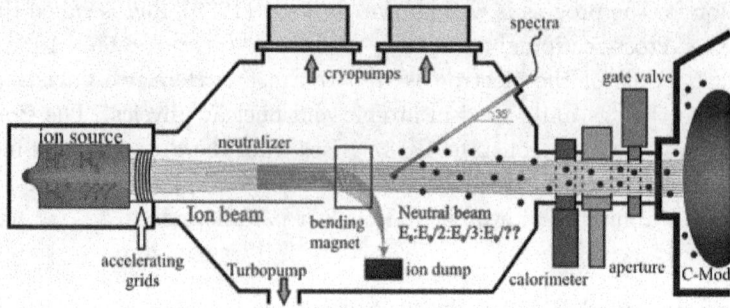

Fig. 8.1   Schematic of a Diagnostic Neutral Beam (DNB) system that is a key diagnostic system for the plasma ions and impurities. The ion source on the left are accelerated with $V = 50\,\mathrm{kV}$ with current of $I = 7\,\mathrm{A}$ for about 1 sec. Accelerated ions from the left enter the neutralizing chamber and then the remaining ions are turned away by the bending magnetic. The remaining neutral atoms, with the energies shown in the diagram, exit right on through the calorimeter and the gate into the tokamak, shown here as C-Mod [Bespamyanov, *et al.* (2006); Bespamyanov, *et al.* (2012)].

drawn from the full range of modern diagnostic techniques, including lasers, X-rays, neutron cameras, impurity monitors, particle spectrometers, radiation bolometers, pressure and gas analysis, and optical fibers.

There are variations in the name and abbreviations used in the fusion literature for this technique of using charge exchange atomic collisions to create radiative ions from the fully stripped working gas ions that are otherwise invisible in the fully ionized plasmas. Once the working gas stripped ion of the working gas (hydrogen, deuterium, helium) has picked up an electron from the injected neutral beam, spectroscopic analysis of the otherwise invisible fully stripped ions yields temperature, velocity vector and density information about the plasma ions. The names include charge exchange recombination spectroscopy abbreviated both CXRS, CERS and CRS. The neutral beam is called the Diagnostic Neutral Beam (DNB) and is used for beam emission spectroscopy called BES and the motional Stark effect called MSE.

A typical DNB is hydrogen neutrals at $50\,\mathrm{keV}$ with lower density half energy component at $25\,\mathrm{KeV}$ injected at various angles depending on the desired plasma measurements. For example in the case of helium plasma the hydrogen neutral beam may produce He^+ ions, equivalently in spectroscopic notation $H\,II$ ions in an excited state that would immediately emit $468\,\mathrm{nm}$ photons in a transition to the ground state. The emission spectrum is complex but yields detailed information about the work gas ions with minimal perturbation to the ion distributions.

8.3 Charge Exchange Recombination Spectroscopy

The charge exchange recombination spectroscopy, or CXRS method, yields information on the light impurity ions in the plasma. The CXRS data is used in Futatani,
et al. (2010) and Rowan, *et al.* (2008) the inward transport of the low mass-low-Z
impurity ions and outward transport of the working gas ions in Alcator C-Mod data.
Since the lighter impurities are fully stripped (fully ionized) in the hot plasma, one
injects a neutral beam H^o which undergoes charge exchange to become H^+ and
transfers an electron to the thermal plasma ion. The newly formed neutral, or partially ionized, thermal plasma atom is in an excited state then emits the measured
optical photons. In Fig. 8.2 the plasma is He and the DNB of H atoms produces
the He^{+1} ion in an excited state. In step 3 the excited He^{+1} ion decays to the
lower energy state of the He^{+1} ion emitting the sharp line photon indicated by $+\gamma$.
Figure 8.3 shows the gamma ray or UV line radiation collected in the "collection
optics" which will have a spectra distribution of emission lines.

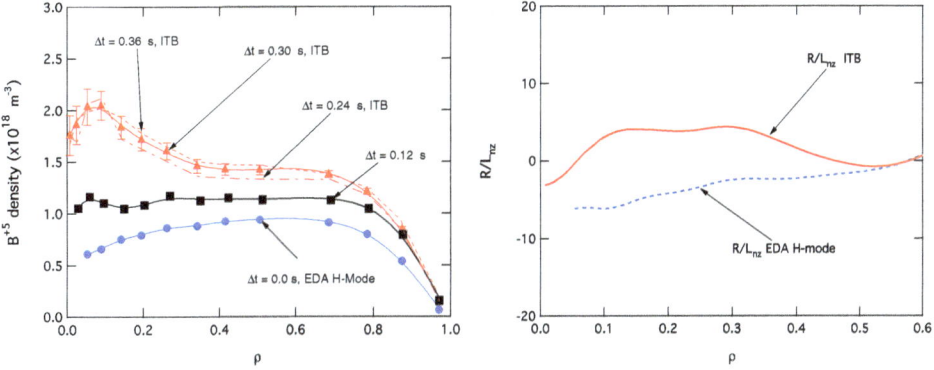

Fig. 8.2 (a) Time evolution of the boron density profile in the C-Mod discharge with 3.5-MW
ICRF heating, from the H-mode (blue circles), transition (black squares), to ITB (red triangles);
(b) boron density gradient profile in the H-mode (blue dash) and ITB (red solid) [Rowan, *et al.*
(2008)].

The measured optical emission, $\varepsilon(\mathbf{r}, \Omega, \lambda)$, is given by

$$\varepsilon(\mathbf{r},\Omega,\lambda)dV\,d\Omega\,d\lambda = \frac{1}{4\pi}\int\int f_n(\mathbf{r},\mathbf{v}_n)f(\mathbf{r},\mathbf{v})q_{em}(|\mathbf{v}_r|)\epsilon(\mathbf{v},\widehat{\mathbf{s}},\Omega,\lambda)d^3\mathbf{v}d^3\mathbf{v}_n \quad (8.1)$$

where ϵ is the atomic line emission model [Rowan, *et al.* (2008); Fiore, *et al.* (2004)].

Atomic line emission function has the form

$$\epsilon(\mathbf{v}, \hat{\mathbf{s}}, \mathbf{B}, \lambda) = \sum_n a_n(\Omega) \delta \left[\lambda - \lambda_n \left(1 + \frac{\mathbf{v} \cdot \hat{\mathbf{s}}}{c} \right) \right] \tag{8.2}$$

where $\hat{\mathbf{s}}$ is defined by the observational direction of the emitted photons.

STRAHL is a comprehensive diagnostic code frequently used for analysis of the optical and UV emissions for the optical emissions from the partially ionized ions in the plasma. The spectral analysis enables one to determine the density and temperature of the plasma.

STRAHL is a 1D transport code which solves coupled equations for the density of species $\langle n_s \rangle$ given the source g_s and the flux I_s

$$\frac{\partial \langle n_s \rangle}{\partial t} = - \left(\frac{\partial V}{\partial \rho} \right)^{-1} \frac{\partial}{\partial \rho} \left(\frac{\partial V}{\partial \rho} \langle \Gamma_s^\rho \rangle \right) + \langle g_s \rangle \tag{8.3}$$

where the surfaced-averaged particle flux is given by the

$$\langle \Gamma^r \rangle = -D^r(r) \frac{\partial n(r)}{\partial r} + v^r(r) n(r) \tag{8.4}$$

for the flux geometry defined in a subroutine.

The D and v parameters of the transport models are determined by minimizing the deviation of the measured data from the model data n_{pred} by the G functional given by

$$G[D(r), v(r)] = \frac{1}{N_r N_t} \sum_i^{N_r} \sum_j^{N_t} \left(n_{\text{meas}}(r_i, t_j) - \frac{\langle n_{\text{meas}} \rangle}{\langle n_{\text{pred}} \rangle} n_{\text{pred}}(r_i, t_j) \right)^2 \tag{8.5}$$

One generates a new best fit by adding random value perturbations to the current best fit and repeating the calculation of G. A neural network is used to choose the best parameter set efficiently.

Futatani, *et al.* (2010) and the Alcator C-Mod group show the results show that $D(r)$ and $v(r)$ are consistent with the turbulent transport driven by the drift waves fluctuations which include the dynamics of the impurity ions and their radial gradients. The impurity density starts in the first phase of the plasma formation as peaked in the edge region. However, the drift waves that include the impurities as an active component produce and inward flow of the impurities and outward flux of the working gas ions. The system is higher is complex than the single ions species plasma and is of critical importance in the fusion plasma since the incoming impurity ions dilute the fuel and would greatly reduce the fusion power. A typical impurity transport analysis is given for Alcator C-Mod in Fu, *et al.* (2012) using the CXRS data to follow the distribution of the impurity (Boron) plasma ions. For ITER the possible inward transport of beryllium ions will be a critical issue and would be measured with the CXRS diagnostics.

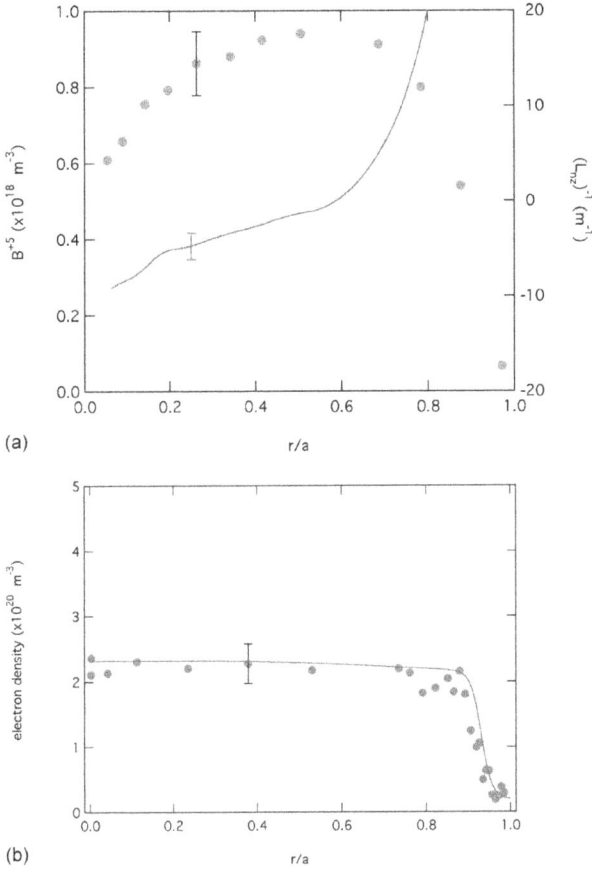

Fig. 8.3   The density profiles and inverse density gradient scale lengths of boron B^{5+} during an H-mode of the Alcator C-Mod tokamak. (a) Measurement of B^{5+} density profile (solid circles) and the inverse gradient scale length L_{nz}^{-1} of B^{5+} density. (b) The measurement of electron density profile (solid circles) and the data fit (solid line). Typical error bars are indicated [Futatani, *et al.* (2010); Fu, *et al.* (2013)].

8.4   Energy Distribution Functions for Electrons, Ions and Alpha Particles

The measurement of the electron density fluctuations down to the small spatial scale of the electron gyroradius is described by Gurchenko, *et al.* (2007). The method is to use the backscattering of ultra-high frequency (UHR) electromagnetic waves off the electron-density fluctuations. This type of backscattering from electron density fluctuations has been used since the 1960s for measuring plasma turbulence in the ionosphere. Ionospheric plasma turbulence is the most severe limiting factor in the fidelity of the global navigational satellite systems (GNSS). In the tokamak the strongest wave scattering is localized to the upper hybrid resonance layer, so that

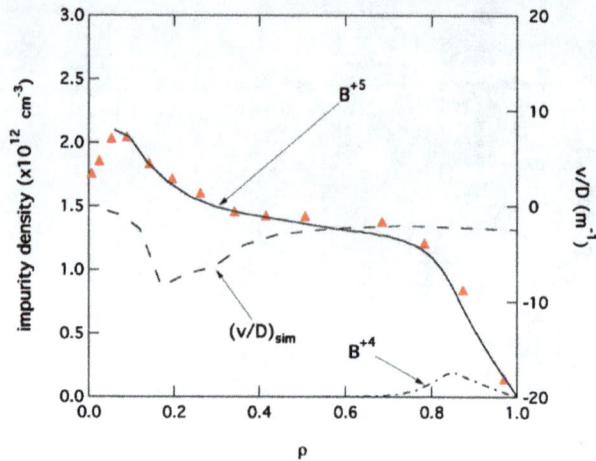

Fig. 8.4 Simulation of B^{5+} density (solid line) and B^{4+} density (dashed-dotted line) compared with the measured B^{5+} density v/D (dashed line) required in the simulation is plotted on the right axis [Futatani, *et al.* (2010); Fu, *et al.* (2013)]

by sweeping the UHR frequency from 50 to 70 GHz, the backscattered signal gives a radial profile of the electron density fluctuations. The scattering experiments performed on the FT-2 (Frascati Tokamak with upgrade 2) give space-time plots of the density fluctuations that validate the ETG turbulence model described in Chapter 4. The electron scattering experiments shows three space-time plots of the scattered power for time interval $t = 30$ to 38 milliseconds over a range of the plasma radius. The linear threshold for the ETG is observed in the scattering data. A sharp increase is observed in the small-scale density fluctuations once the stability parameter $\eta_e = L_n/L_{T_e}$ exceeds the critical value required for the onset of the electron temperature gradient instability. This critical value is well known from the drift wave stability theory [Horton (2012)], Chapter 4. In the Frascati plasma, as in most tokamaks, this critical value is easily exceeded with $\eta_e = L_n/L_{T_e}$ greater than unity. The resulting turbulence produces an anomalous transport of electron thermal energy across the confining magnetic field the rate derived in Chapter 4.4.

8.5   Scattering of High-Frequency Electromagnetic Waves from Plasmas

The electron temperature and the shifts in the electron velocity distribution functions are measured by Thomson scattering experiments. In Thomson scattering a laser or high-frequency RF beam, with the frequency much higher that the electron plasma frequency, is passed through the plasma and recorded in a series of detectors and various positions around the plasma. By the geometry of the transmitted

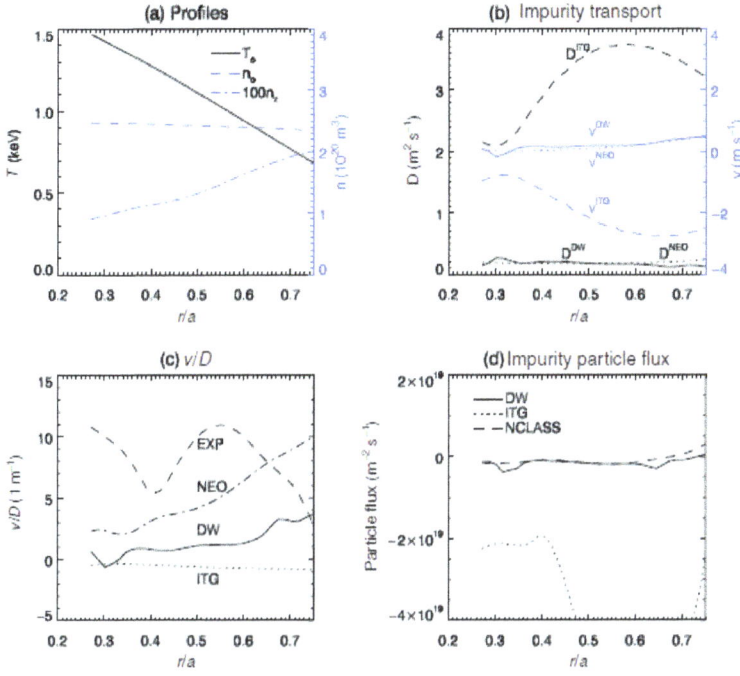

Fig. 8.5 (a) C-Mod shot #1070831028 H-mode (a) plasma density, electron temperature, and Boron density profile, (b) calculated diffusion coefficients (black curves) and convection velocity (blue curves), (c) comparison of V/D with neoclassical and experimental results ($= -1/L_{nz}$) and (d) neoclassical and turbulent particle flux for impurities. DW: impurity drift wave; ITG: ITG mode; NEO: neoclassical results; and EXP: experimental results.

beam and the received beam, and the shape of the frequency spectrum in the receivers, one derives the electron density fluctuation level at the vector wavenumber $\mathbf{k} = \mathbf{k}_{\text{out}} - \mathbf{k}_{\text{in}}$ that scattered the high-frequency input beam at $\mathbf{k}_{\text{in}}$. The scattering cross-section for the process is well known [Jackson (1999)] and is called the Thomson scattering cross-section given by $\sigma_T = (8\pi/3)e^4/m_e^2 c^4 = 2.6 \times 10^{-26} \text{cm}^2$ and the classical radius of the electron is $r_0 = e^2/m_e c^2$. Here, we deviate from the MKS units to the cgs units used in atomic and nuclear physics. The Bohr radius is $a_0 = (137)^2 r_0$, where the constant $1/137$ arises from the dimensionless interaction parameter of quantum electrodynamics $\alpha = e^2/\hbar c \cong 1/137$. For quick reference, Table 8.1 gives a number of atomic and nuclear parameters.

8.5.1 *Scattering of RF waves in the turbulent plasma*

The scattered electromagnetic wave has both a change in vector wavenumber $\delta \mathbf{k}$ and in frequency $\delta \omega$. However, the change in frequency $\delta \omega / \omega_{\text{RF}}$ is small until the electron velocity approaches the speed of light. So for the thermal plasma the

Table 8.2 Drift Wave Turbulence Classification according to the gradient driving mechanisms.

Mode	Abbreviation	Free Energy	λ, ω scale	Driver
Electron Temperature Gradient	ETG	$\dfrac{\nabla T_e}{T_e}$	$k_\perp \rho_s > 1$ $\omega < \omega_{*e}$	temperature gradient and bad curvature
Trapped Electron Mode	TEM	$\dfrac{\nabla T_e}{T_e} + \dfrac{\nabla n}{n}$	$k \perp \rho_s \sim 1$ $\omega < \omega_{*e}$	trapped electron precession resonance
Ion Temperature Gradient	ITG	$\dfrac{\nabla T_i}{T_i}$	$k_\perp \rho_s \sim 0.3$ $\omega < \omega_{*pi}$	temperature gradient and bad curvature
Trapped Ion Mode	TIM	$\dfrac{T_i}{T_e} + \dfrac{T_i}{T_e} + \dfrac{\nabla n}{n}$	$k_\perp \rho_s \sim 1$ $\omega \sim \omega_{*e}$	trapped ion precession resonance

Thomson scattering gives the wavenumber of the fluctuation scattering the incident wave. This method is used to measure the drift wave frequency spectrum as first done in the ATC experiment by Mazzucato and repeated in many experiments to derive the spectral distribution of the fluctuations. When the frequency of the incident electromagnetic wave is in the infrared or visible spectrum the frequency shift is given by the $\delta \boldsymbol{k} \cdot \boldsymbol{v}$ of the electrons. This is the method used to determine the electron temperature of the plasma. Typically, infrared lasers, particularly the CO_2 laser is used for this purpose. The method becomes difficult in regions of low electron density since the scattering cross-section is small – given by the square of the classical radius of the electron r_0 as shown in Table 8.3.

8.6 Magnetic Probes

The magnetic fluctuations in plasmas are important for understanding the MHD waves, particularly the Alfvén waves. The magnetic field probes are simple with a coil of N_t turns of wire on a circle of radius R_t. The coils can be arranged to have three perpendicular axes so as to measure the $\delta B_x, \delta B_y, \delta B_z$ components of the fluctuating magnetic fluxes from the $\delta B \Delta$ area of the magnetic probe. These probes are suitable for the plasma edge near the plasma chamber wall but disturb fusion plasmas too strongly to measure the interior magnetic fields.

For stable plasmas the interior magnetic field fluctuations are of order 10^{-4} of the ambient magnetic field, so mostly these fluctuating magnetic fields go unmeasured. Modern high-tech magnetometers as in cell phones use the Hall effect to measure the three axis magnetic field components as low as nano-tesla. The helicity injection torus called HIT at the University of Washington has a large number of these Hall effect probes for reconstructing the complex merging of the injected magnetic flux ropes into the central toroidal chamber to achieve steady-state toroidal current drive from magnetic helicity injection.

8.7 X-ray Spectra and Electron Cyclotron Emission

Key diagnostics measuring the electron density and momentum distributions are the spectrometers for the high-frequency electromagnetic waves emitted during electron acceleration. Historically, the radiation is called Bremsstrahlung – German for braking radiation – and was developed early by high-energy physicists to infer the distribution of electrons in astrophysics and High-Energy Physics Accelerators [Janossy (1948)]. Cosmic rays is the named used for highly relativistic-charged particles and was shown to be the source of the gamma-rays [γ rays] first measured by Wilson (1927), who invented the cloud chamber.

Instruments monitoring the Bremsstrahlung and Synchrotron radiation emitted by the tokamak plasma have been essential tools and will be used in ITER. Bremsstrahlung is one method used to measure the Z_{eff} when the electron density and temperature are known. Electron cyclotron emission (ECE) is used to measure the electron temperature and its profile along with Thomson scattering of a diagnostic laser beam. These diagnostics give the vital information about the electrons starting to "runaway," as it is called, when the higher energy and thus less collisional electrons that carry the toroidal plasma current become too strongly accelerated by the toroidal electric field in the plasma. In the early tokamak experiments the formation of runaway distributions was a serious problem that lead to the electron beams burning holes through the vacuum chamber walls in some experiments. Now, the radiation is monitored in real-time control systems to keep the electron currents at the desired level preventing the formation of runaway discharges. Using the RF waves to control the plasma current eliminates the runaway problem driven by Ohmic heating.

8.8 Langmuir Probes

In the edge plasma below $10\,\text{eV}$ in temperature and in the divertor chambers the characteristics of the plasma are measured with arrays of Langmuir probes. By inserting permanently, or for higher temperature plasmas using fast reciprocating probes, one can record the current drawn by the tip of the probe as a function of the bias voltage applied to the probe tip. The voltage on the probe tip is measured with respect to the walls of the vacuum chamber that confines the plasma. When the tip is biased sufficiently negative the electrons are reflected and the probe current is the ion saturation current given by $I_{\text{sat}} = en_e v_{\text{sat}} \times A_{\text{probe}}$ where sufficiently negative means a negative voltage comparable to the electron temperature T_e. Then as the probe tip voltage is swept to the corresponding positive values, the probe current decreases and goes through zero at the point called the floating potential given approximately by $V_{\text{sat}} = V_{\text{plasma}} + 4.5\,T_e/e$. Then as the probe tip swings further positive the probe current runs up to a large value and saturates at the electron saturation current where current density is approximately $j_{el} = -en_e(T_e/m_e)^{1/2}$ and thus an order of magnitude larger than the ion saturation current.

The full probe formula is simple for the unmagnetized plasma but for the magnetized plasma the formula is sufficiently complicated that we refer the reader to Hutchinson (2002). The complication arises in the magnetized plasma in that the charges move freely along the magnetic field but are restrained by their gyroradius in the motion across the magnetic field. Thus, a pre-sheath region of length L is large compared to the mean free-path of the ions so that the analysis becomes complicated.

The physical origin of the pre-sheath in the magnetized plasma is clear and immediately suggests that one can use the two different ion saturation currents collected on the opposite sides of a double-faced probe to determine the net flow velocity of the plasma along the magnetic field. This probe is important for analysis of the plasma flows in the SOL of large tokamaks and is called the Mach probe.

8.8.1 *Pellet injection as a diagnostic probe*

Pellet injection is used to fuel plasmas particularly for raising the density to its maximum allowed values and the emissions from the injected pellets are important for understanding the dynamics of impurities in the plasma. Lithium atoms have a convenient emission line in the visible spectrum at 670 nm which splits into a two distinct lines in a strong magnetic field by the Zeeman effect. Injection of pellets with lithium atoms is used to track and measure the transport of the lithium atoms during the progressive ionization of the injected pellet [Terry, *et al.* (2005)] entering the hot plasma.

A variation is to use the Stark effect on the pellet injected atoms to measure the local electric field in the plasma. This method is used both for transport of impurities in the plasma and for the measurement of the RF electric fields during intense RF heating of plasmas.

A clever version of this method to trace the transport properties of diffusion and convection of the high-Z metal impurities from the stainless steel chamber walls. An example of this diagnostic is used in test pellet injection experiments with specially designed pellets with three tracer elements center around the metals of Cr and Fe that are the impurities that arise from the erosion of the chamber walls as performed by Sudo, *et al.* (2012). These high-Z impurities are particularly dangerous in terms of rapidly degrading the performance of a fusion reactor. Thus, the diagnostic uses high-speed injection of encapsulated pellets with known amounts of V, Mn and Co which surround the elements Cr and Fe in the periodic table. The tracer pellets deposit 10^{17} of these tracer atoms in the plasma. The spectroscopy allows the instruments to follow the paths of these atoms in their ionized states with

Table 8.3   Nuclear Reactions.

Neutron Sources

$$D + D \rightarrow \mathrm{He}^3 + n$$
$$D + T \rightarrow \mathrm{He}^4 + n$$
$$\mathrm{Li}^7 + d \rightarrow \mathrm{Be}^8 + n$$
$$\mathrm{Be}^9 + d \rightarrow \mathrm{Be}^{10} + n$$

Neutron Detection

1) absorption of neutron with emission of fast charged particle
$$B^{10} + n \rightarrow \mathrm{Li} + \alpha$$
2) absorption by nucleus to form radio nucleus
3) absorption with fission of compound nucleus

the ions that reach the core plasma still retaining some core electrons that produce discrete emission lines. By combining the data with the high-performance spectral simulation code (STRAHL) the line emission data is converted into the density and transport fluxes of these heavy ions that have the same properties as the metal wall impurities. In Alcator C-Mod a large number of intrinsic impurities (B, F, Fe, Mo and W) were studied along with the seeded impurities with gas puffing of He, N, Ne, and Ar and with laser blow-off experiments for Al, Ca and Ni. The plasma facing metals were Mo and W. The research shows that the transport of the metal impurities to the core plasma is a strong function of plasma density and temperature consistent with drift wave turbulent impurity transport analysis [Fu, *et al.* (2012)].

8.8.2   *Alpha particle and neutron detectors*

In the future burning plasmas, the D-D and especially D-T plasmas with sufficient temperature to have a significant neutron and alpha particle emission rate [$T_i > 10\mathrm{KeV}$] other diagnostic instruments are used. Some of the highest ion temperature JET and TFTR plasmas were able to use the alpha particle spectra and neutron spectra for plasma diagnostics.

The alpha particle spectrum is measure with spectrometers in the wall of the vacuum chamber. These mass-energy spectrometers have window (a slit) through which the high-energy alpha particles enter into a box with crossed electric and magnetic fields. There are two mutually perpendicular scintillation counters that record the hits of the alpha particles. This instrument yields precise information on the energy and velocity spectrum of the escaping alpha particles.

The neutron fluxes are measured by using the electromagnetic emissions as the neutrons hit diamond detectors embedded in the chamber walls. The first detailed measurements of the fusion neutrons were made in TFTR [Nazikian, *et al.* (1997)].

8.9   Gas Puff Imagining and Phase Contrast Imaging

Gas puff imaging (GPI) diagnostic is a technique used to measure the structures in the plasma turbulence and was used in Alcator C-Mod plasmas to understand the vortex and blob structures that are emitted in the edge turbulence where the sharp density gradients is created by the magnetic separatrix [Terry, *et al.* (2005)]. The technique gives two dimensional images directly overcoming the uncertainties introduced by line-integrated optical measurements. With GPI using emission lines of D_α and He I emission lines clear pictures of the vortex structures often called "blobs" of higher density plasma from mm to cm sizes are followed in time. These features were known to exist earlier and were called by the generic name of QCMs for quasi-coherent modes.

With the GPI and phase contrast imaging diagnostics their structure has become clear and is now known to be the vortex structures found in the edge turbulence simulations of the scrape off layer SOL plasmas. The diagnostic in the GPI camera gives the spectral resolution in frequency f and wavenumber k_θ for with high time resolution recorded as movies of the structures being emitted from just inside the magnetic separatrix into the SOL plasma. The data from the inside high-field region of the torus shows much weaker vortex structures than those occurring in the low-field outboard plane as predicted by the drift waves driven by the product of the density gradient and the toroidal magnetic curvature vector.

In future burning plasmas, the D-D and D-T plasmas with sufficient temperature to have a significant neutron and alpha particle emission rate $[T_i > 10\text{KeV}]$ additional diagnostic instruments will become useful. Some of the highest ion temperature JET and TFTR plasmas were able to use the alpha particle spectra and neutron spectra for plasma diagnostics. The information from the phase-contrast imaging (PCI) and the beam emission spectroscopy (BES) diagnostics combine to give a wide range of wavenumbers and structure characteristics for the plasma turbulence in ITER.

References

Bespamyanov, I. O., *et al.* (2006). *Rev. Sci. Instrum.* **77**, p. 10F123.

Bespamyanov, I. O., *et al.* (2012). *Comput. Phys. Comm.* **183** p. 669-676.

Dong, J. Q., *et al.* (2013). (Iop Publishing and International Atomic Energy Agency, Nuclear Fusion) *Nucl. Fusion* **53**, 027001 (11pp), doi:10.1088/0029-5515/53/2/027001, APTWG: 2nd Asia-Pacific Transport Working Group Meeting.

Fiore, C. L., Bonoli, P. T., Ernst, D. R., Hubbard, A. E., Greenwald, M. J., Lynn, A., Marmar, E. S., Phillips, P., Redi, M. H., Rice, J. E., Wolfe, S. M., Wukitch, S. J., and Zhurovich, K. (2004). Control of internal transport barriers on Alcator C-Mod, *Phys. Plasmas* **11**, p. 2480, http://dx.doi.org/10.1063/1.1652785.

Fu, S. R., Horton, W., Bespamyanov, I. O., Rowan, W. L., Benkadda, S., Fiore, C. L., Futatani, S., and Liao, K. T. (2013), Turbulent impurity transport modeling for Alcator C-Mod, *J. Plasma Phys.* **79**, part 5, pp. 837846. (Cambridge University Press, 2013) doi:10.1017/S0022377813000548837.

Fu, X., Horton, W., Xiao, Y., Lin, Z., Sen, A. K., and Sokolov, V. (2012). Validation of electron temperature gradient turbulence in the Columbia Linear Machine, *Phys. Plasmas* **19**, pp. 032303-032310, http://dx.doi.org/10.1063/1.3686148.

Futatani, S., Horton, W., Benkadda, S., Bespamyanov, I. O., and Rowan, W. L. (2010). Fluid models of impurity transport via drift wave turbulence, *Phys. Plasmas* **17**, p. 072512, doi:10.1063/1.3459062.

Gurchenko, A. D., Gusakov, E. Z., Altukhov, A. B., Stepanov, A. Yu., Esipov, L. A., Kantor, M. Yu, Kouprienko, D. V., Dyachenko, V. V., and Lashkul, S. I. (2007), Observation of the ETG mode component of tokamak plasma turbulence by the UHR backscattering diagnostics, *Nucl. Fusion* **47**, pp. 245-250 (2007), http://iopscience.iop.org/0029-5515/47/4/001.

Horton, W. (2012). *Turbulent Transport in Magnetized Plasmas* (World Scientific) ISBN:978-981-4383-53-0.

Hutchinson, I. H. (2002). *Principles of Plasma Diagnostics*, 2nd Edition, Chapter pp. 74-89 (Cambridge University Press, Cambridge, UK) ISBN:0-521-67574-X.

Jackson, J. D. (1999). *Classical Electrodynamics* (Wiley, New York) 3rd Ed., pp. 708-724.

Janossy, L. (1948). *Cosmic Rays* (Oxford Press).

Liao, K. T. (2014). Helium Charge Exchange Spectroscopy on Alcator C-Mod Tokamak. PhD Thesis. The University of Texas at Austin.

Nazikian, R., *et al.* (1997). *Fusion Energy*, International Atomic Energy Agency, p. 281.

Peacock, N. J., Robinson, D. C., Forrest, M. J., Wilcock, P. D., and Sannikov, V. V. (1969). Measurement of the electron temperature by Thomson scattering in tokamak T3, *Nature* **224**, pp. 488-490, doi:10.1038/224488a0.

Rowan, W. L., Bespamyatnov, Igor O., and Fiore, C. L. (2008). Light impurity transport at an internal transport barrier in Alcator C-Mod, *Nucl. Fusion* **48**, 10, p. 105005, http://stacks.iop.org/0029-5515/48/i=10/a=105005.

Sudo, S., Tamura, N., Suzuki, C., Muto, S., Funaba, H., and LHD Group (2012). Multiple-tracer TESPEL injection for studying impurity behavior in a magnetically confined plasma, *Nucl. Fusion* **52**, p. 063012, doi 10.1088/0029-5515/52/6/063012.

Terry, J. L., Basse, N. P., Cziegler, I., Greenwald, M., Grulke, O., LaBombard, B., Zweben, S. J., Edlund, E. M., Hughes, J. W., Lin, L., Lin, Y., Porkolab, M., Sampsell, M., Veto, B., and Wukitch, S. J. (2005). Transport phenomena in the edge of Alcator C-Mod plasmas, *Nucl. Fusion* **45**, p. 1321, doi 10.1088/0029-5515/45/11/013.

Wilson, C. T. R. (1927). On the cloud method of making visible ions and the tracks of ionizing particles (*Nobel Lecture*, December 12, 1927).

Chapter 9

Plasma Facing Components and Plasma-Wall Interaction Physics

The WEST experiment is designed to run steady with a fusion grade plasma for testing the physics models of the plasma facing components, defined as PFCs in the literature. The behavior of the PFCs in the ITER experiment is area of concern as relatively little is known about the problem for the long pulse durations of 500 s under the conditions of multi-KeV plasma core temperatures. The nominal design specification for the integrity of the beryllium coated stainless steel walls in ITER is $10\,\mathrm{MW/m^2}$ for thousands of discharges.

The Southwestern Institute of Physics in Chengdu, China, as a partner in the WEST project, is building additional toroidal coils that will produce the divertor poloidal magnetic field. The experiment will have the lower null-divertor magnetic configuration as an exact model of the separate flux surface in the ITER. Information for the status of the WEST project can be found in Newsletters at the website `http://west.cea.fr`. Research and industrial participants in the construction and design include both Europe (F4E) and Japan (JADA)whose mission it is to build ITER. The machine will have 48 stainless steel panels for the vacuum vessel and 480 magnetic sensors built into WEST vessel walls. The WEST website, `http://west.cea.fr`, at the Institute for Magnetic Fusion Research (IRFM) at CEA Cadarache has the current status of the experiment.

Since the same technology planned for ITER will be used in building the plasma-facing walls (PFCs) of WEST, the early 2015-2018 operation of WEST allows some assurances that the specifications for the materials and the fabrication of the chamber walls and divertor chamber for ITER will be fulfilled successfully.

Dense sets of flush-mounted Langmuir probes are built into the plasma facing components to provide detailed data for understanding the stresses and temperature limits for blistering and bubble formation on the PFC surfaces. Those components in ITER that will be subjected to intense plasma bombardment including parts to the divertor chamber, the divertor dome, neighboring components and the RF antennas. These components will be actively cooled by forcing water through the core of large arrays of rectangular tungsten blocks as shown in Fig. 9.1. These PFCs are long rectangular tubes with cross-sections of about $3\,\mathrm{cm} \times 3\,\mathrm{cm}$ and are called "monoblocks". The power handling limits of the tungsten monoblocks will

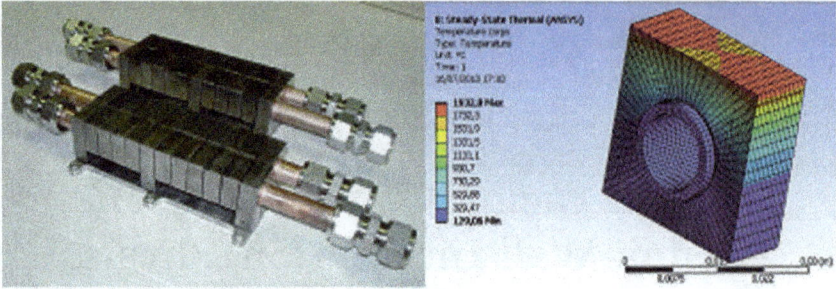

Fig. 9.1   Plasma wall architecture for WEST. (a) Tungsten monoblocks with water pipes for cooling the first wall of the superconducting steady-state machine. (b) Simulation of the temperature
distributions on a monoblock approaching the critical condition for blistering (M. Grossman, CEA,
2014).

be determined on WEST. The name WEST derives from these features: Tungsten
(W) Environment (E) Steady state (S) Tokamak (T).

WEST is under construction at the time of writing and will be a full model of
the steady-state operation of the $500\,\mathrm{m}^3$ ITER machine in a $15\,\mathrm{m}^3$ superconducting
steady state tokamak. The principal heating and current drive for the steady state
operation is provided by the two high-power lower-hybrid antennas described in
Chapter 4 and Fig. 4.1. These RF antennas were successful in producing a steady
fusion grade plasmas for periods from several minutes up to six minutes in the pro-
cessor – Tore Supra – with circular cross-section and conventional plasma limiters.
WEST will have the divertor structure shown in Fig. 9.2.

9.1   Crystallization and Melting Limits

As the temperature of the tungsten monoblocks and other tungsten parts of the machine rises there are two critical temperatures. The recrystallization temperature
T_{rec} =1000-1200 C and the melting temperature $T_{\mathrm{melt}} = 3700\,\mathrm{K}$ for the tungsten
coated surfaces. The machines will be operated so as to allow the wall temperature
rise to and above the recrystallization temperature but stay well below the melting
temperature for the modeling of the ITER discharges. There may rarely be unexpected excursions of the plasma temperature that would raise the PFC components
to above their temperature limits but the plasma control systems based on real time
input-out controls is designed to forecast the probability of such an excursion and
then to trigger a massive inert gas (such as argon) injection into the chamber to
immediately cool the edge plasma.

In the final testing in WEST the heat loads on the PFCs would be allowed
to run up to higher values for short periods to explore the regimes where severe
wall damage occurs. Such experiments are used to determine the lifetime of the
components.

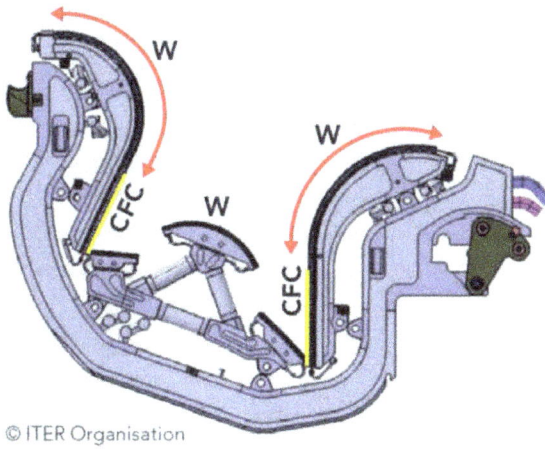

Fig. 9.2   Architecture of the divertor structure for ITER with Tungsten W walls covered with the monoblocks shown in Fig. 9.1 where the first wall receives the most intense plasma heating and covered with carbon-facing components where the plasma wall-interactions are less intense (M. Grossman, CEA, 2014).

The solid state aspect of the tungsten damage in WEST is in the regime called chemical surface damage where the atomic bounds are broken by the atomic vibrations from the high temperature of the atoms in the lattice.

The second wall damage that will be tested in ITER is the nuclear wall damage of the tungsten that arises from the $14\,\mathrm{MeV}$ neutrons knocking out the wall atoms as they recoil from the neutron collisions. The number of atoms displaced in this manner strictly limits the integrity of the walls.

The nuclear cross-sections for the neutrons "knocking" out atoms in solid-state materials is described by Halliday (1950) (pp. 224-253) with the subsection on the Boron detector based on $B^{10} + n \rightarrow Li^7 + \alpha$ being particularly relevant to fusion machines. The neutron cross-sections for both knock-on collisions and other elastic collisions are of order 10 to 100 barns where 1 barn is $10^{-24}\,\mathrm{cm}^2$. However, there are complex nuclear resonances for the collisions as used in the B^{10} isotope for the neutron detectors.

Heavier nuclei as in U^{235} and U^{238} have very large resonance cross-sections reaching 1,000 barns for slow or thermal neutrons. The wave properties of neutrons produce wave diffraction [Born and Wolf (1959)] that allows diagnostic instruments similar to those used for X-ray detectors.

Graphite is used as a way to thermalize neutrons, since neutron-proton collisions are most effective in transferring energy from the neutrons to the blanket material. ITER has a neutron blanket or shield between inner metallic wall facing the plasma and the outer magnetic field coils. This shield is necessary to protect the outer superconductors from the neutron flux. The shield has regions where samples

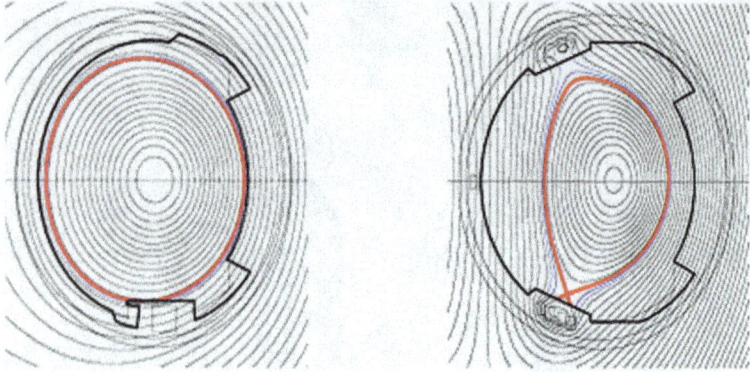

Fig. 9.3   The two types of magnetic field geometries used in tokamaks. The left figure, 9.3a, gives the classical closed flux surfaces as used in Tore Supra and TFTR for example and the poloidal flux function. The right figure, 9.3b, gives the single null magnetic divertor magnetic configuration as used in JET and JT-60U and will be used in ITER. Figure 9.3b shows the X-point that divides interior hot confined plasma from the exterior plasma that flows freely to the divertor chamber. Sharp density and temperature gradients develop across the last closed flux surface (M. Grossman, CEA, 2014).

of materials can be tested for the future fusion reactor blankets that convert the neutron energy into thermal energy.

For tungsten walls the number of displacements per atom, known as DPAs in the literature, from the neutron flux is thought to be limited to 100 to 150 for the useful lifetime of a fusion reactor walls. The cross-section σ_n for the neutron to knock a tungsten atom out of the lattice is very small since the neutron has no electric charge and the collision cross-section is directly given by the area of the nucleus of the tungsten atom. The charged alpha particles are largely deflected by the magnetic field so their flux to the wall is much lower, however the cross-section for the inelastic collision of alphas with the tungsten atoms is larger. WEST will not have a significant neutron flux so the testing for nuclear wall damage is to be carried out in a different machines dedicated to this reactor wall lifetime research.

Surface melting experiments are reported from IFRM Cadarache and from the TEXTOR laboratory at Julich in Germany. These experiments show under what conditions the cracks that develop on the surfaces of the monoblock tubes and the time taken for the cracks to spread to the core of the tubes [Dejarnac, *et al.* (2014); Gunn, *et al.* (2011)].

The core plasma temperature in a fusion reactor may easily run up higher than 15 KeV. For example, the core temperature during transients discharges in the JET and JT-60U machines reached values up to 30 to 40 KeV. Some of this hot core plasma will cross the magnetic separatrix and flow in the scrape-off layer near to and perhaps touching the first metallic wall. These temperatures are well above the melting point temperature T_m of the metals.

The metal with the highest melting temperature is tungsten W which melts at 3653 K or 3380 C. Before reaching the melting temperature, however, the walls become unusable for the confinement of the plasma. Molybdenum has been used for years on the walls of the Alcator-C Mod with a coating of boron (melting point 2570 K). The boron coating of the walls significantly improved the performance of Alcator.

The melting point of Mo and carbon C are approximately 2900 K. In JET the original carbon-coated surface walls proved unable to survive the intense plasma heating. Subsequently, Dietz (1990) and the JET Team changed the wall coatings to beryllium, proving the superior plasma behavior with Be coated walls.

9.2   Wall Erosion Due to Evaporation

When solids are heated to high temperatures but well below their melting point T_m temperature, there is an erosion and blistering of the surface material. The atoms have sufficient kinetic energy to overcome the surface chemical binding energies and thus evaporate from the surface in clouds. The rate of the evaporation is characterized by the increase in the pressure at the surface defined as the vapor pressure. Vapor pressure for metals increases with temperature as given by the Boltzmann law

$$p_{\text{vapor}} = p_o \exp\left[-\frac{\delta H}{k_B T}\right],$$
(9.1)

where δH is the atomic surface binding energy of wall material which is typically stainless steel.

As the surface temperature increases, clouds of surface material are blown off into the neighboring plasma, much as dust devils blow off dust on hot sandy desert floors from solar heating of the SiO_2 sand grains.

9.3   Wall Blistering Below the Melting Temperature

A second process is blistering of the surface material from the bombardment of the surface by the hydrogen ions in the plasma. The metals for walls divide into two groups: the first group is V, Nb, Ta, Ti and Zr, which have positive heat of solutions for hydrogen and thus high solubility. The second group is Mo, W, Cu and Be, which have negative heat solution or heat of solubility in hydrogen. These properties control the degree of blistering of the material as the wall temperature rises.

The blisters will be measured in the WEST tokamak. This machine will allow extensive studies of the properties of candidate wall metals for plasma with confinement electron temperatures from 1 to 2 KeV.

A typical blister on the metal surface may have a height $h = 10\,\mu\text{m}$ and diameter $d = 20\,\mu\text{m}$ with a shell thickness of a few microns. The pressure inside the blister is much greater than atmospheric pressure.

The special wall insulating materials and the protective wall blocks of tungsten W, called monoblocks, shown in Fig. 9.1, being designed for ITER, will be tested in WEST. These long-thin wall-protecting blocks about $2\,\mathrm{cm} \times 2\,\mathrm{cm}$ in cross-section have a water cooling hole down their center. The monoblocks will line the walls where the intense plasma bombardment is measured. The blocks can then be removed, studied in detail for blistering and erosion and replaced with newly designed blocks well before the final wall assembly is carried out for ITER. The proceedings of a June 2014 International Workshop devoted to the plans for WEST are available at `http://west.cea.fr./workshop2014`.

The Eighth IAEA Technical Meeting on Steady-State Operation of Magnetic Fusion Devices; 26-29 May 2015, Nara, Japan identified the following key issues for future research for ITER and the following DEMO fusion reactors: (1) superconducting devices, (2) long-pulse operation, (3) steady-state fusion-plasma wall interactions and (4) particle control and power exhaust.

9.4   Radiation Limits for Fusion Reactors

The material wall power flux limit of $10\,\mathrm{MW/m^2}$ implies the need for a substantial fraction of the escaping power to be in the form of radiation power. For the reference parameters of ITER operation at $n_e T \tau_E = 10^{21}\mathrm{KeV\text{-}s/m^3}$ and the reference confinement time of $\tau_E = 5\,\mathrm{s}$ gives the radiated power $P_{\mathrm{rad}} \cong 0.1\,\mathrm{MW/m^2}$ over walls of the $500\,\mathrm{m^3}$ plasma volume.

For the nominal limit of wall loading of $10\,\mathrm{MW/m^2}$ in Sec. 9.1 as the reference value, we can then determine the expected radiation on the walls from Bremsstrahlung and impurity line radiation. Following the calculations of Mirnov (2009), the radiated power load on the first wall is 0.5 to $1.0\,\mathrm{MW/m^2}$. Kotschenreuther, *et al.* (2008) recommend a design forcing the machine to operate with 80% of the power radiated.

The plasma power crossing the separatrix and striking the divertor plates must be reduced by the radiated power to critical limits for the survival in the steady state of the divertor chamber walls and the divertor dome. When the divertor chamber is too small in volume, then most of the power from the plasma crossing into the open field line regions needs to be radiated. The leading demonstrated method for providing the maximum protection to the divertor chamber in the tokamak is to raise the impurity radiation by the injection of an impurity gas (argon or neon) with n_{imp} level approaching that of the low electron density $n_{\mathrm{imp}} \lesssim 10^{19}\mathrm{m^{-3}}$ in the scrape-off layer (SOL) as described in Apicella, *et al.* (2009).

9.5   International Fusion Materials Irradiation Facility (IFMIF)

Neutrons from fusion reactions have a higher kinetic energy than those from fission reactions. This makes it difficult to estimate the rate of damage and the lifetime of the metal walls in the future fusion reactor.

The displacement of the atoms accumulates with time so the number of allowed d.p.a. effectively determines the lifetime of the walls in the fusion reactor. Current research suggest that the expected lifetime for the ITER-like walls is 125 d.p.a. Then with this limit known, one still needs the results of the ITER experiments to determine how long this amounts to in years for the mean replacement time for the reactor walls. Clearly, wall replacement would require a major time and expense so the cost of the fusion power in Euros per kw.hr is a steep function of the reactor wall lifetimes.

To answer these questions on the nuclear and atomic physics of the walls as soon as possible, a new project called the International Fusion Materials Irradiation Facility was started at the Rokkasho ITER site in Japan. At this facility, the Rokkasho Laboratory, the first of several joint research projects between Japan and the European Union (EU) is being carried out in support of fusion science and fusion power development. This agreement for a joint EU and Japanese research program is called the Broader Approach for Fusion Energy Research. The questions of the lifetime of the walls of the tokamak is clearly an issue that applies across the range of all fusion power systems. The Broader Approach activities agreement was signed on 1 June 2007 and is working to support the final design of the ITER walls and neutron blankets.

There are two later projects in addition to the Fusion Materials Test Facility. The second project is to build a follow-up to the successful JT-60U machine called the JT-60SA where informally the fusion community understands that SA denotes "super advanced". The new fusion reactor research establishment of an International Fusion Energy Research Center is called IFERC.

The information derived from the broader approach (BA) research is largely complementary that we will derive from the ITER machine which measures plasma confinement in response to high-power auxiliary heating. The information from the BA is needed to proceed with the next level machine called DEMO and for the design of any other fusion reactor machine based on the large helical device (LHD), a mirror reactor or a reversed field pinch machine. There have been numerous BA workshops and details plans are advancing rapidly as found on the website http://www.ba-fusion/org.

The core plasma temperature in a fusion reactor must run higher than 15 keV. Temperatures have reached values up to 30 to 40 KeV as transients in the JET and JT-60U discharges. When solids, as in the metal walls of the confinement vessel, are heated by these plasma temperatures they rapidly melt. The metal with the highest melting temperature is tungsten W, which melts at 3653 K or 3380 C.

Before reaching the melting temperature, however, the walls become unusable for confinement of the plasma. Molybdenum has been used for a decade on the walls of the Alcator-C Mod with a coating of Boron. In JET the original carbon walls proved unable to survive the intense plasma heating and the wall surface coating material was changed to Beryllium, resulting in a marked improvement in the purity of the plasma. In Table 8.1 we give the melting temperatures of a number of elements and material including a common stainless steel (SS304 with $T_m = 1400\,\text{K}$) and Boron (T_m (boron) $= 2573\,\text{K}$).

9.6   Lifetime of Wall and Divertor Elements

Impurity transport observations from the ASDEX upgrade machine and JET are discussed in relation to the core particle transport in Gruber, *et al.* (1999), and for the Joint European Torus in Pamela, *et al.* (2003). These articles state that a robust experimental behavior is observed in both the devices and that the related parametric dependences for the plasma are identified. The observations are compared with theoretical and simulation results derived from the core turbulent particle transport to and from the wall.

We see that the path of the impurities from the chamber wall, labeled plasma-facing components in Fig. 9.3, has three paths. The most critical path is the component labeled Accumulation that goes into the core plasma. The second component of higher density remains in the low-density scrape-off layer plasma and remains there to interact again with the walls, and flows to the divertor chamber to be flushed from the machine. There is a third component that returns immediately to the first wall called the plasma-facing components, to interact again with the wall and its coating. This component must be kept small and is perhaps the most difficult component to control.

Impurity transport studies in the high density core of the Large Helical Device show agreement with the neoclassical collision dominated model with positive $E_r(r)$ expelling impurity ions in low-collisionality regime, but in mid-collisionality regime, the impurity transport data shows impurity accumulation. The positive $E_r(r)$ regime is called the electron root of the ambipolar condition. In tokamaks the usual regime is a negative $E_r(r)$ from the combination of collisional and turbulence transport and is called the ion root of the ambipolar condition. In this regime of an inward pointing radial electric field the plasma rotates in the electron diamagnetic direction and the impurity ions are pushed into the plasma until their profile becomes peaked in the core and the turbulent diffusion rises to produce a balance with a small residual impurity ion radial flux [Nakamura, *et al.* (2014)].

The problem of using ITER to determine the lifetime of proposed fusion reactor walls was discovered by the INTOR team and described in Stacey (2010). At some point the INTOR design team group realized the conflict between the demands of a machine designed like ITER to achieve break-even or high-Q plasmas called "fusion power-relevant plasmas") and a different type of design to produce "reactor relevant engineering technology" [Stacey (2010)], p. 78-80. This issue was not solved and the conclusion was that long run times at high neutron flux levels would be required from a different machine. This branch in research directions has now been accepted in the bilateral agreement between EU and Japan to build a new machine designed to test the life time of proposed wall components. The new project called IFMIF is described in Chapter 10.

The physics of the problem is that determining the life time of the walls requires a higher neutron fluence (defined by the neutron flux times the exposure time) which is estimated to require running times much longer that would be affordable or practical with the "fusion power-relevant" ITER-like machine. The exposure time for testing wall designs was estimated to $4\,\mathrm{MW/m^2}$ for a number of years. The INTOR group concluded that it would not be feasible to build a "dual-purpose" machine to test both these goals. One needs a high fluence neutron source without regard to the fusion power amplification to test the engineering of the walls.

9.7   Surface Quantum Physics

The surface plasma-wall interaction is governed by quantum surface physics and thus is not amenable to direct engineering control. The quantum states at the surface depend on the surface materials. Gaining knowledge of the plasma-surface interactions is an active area of research at the time of writing. Special laboratory experiments are underway at the Dusty Plasma Laboratory at Aix-Marseilles University and at University of Auburn (Prof. Edward Thomas, in University of Auburn, etjr@physics.auburn.edu) to develop a predictive understanding of the plasma-surface interactions.

In the history of the plasma confinement experiments the first wall coating materials were the carbon (C) materials used in spacecraft re-entry shields. The plasma experiments showed what were called the "carbon blooms" as the confinement temperatures increased. Then the plasma-facing material lining the metal chamber walls was changed to boron (B) and then finally to beryllium (Be). The beryllium-coated walls introduced by Dietz, *et al.* (1990) in JET have shown the best performance and thus were chosen for the wall-lining element of the PFC for the ITER machine. Recalling that the order of the periodic table is H, He, Li, Be, B and C, we see that the only remaining lighter element to use for the wall-lining surface is lithium. Liquid lithium walls have a flowing liquid metal coating of lithium and are an active area of current experiments in the NSTX tokamak [Kaye, *et al.* (2007)].

Table 8.1 gives the melting temperatures of a number of elements and material used in toroidal plasma experiments, including a common stainless steel (SS304 with $T_m = 1400\,\mathrm{K}$) and a number of metals used in nuclear reactors.

References

Apicella, M. L., Lazarev, V., Lybulinski, Il., Massitelli, G., Mirnov, S., Vertkov, A. (2009). Lithium capillary porous system behavior as PFM in FTU tokamak experiments, *J. Nucl. Mater.* **386-388**, pp. 821-823, doi:10.1016/j.jnucmat.2008.12.238.

Born, Max, and Wolf, Emil (1959). *Principles of Optics-Electromagnetic Theory of Propagation, Interference and Diffraction of Light* (Pergamon Press, London, 1959).

Dejarnac, R., Podolnik, A., Komm, M., Arnoux, G., Coenen, J. W., Devaux, S., Frassinetti, L., Gunn, J. P., Matthews, G. F., Pitts, R. A., and JET-EFDA Contributors. (2014). Numerical evaluation of heat flux and surface temperature on a misaligned JET divertor W lamella during ELMs, *Nucl. Fusion* **54**, 12R, p. 123011, doi:10.1088/0029-5515/54/12/123011.

Dietz, K. J., and The JET Team (1990). Effect of beryllium on plasma performance in JET, *Plasma Phys. Control. Fusion* **32**, pp. 837-852(1990), http://dx.doi.org/10.1088/0741-3335/32/11/002).

Gruber, O., Bosch, H.-S., Günter, S., Herrmann, A., Kallenbach, A., Kaufmann, K. Krieger, Lackner, K., Mertens, V., Neu, R., Ryter, F., Schweinzer, J., Stäbler, A., Suttrop, W., Wolf, R., Asmussen, K., Bard, A., Becker, G., Behler, K., Behringer, K., Bergmann, A., Bessenrodt-Weberpals, M., Borrass, K., Braamsa, B., Brambilla, M., Brandenburgb, R., Braun, F., Brinkschulte, H., Brückner, R., Brüsehaber, B., Büchl, K., Buhler, A., Callaghanc, H. P., Carlson, A., Coster, D. P., Cupidod, L., de Peña Hempel, S., Dorn, C., Drube, R., Dux, R., Egorove, S., Engelhardt, W., Fahrbach, H.-U., Fantzf, U., Feist, H.-U., Franzen, P., Fuchs, J. C., Fussmann, G., Gafert, J., Gantenbeing, G., Gehre, O., Geier, A., Gernhardt, J., Gubanka, E., Gude, A., Haas, G., Hallatschek, K., Hartmann, D., Heinemann, B., Herppich, G., Herrmann, W., Hofmeister, F., Holzhauerg, E., Jacobi, D., Kakoulidish, M., Karakatsanish, N., Kardaun, O., Khutoretskii, A., Kollotzek, H., Kötterl, S., Kraus, W., Kurzan, B., Kyriakakish, G., Lang, P. T., Lang, R. S., Laux, M., Lengyel, L. L., Leuterer, F., Lorenz, A., Maier, H., Mansod, M., Maraschek, M., Markoulakih, M., Mast, K.-F., McCarthyc, P. J., Meisel, D., Meister, H., Merkel, R., Meskatg, J. P., Müller, H. W., Münich, M., Murmann, H., Napiontek, B., Neu, G., Neuhauser, J., Niethammer, M., Noterdaeme, J.-M., Pautasso, G., Peeters, A. G., Pereverzev, G., Pinches, S., Raupp, G., Reinmüller, K., Riedl, R., Rohde, V., Röhr, H., Roth, J., Salzmann, H., Sandmann, W., Schilling, H.-B., Schlögl, D., Schmidtmann, K., Schneider, H., Schneider, R., Schneider, W., Schramm, G., Schweizer, S., Schwörer, R. R., Scott, B. D., Seidel, U., Serrad, F., Sesnic, S., Sihler, C., Silvad, A., Speth, E., Steuer, K.-H., Stober, J., Streibl, B., Thoma, A., Treutterer, W., Troppmann, M., Tsoish, N., Ullrich, W., Ulrich, M., Varelad, P., Verbeek, H., Vollmer, O., Wedler,

H., Weinlich, M., Wenzel, U., Wesner, F., Wunderlich, R., Xantopoulosh, N., Yuj, Q., Zasche, D., Zehetbauer, T., Zehrfeld, H.-P., Zohmg, H., and Zouhar, M. (1999). Overview of ASDEX Upgrade Results, *Nucl. Fusion* **39**, p. 1321, doi:10.1088/0029-5515/39/9Y/309.

Gunn, J. P., Pascal, J.-Y., Saint-Laurent, F., and Gil, C. (2011). Electric Probes in Tokamaks: Experience in Tore Supra, Contributor *Plasma Phys.* **51**, pp. 256-263, doi: 10.1002/ctpp.201000074.

Halliday, D. (1950). *Neutron production, detection and interaction with matter*, Ch. 6 pp. 214-254, (John Wiley and Sons, 1950).

Kaye, S. M., Levinton, F. M., Stutman, D., Tritz, K., Yuh, H., Bell, M. G., Bell, R. E., Domier, C. W., Gates, D., Horton, W., Kim, J.-Y. LeBlanc, B. P., Luhman, Jr., N. C., Maingi, R., Mazzucato, E., Menard, J. E., Mikkelsen, D., Mueller, D., Park, H., Rewoldt, G., Sabbagh, S. A., Smith, D. R., and Wang, W. (2007). Confinement and Transport in the National Spherical Torus Experiment, *Nucl. Fusion* **47**, p. 499, doi:10.1088/0029-5515/47/7/001.

Kotschenreuther, M., Valanju, P., Mahajan, S., Zheng, L. J., Pearlstein, L. D., Bulmer, R. H., Canik, J., and Maingi, R. (2008). The super x divertor (sxd) and high power density experiment (hpdx) (22nd International Atomic Energy Agency Fusion).

Mirnov, S. J. (2009). Plasma-wall interactions and plasma behaviour in fusion devices with liquid lithium plasma facing components, *Nucl. Mater.*, **390-391**, pp. 876-865, doi:10.1016/j.jnucmat.2009.01.228.

Nakamura, Y., Kobayashi, M., Yoshimura, S., Tamura, N., Yoshinuma, M., Tanaka, K., Suzuki, C., Peterson, B. J., Sakamoto, R., Morisaki, T., and the LHD Experiment Group. (2014). Impurity shielding criteria for steady state hydrogen plasmas in the LHD, a heliotron-type device, *Plasma Phys. Control. Fusion* **56** p. 075014, doi:10.1088/0741-3335/56/7/075014.

Pamela, J., Solano, E. R., and JET EFDA Contributors. (2003). Overview of JET results, *Nucl. Fusion* **43**, p. 1540, doi:10.1088/0029-5515/43/12/002.

Stacey, W. M. (2910). *The Quest for a Fusion Energy Reactor*, pp. 78-80 (Oxford Press, 2010), ISBN:978-0-19-973384-2.

Chapter 10

The Broader Approach and Tritium Breeding Blankets

The major portion of the energy and power – [14.1 MeV/17.6 MeV = 80%] – produced in the fusion reactor is from the 14.1 MeV neutrons emitted from the D+T nuclear reactions. In this regard the fusion reactor is similar to the fission reactor in which 1-2 MeV neutrons provide the nuclear energy and the power is from the fission nuclear reactions. Both devices require what is a called the "blanket" whose function is to convert the kinetic energy in the neutrons into thermal energy for driving the electric power-producing turbines. Neutrons easily penetrate most materials and activate some nuclei making the surrounding materials radioactive. Thus, extensive shielding is required in both fission and fusion nuclear reactors. The shielding material is called the blanket. Currently the world has 439 fission power reactors with 69 new reactors under construction as described by the IAEA Director General Yukiya Amano (`https://www.iaea.org/newscenter/focus/nuclear-power`). The IAEA declares that the emphasis now for research is on the back-end of the fission nuclear power system, meaning the disposing of the nuclear waste materials. Fusion reactors are one method being considered for aiding the deactivation of nuclear waste from fission reactors.

10.1 Neutron Blanket and Breeding Tritium

The blanket in the fusion reactor will contain lithium (Li) to use the neutron-on-lithium reactions to produce the tritium fuel. Helium is the other product of the $n + \text{Li}$ nuclear reactions. This generation of tritium is called the breeder function of the thick blanket surrounding the regions where the neutron flux penetrates the first vacuum chamber walls. Owing to the high energy of the fusion neutrons, the blanket will be of order one meter thick and just outside the first metallic vacuum chamber wall. This first vacuum chamber wall in ITER is a double wall with an internal metal rib to give the wall great structural strength. Beyond the blanket in the reactor there will be a final neutron shield to stop any remaining neutrons. Thus the ITER project created the need for major new nuclear engineering projects to understand and design the energy-exchanging layers and tritium breeding functions of the thick blankets just outside the first metal wall of the plasma chamber.

Both fission and fusion reactors require blankets to convert the kinetic energy of the neutrons into thermal energy to drive the electric power generators. The lithium blanket may serve three functions of (1) breeding tritium, (2) slowing down the neutrons to transform their kinetic energy into thermal energy and (3) to research, with the addition of fluorine (F), the formation of a molten salt (2Li F.BaF$_2$) for the efficient heat transfer fluid for powering the steam turbines. This thermal energy transfer function arises from the important properties of liquid lithium and its high specific heat capacity of $C_v = 4200 \, \text{J/kgK}$.

10.2   The Broader Approach and IFMIF

To answer these questions on the nuclear and atomic physics of the walls as soon as possible, a new project called the International Fusion Materials Irradiation Facility (IFMIF), was started at the Rokkasho Research Laboratory in Japan. This facility is the Rokkasho adjunct to the ITER project and is the first of several joint research projects between Japan and the European Union (EU). The projects are being carried out to strengthen and hasten the development of fusion engineering and science for the development of fusion power. One argument for this "Broader Approach and IFMIF" is that the results are needed to move developed countries away from carbon fuels more rapidly.

This bilateral agreement for joint EU and Japanese research is called the Broader Approach for Fusion Energy Research. The questions of the lifetime of the tokamak walls is clearly an issue that impacts the cost and the viability of all fusion power systems. The Broader Approach (BA) agreement was signed on 1 June 2007 and calls for three projects in support of fusion energy research support complementary to those pursued in the ITER project. The BA projects are required for the development of fusion powered electric generators.

The first major project supports the physics issues for the walls and blankets for fusion reactors. This project calls for using a high-energy linear accelerator for testing the nuclear physics of the wall materials and blankets. The Broader Approach agreement has two later projects. The second project is to build a follow-up to the successful JT-60U machine called the JT-60SA where informally the fusion community understands that SA denotes "super advanced". The JT-60S machine would research advanced high-performance steady-state tokamaks. The third project is the establishment of an International Fusion Energy Research Center (IFERC).

The information derived from the Broader Approach (BA) research is largely complementary to that derived from the ITER machine. The information from the BA is needed to proceed with the next level machine called DEMO and for the design of other possible fusion reactor machines like, for example, the Large Helical Device, a fusion neutron source for nuclear science, or the more advanced reversed field pinch (RFP) toroidal fusion machine. There have been numerous BA workshops and detailed plans are advancing rapidly and can be found on the website `http://www.ba-fusion/org`.

10.3   Neutron Shielding, the Cryostat and the Cooling Systems

The tests of the first wall radiation damage and determination of the life-time of wall materials is a key project for all fusion power machines. The neutrons at 14 MeV with no electric charge will move through the first wall. A thick shield called the "blanket" is placed behind the first wall to absorb the neutrons. The first metallic wall is made of stainless steel with a coating of tungsten $W(Z = 74, A = 184)$ and beryllium $Be(Z = 4, A = 9)$. The tungsten surfaces are used on the surfaces that receive the high-energy plasma flux owing to its high melting temperature. When a major disruption occurs, the plasma striking the surface has a temperature equivalent to the solar core temperatures. This wall life-time problem is a key research area and will be addressed by additional experiments at an organization called the International Materials Irradiation Facility with the acronym IFMIF (`http:www.ifmif.org`).

Fig. 10.1   The ITER blanket modules provide shielding from the high thermal loads from within the vacuum vessel and the high-energy neutrons produced by the fusion reactions. In later experiments various modules will be used to research the best tritium breeding designs. The blanket covers the interior surfaces of the vacuum vessel, providing shielding to the vessel and the superconducting magnets from the heat and neutron fluxes of the fusion reactions.

Shielding blankets, as shown in Fig. 10.1 of several designs will be installed on the plasma chamber wall in ITER to find the best method to convert the neutron 14 MeV energy into usable thermal energy for driving the steam turbines producing the electric power. The best-performing and longest-lasting wall materials will be incorporated in the future electric power reactor called DEMO. DEMO will be the first model of a fusion power reactor driving the electric power generating turbine. With DEMO the cost of electric power in kilowatts per Euro can be estimated.

Currently, the walls of ITER are to be double layers of tungsten with water cooling flowing between the walls to keep the wall temperature below the melting temperature of approximately 3500 K. Tungsten (W_{74}^{184}) has the highest melting temperature of any metal and is widely used for similar heat-withstanding applications as filaments in electric lights. The inner tungsten surfaces of the vessel are to be covered with beryllium as the material that is known to keep the impurity content low from the plasma-first wall interactions, as described in Chapter 9. The tungsten is typically a layer deposited on stainless steel for the structural strength of the double layered wall confining the plasma.

The blanket that slows the neutrons down to thermal temperature and transforms their energy into heat for driving the steam turbines contains lithium [with the two isotopes Li_7^3 and Li_6^3]. Lithium is in high demand for the electronic industry for batteries. The lithium atoms hit by the MeV energy neutrons undergo a nuclear reaction transforming the lithium into tritium plus helium. There are two isotopes of lithium: the Li_6^3 with three neutrons (and three protons) yields tritium and helium. The lithium isotope with four neutrons yields the same tritium and helium plus a moderate energy neutron. This extra neutron is useful for most neutron blanket designs to make up for lost neutrons. This set of reactions is called the Breeder reaction, in that it takes the spent tritium fuel and transforms this waste nuclear products into fresh fuel, namely tritium.

The outer wall is a vacuum-tight toroidal chamber called the cryostat made from 3,800 tons of stainless steel. The cryostat chamber would be about 8,500 cubic meters in volume. The structure is reinforced with many ribs to withstand the forces from the atmospheric pressure over this large surface area enclosing the vacuum. Atmospheric pressure is approximately $100,000$ Pa, and inside the vacuum chamber pressure will be at or less than 10^{-4} Pa. As a reminder, the force on a chamber is the pressure difference between the outside and the inside multiplied by the surface area giving 10^8 N.

For a fusion power reactor, as for ITER on a smaller scale, there must be an elaborate cooling system with large cooling towers and a pipeline supplying water from a large water reservoir. For ITER the water reservoir is from the nearby canal that runs along the Durance river close (within 5 km) to the ITER facility. The ITER system will dissipate some 500 MW of thermal heat during the operations.

For historical perspective, we include in the Table 10.1, the news announcement in November 1991 that the Joint European Torus (JET) with the first JET results of the deuterium-tritium shots, namely shot number 26148 that produced a neutron yield of 6.3×10^{17} neutron per second which gave the equivalent to 2.2 MJ of energy in a pulse of about one second. This was one of the early trace tritium shots where the tritium to deuterium density ratio was about 10%.

Again, in 1997 the record shots in JET producing 16 MW of fusion power are described in Chapter 1, showing much higher neutron yields and essentially reaching the break-even condition for a short period of time with fusion power comparable to injected power of 24 MW (www.fusionforefrontenergy.european). The 1999 publication documenting the record JET shots is Keilhacker, *et al.* (1999) from the

Table 10.1 JET discharge #26148 in 1991 with 10% trace tritium producing 6×10^{17} n/s equivalent to 1.8 MW of fusion power.

I_p	$=$	3.1 MA
B_T	$=$	2.8 T
P_{INJ}	$=$	1.53 MW
$n_e(0)$	$=$	3.6×10^{19} m^{-3}
$n_D(0) + n_T(0)$	$=$	2.4×10^{19} m^{-3}
$\overline{Z}_{\text{eff}}$	$\sim$	2.7
$T_i(0)$	$=$	18 keV
$T_e(0)$	$=$	10 keV
W_{dia}	$=$	9.1 MJ
W_{dia}	$=$	6 MW
τ_E	$\sim$	1 s
$(n_D + n_T)T_i\tau_E$	$=$	4.3×10^{20} m^{-3}keV s

JET science team in the 1999 journal *Nuclear Fusion*. The record shot 42976 in 1997 produced approximately 22 MJ of energy from a peak fusion power of 16 MW over a 4 second period. The ion temperature reached 40 keV and was driven by both neutral beam heating and ion cyclotron heating as described in detail in Chapter 1. The plasma confinement was in an ELM-free H-mode.

JET received a five-year renewal in 2014 to extend its search for achieving a net fusion power greater than the injected power, called $Q_{\text{fus}} > 1$, with deuterium and tritium plasmas.

Thus, with the coming of the future D-T shots in ITER the international physics and engineering community fully anticipate to increase the fusion power by an order of magnitude and to convincingly demonstrate the feasibility of building a functional first nuclear fusion powered electric generator. The number of years required to reach this state in ITER is not predictable but the achievement is certain.

10.4 Steady-State High Beta-High Fluence Machine

Before construction of the demonstration electric power plant (DEMO), there needs to be a fusion machine running at moderate power gain factors with $Q \leq 5$ that runs steady state for periods of weeks. The machine would be used to find the optimal neutron blankets for DEMO with a series of blanket designs that would receive neutron fluxes from 1-5 MW/m^2 for periods of weeks. This would allow testing research on the power extraction and the material nuclear science for the lifetime of machine components before finalizing the design of the demonstration fusion power reactor.

This next-step machine described by Stambaugh, *et al.* (2011) would have a neutron fluence of order 5 MW-yr/m^2 over several years of operation. The blankets would be optimized both for tritium production and for conversion of the neutron kinetic energy into thermal energy for the production of electric power. The experiments would validate the designs for the life-time of the irradiated materials, the life-time of the magnetic coils and the diagnostic systems.

The intermediate-step tokamak would give valuable information for all magnetic confinement fusion machines for the erosion of the plasma facing components with ten times the fluence of ITER. Questions concerning the retention of tritium in machine components would be answered and the life-time of a myriad of components inside the reactor chamber would be determined.

To achieve the weeks of steady-state operation, the real-time diagnostic-control systems would be optimized and validated. The exhaust power handling would be optimized for steady-state operation. Here the resistive wall modes and ELM control systems would be thoroughly optimized.

With respect to ITER, which was designed to achieve the highest performance over periods of order five minutes, the intermediate step machine would be designed to search for the most reliable steady-state fusion power with the required production and control of kilograms of tritium and the conversion of the neutron power into electric power for periods of weeks. The difference in the design philosophy may be likened to that of the Grand Prix race car versus the fast, long-distance bus or truck [Chan, *et al.* (2015)].

10.5   Radiation Diagnostics, Neutron and Hard X-ray Radiation Monitoring and Remote Handling for Maintenance

The dominant competition for nuclear fusion in the laboratory over the past 40 years has been what is called Inertial Confinement of plasmas created in laser-driven implosions followed by explosions of ball-bearing sized capsules of deuterium and tritium. The most successful inertia confinement experiments are at the National Ignition Facility (NIF) at the Lawrence Livermore National Laboratory in California. Currently, the laser ignition experiment runs with 192 laser beams focused into a hollow cylinder called by its German name, the Hohlraum, for a hollow room. The pellet of D-T is isotropically compressed (crushed) by the force of the radiation pressure in the Hohlraum created by the reflecting laser beams. The radiation pressure compresses the small, sphere to solid-state density. Currently, the lasers deliver 3 MJ of energy in a few nanoseconds to the D-T pellet in a spherically-symmetric radiation field, created by the Hohlraum reflections of the incident laser beams. There are other inertial confinement fusion experiments in Japan and at the University of Rochester that do not use the Hohlraum, but directly focus the laser beams on the D-T target. This approach is called direct drive, which as yet, has not matched the yield of the indirect-drive ignition experiments. At the time of writing the NIFS experiments are short of the design and simulation expectations.

Consequently, in 2014 the Department Energy changed the priorities of the NIFS experiment to have nuclear weapons design as top priority and nuclear fusion research as second priority in the next few years. The full report of this is found in

"Progress toward ignition on the National Ignition Facility" [Hinkel, *et al.* (2013)].

In closing, we note that further background information on the engineering design of the ITER tokamak is given in Chapter 10 (pp. 113-126) of *The Fusion Quest* by T. K. Fowler (ISBN:0-8018-5456-3) with interesting accounts of the decisions made by the designers of the most successful fusion machine JET.

Now with the success of JET the aim became to build a machine that would achieve ignition and sufficient transformer flux (1000 volt-sec) to run for about 1000/s producing 1000 MW of fusion power. This objective appeared out of reach to the Conceptual Design Activity team led by Boris Kadomtsev, with leaders including Ken Tomabechi, who directed design activity from the research center in Munich during the early ITER design periods. Further details of the debates and compromises made in designing the ITER machine are described in detail from the USA perspective by Steve Dean in the book *Search for the Ultimate Energy Source* (Chs. 9 and 10, pp. 149-192) and the reaction of the private sector industrial participants, both in high-tech industry and the nuclear power industry. Dean also gives a detailed account of the political ups and downs from the oil and gas energy crisis period in the 1970s to the change to the environmental crisis with respect to carbon pollution of the 2010s governing the decisions of the Bush and Obama administrations. Finally, we note that the early disagreements between the environmentalists and the electric power industry in the USA are covered in detail in the book *Energy* [Holdren and Herrera (1971)]. While nuclear safety is not an issue with fusion power, there remains the issue of the reactor cooling water discharge warming the lakes, and other environmental problems generic to all large electric power generating stations by the laws of thermodynamics.

References

Chan, V. S., Costley, A. E., Wan, B. N., Garofalo, A. M., and Leuer, J. A. (2015). Evaluation of CFETR as a Fusion Nuclear Science Facility using multiple system codes, *Nucl. Fusion* **55** p. 023017, doi:10.1088/0029-5515/55/2/023017.

Dean, S. O. (2013). *Search for the Ultimate Energy Source: A History of the U.S. Fusion Energy Program* (Springer Science & Business Media) Chapters 9-10, pp. 149-192.

Fowler, T. K. (1997). *The Fusion Quest* (Johns Hopkins University Press, Baltimore, 1997) ISBN:0-8018-5456-3.

Holdren, J., and Herrera, P. (1971), Energy, A Crisis in Power, Sierra Club (1971), ISBN:87156-055-0.

Hinkel, D. E., Edwards, M. J., Amendt, P. A., Benedetti, R., Hopkins, L. Berzak, Bleuel, D., Boehly, T. R., Bradley, D. K., Caggiano, J. A., Callahan, D. A., Celliers, P. M., Cerjan, C. J., Clark, D., Collins, G. W., Dewald, E. L., Dittrich, T. R., Divol, L., Dixit, S. N., Doeppner, T. Edgell, D., Eggert, J., Farley, D., Frenje, J. A., Glebov, V., Glenn, S. M., Haan, S. W., Hamza, A., Hammel, B. A., Haynam, C. A., Hammer, J. H., Heeter, R. F., Herrmann, H. W., Ho, D., Hurricane, O., Izumi, N., Johnson, M. Gatu, Jones, O. S., Kalantar, D. H., Kauffman, R. L., Kilkenny, J. D., Kline, J. L., Knauer, J P. J., Koch, A., Kritcher, A. Kyrala, G. A., LaFortune, K., Landen, O. L., Lasinski, B. F., Ma, T., Mackinnon, A. J., Macphee, A. J., Mapoles, E., Milovich, J. L., Moody, J. D., Meeker, D., Meezan, N. B., Michel, P., Moore, A. S., Munro, D. H., Nikroo, A., Olson, R. E., Opachich, K., Pak, A., Parham, T., Patel, P., Park, H-S., Petrasso, R. P., Ralph, J., Regan, S. P., Remington, B. A., Rinderknecht, H. G., Robey, H. F., Rosen, M. D., Ross, J. S., Rygg, R., Salmonson, J. D., Sangster, T. C., Schneider, M. B., Smalyuk, V., Spears, B. K., Springer, P. T., Storm, E., Strozzi, D. J., Suter, L. J., Thomas, C. A., Town, R. P. J., Williams, E. A., Weber, S. V., Wegner, P. J., Wilson, D. C., Widmann, K., Yeamans, C., Zylstra, A., Lindl, J. D., Atherton, L. J., Hsing, W. W., MacGowan, B. J., Van Wonterghem, B. M., and Moses, E. I. (2013). Progress toward ignition at the National Ignition Facility, *Plasma Phys. Control. Fusion* **55**, p. 124015, doi:10.1088/0741-3335/55/12/124015.

Keilhacker, M., Gibson, A., Gormezano, C., Lomas, P.J., Thomas, P. R., Watkins, M. L., Andrew, P., Balet, B., Borba, D., Challis, C. D., Coffey, I., Cottrell, G. A., De Esch, H. P. L., Deliyanakis, N., Fasoli, A., Gowers, C. W., Guo, H. Y., Huysmans, G. T. A., Jones, T. T. C., Kerner, W., König, R. W. T., Loughlin, M. J., Maas, A., Marcus, F. B., Nave, M. F. F., Rimini, F. G., Sadler, G. J., Sharapov, S. E., Sips, G., Smeulders, P., Söldner, F. X., Taroni, A., Tubbing, B. J. D., von Hellermann, M. G., Ward, D. J., and JET Team. (1999). High-fusion performance from deuterium-tritium plasmas in JET, *Nucl. Fusion* **39**, p. 209, doi:10.1088/0029-5515/39/2/306.

Stambaugh, R. D., Chan, V. S., Garofalo, A. M., Sawan, M., Humphreys, D. A., Lao, L. L., Leuer, J. A., Petrie, T. W., Prater, R., Snyder, P. B., Smith, J. P., and Wong, C. P. C. (2011). Fusion Nuclear Science Facility Candidates, *Fusion Sci. Tech.* **59**, 2, pp. 279-307.

Glossary Index

General Index